International Political Economy Series

Series Editor
Timothy M. Shaw, University of Massachusetts Boston, Boston
USA;
Emeritus Professor, University of London, London, UK

The global political economy is in flux as a series of cumulative crises impacts its organization and governance. The IPE series has tracked its development in both analysis and structure over the last three decades. It has always had a concentration on the global South. Now the South increasingly challenges the North as the centre of development, also reflected in a growing number of submissions and publications on indebted Eurozone economies in Southern Europe. An indispensable resource for scholars and researchers, the series examines a variety of capitalisms and connections by focusing on emerging economies, companies and sectors, debates and policies. It informs diverse policy communities as the established trans-Atlantic North declines and 'the rest', especially the BRICS, rise. NOW INDEXED ON SCOPUS!

More information about this series at
http://www.palgrave.com/gp/series/13996

Leila Simona Talani

The International Political Economy of Migration in the Globalization Era

palgrave
macmillan

Leila Simona Talani
European and International Studies Department
King's College London
London, UK

ISSN 2662-2483 ISSN 2662-2491 (electronic)
International Political Economy Series
ISBN 978-3-030-79320-3 ISBN 978-3-030-79321-0 (eBook)
https://doi.org/10.1007/978-3-030-79321-0

This Palgrave Macmillan imprint is published by the registered company Springer Nature Switzerland AG
The registered company address is: Gewerbestrasse 11, 6330 Cham, Switzerland

A Gabriellone ed alla mia mamma

Acknowledgements

The author is very grateful to Prof. Frieden, Prof. Freeman and Prof. Alesina for having accepted to talk with the author about the topics discussed in the book in the course of her period as Pierre Keller visiting Professor at the Harvard Kennedy School in 2018.

She would also like to thank Prof. Rodrik for permitting her to attend his Harvard module in the same period and comment on its contents. Finally, the author is grateful to Harvard Kennedy School and Harvard University for allowing her to participate in various seminars and workshops where she could gather the insights of scholars such as Prof. Putnam, Prof. Borjas, Prof. Inglehart and Prof. Norris on the issues that concern this book.

Introduction

The Theoretical Rationale

This book is concerned with the analysis of the impact of globalisation on international migration from a distinct international political economy perspective. It confronts theoretical debates from the different international political economy (IPE) approaches and elaborates on the implications of different theories in policy-making and political realms. Here migration is examined as an integral part of the global political economy that is structurally connected to the process of globalisation, although the definition of globalisation itself is a subject of enquiry. Within this context, one of the most debated questions is whether, in the age of globalisation, the state can regulate international migration? Or whether this is now the task of international organisations? Or perhaps international migration simply cannot be governed? This is known in the literature as the 'policy gap' or 'gap hypothesis' between migration policies and their results. Its discussion provides different conclusions according to which theoretical perspective is adopted (de Haas and Czaika 2013; Guiraudon and Lahav 2007; Boswell and Geddes 2011). So what is the international political economy of migration and which theories are referred to in this book?

The field of migration studies has been enriched by the contribution of a variety of different academic disciplines and perspectives (Brettell and Hollifield 2008). However, international political economy scholars have only recently come to systematically consider the various issues deriving in the global political economy from the increase of mass migration and brain drain (Boswell and Geddes 2011). This book mainly focuses on questions relating to immigration policy, i.e. to the entry and exit of migrants whether regularly or, importantly, irregularly. Indeed, one of the most relevant questions to address is whether irregular migration is the inevitable, perhaps for someone even welcome, outcome of the increased securitisation of international migration in the global context. This leads the author to attempt a political economy of irregular migration, still relying on the contributions of the different schools of international political economy. Less attention is paid here to the integration of migrants, what is called 'immigrant policy' (Hammar 1985), although this book tackles the issue within the context of the role played by social capital in favouring or hindering it. Moreover, this book is mainly concerned with discussing economic migration and less political migration. Still, EU asylum and refugee policy are addressed in the book, emphasising the impact of the EU refugee/migration crisis of 2014/15 on the militarisation of the EU border and the externalisation of the EU's approach to migration.

It is worth briefly recalling here that the IPE debate is traditionally characterised by two overlapping dichotomies. On the one hand, there is the debate between the two mainstream approaches to international migration, namely the liberal-institutionalist and realist ones. On the other hand, the mainstream is challenged by the critical perspectives summarised in the so-called 'globalisation thesis' (Brettell and Hollifield 2008). With respect to migration, realists and liberal-institutionalists tend to stress the role of national or international institutions, especially in relation to their ability to govern the phenomenon of international migration (Czaika and de Haas 2013; De Haas 2011; Brettell and Hollifield 2008; Boswell and Geddes 2011; Guiraudon and Lahav 2007). While realists insist that the state, in the globalisation era, is still perfectly able to implement an effective immigration policy, liberal-institutionalists posit that the only possibility remaining to govern migration is to rely on international organisations and international bodies of law. However, they both coincide in believing that international migration can be contained and governed in the age of globalisation. On the contrary, the globalisation thesis stresses the structural nature of international migratory flows in the global political economy and, therefore, its inevitability. It then concludes that the phenomenon cannot be controlled by neither national nor international or supranational institutions. Within the context of the above mentioned broad IPE debate on migration, this book tackles some of the principal concerns of the current state of the art: the paradox of securitisation of migration and the increase of insecurity; the globalisation backlash and the growing hostility towards international migrants; the tension between regular and irregular migration and the political economy of migrants' insertion in the labour markets of receiving countries, just to cite some of the topics addressed in the book.

More specifically, the first two chapters of the book are aimed at identifying the changing nature of international migration in the globalisation era. In relation to this, the aim is to assess the problem of migration, both regular and irregular, in the global political economy; and, second, to verify the connections between globalisation, marginalisation and national and regional policies to tackle both mass immigration and brain drain from the global south. In this section, including Chapters 1 and 2, the focus is on questions such as:

- How is the process of globalisation related to the increase of migratory flows from the global south to the global north?
- Is it still possible for the state to contain and regulate migration in the globalisation era?
- Is there a scope for global governance in the field of migration?
- Is there tension between regular and irregular migration?

Summing up, this section comprises an enquiry into how the main IPE approaches have tackled the question of international migration in the age of globalisation. The section will conclude by proposing an integrated vision of the impact of globalisation on migration including the economic, social and political dimensions. In particular, it identifies the three paradoxes which will constitute the framework around which the analysis of the book is constructed: the paradox of marginalisation within globalisation; the paradox of regionalisation within globalisation and the paradox of securitisation within globalisation.

The following section addresses the three main paradoxes of globalisation with respect to migration. It starts with Chapter 3 which proposes an analysis of the role played by marginalisation within globalisation in migratory flows from the MENA area to the EU. In Chapter 4, the paradox of regionalisation within globalisation is tackled both from the theoretical and practical point of view. The analysis will be effected by comparing the dynamics of migration from Mexico to the US and from the MENA area to the EU. Special attention will be devoted to the consequences of migration on sending countries in terms of both brain drain and remittances. In Chapter 5, the so-called paradox of securitisation is discussed with reference to recent developments in the EU migratory regime.

The following two chapters are dedicated to a systematic analysis of irregular migration and to the consequences of migration on the labour markets of receiving countries. In Chapter 6, the author presents the dynamics of irregular migration in theory and in practice. The chapter includes a discussion of the role of the underground economy as a potent pull factor for irregular migrants, especially in the case of southern European countries.

In Chapter 7, the insertion of migrants into the labour markets of receiving countries is analysed with the aim of identifying instances of job competition with the local population. The chapter will proceed by assessing whether regular migrants are a burden or an asset for hosting societies.

Chapter 8 will discuss the relationship between social capital, Islamophobia and terrorism to verify the extent to which the paradox of securitisation is creating more insecurity. This will entail tackling the question of Islamophobia and the dynamics of integration of especially Muslim migrants into host societies, both before and after the events of September the 11th. Special attention will be paid to the role of social capital in facilitating or hindering integration of Muslim migrants.

Chapter 9 deals with the relationship between populism and migration. It asks about the nature of populism and whether it is inherently xenophobic and anti-migrant or acquires similar characteristics in specific circumstances such as, for example, when it is in power. The analysis is predicated around the case of the Italian coalition government of the Five Star Movement and the Lega as this is a good example of anti-migrant policies adopted by populist parties while in government.

Finally, the conclusions are concerned with appraising to what extent current migratory dynamics contribute to the dark side of globalisation.

From a methodological point of view, the book heavily relies on research made by the author on the subject of migration. In particular, the author has discussed the topics addressed in the book with leading scholars, such as Frieden, Freeman and Alesina in the course of her visiting Professorship at the Harvard Kennedy School in 2018. During this period, the author was also able to gather the insights of scholars such as Rodrick, Norris, Inglehart, Borjas and Putnam, all quoted in the book, by attending their Harvard modules and seminars. The rest of the research was conducted by the author in the field, as is the case for the survey on motivations at the point of departure carried out by the author in Morocco, for the data on migration from the MENA area or the evaluation of projects of the EU Commission on the underground economy and migration. This primary research has been duly triangulated with secondary sources and documents to obtain more reliable conclusions.

The Book Outline

Chapter 1: Ipe and Migration: The Role of the State

In this chapter, the author addresses the theoretical debate in IPE relating to the role of the state in immigration policy. The analysis rotates around the so-called "policy gap" or “gap hypothesis”. This refers to the debate in migration studies on the persistence of a ‘gap’ between the aims identified by the state in pursuing immigration policy and its outcome. The debate is a theoretical one as the conclusions reached by the different IPE approaches do not coincide. In the neo-realist analysis, for example, the state is still perfectly able to govern migratory phenomena and therefore the policy gap is, at best, a rhetorical one. Here, the state claims to be willing to stop migration all together for political reasons, while in reality the intended objective is to allow irregular migration.

On the contrary, the other mainstream IPE approach, neo-institutionalism, posits that the national level of governance is unable to control migration in the age of globalisation, thus it needs to be substituted by the supranational level. This leads neo-institutionalists to discuss the characteristics of existing and future regional and global migratory regimes. Moreover, neo-institutionalists, such as Keohane, would emphasise the role played by coalitions of interests in favouring international migration in the most powerful liberal states eventually leading to regional and global governance solutions.

Finally, neoliberal complex interdependence theory would predict that the existence of an international regulatory system or institution devoted to controlling and regulating migratory flows would make it easier for the state to open their borders, both to trade and to migration.

Chapter 2: The three paradoxes of globalisation and migration

In Chapter 2, the analysis moves to the discussion of the so-called "globalisation thesis" as proposed by the transnationalist IPE approach and, more specifically, to the impact of globalisation on migration.

The definition of globalisation provided by transnationalists is a qualitative one. According to this definition, at the root of the transformations brought about by globalisation is technological change. The latter modifies substantially both the productive and the financial structure. What is important to assess the impact of globalisation on migration are the structural changes produced by technological progress on the productive structure and, in particular the geographical reallocation of production and the New Global Division of Labour and Power. These happen through Foreign Direct Investments (FDIs), the adoption of a policy of Mergers and Acquisitions (M&A) or by establishing Export Processing Zones in the global south.[1] This restructuring of the productive processes leads to changes in the labour structure with the substantial reallocation of labour-intensive production in developing countries. On the other hand, however, the same structural dynamics increase reliance on distant work as well as increasing labour mobility under the form of both mass migration and brain drain.

Thus, the globalisation thesis explains the most recent migratory flows as a consequence of structural transformations due to technological progress. More specifically, the contribution of scholars, such as Dicken, Mittlemann or Overbeek, allow the author of this book to identify three paradoxes of globalisation that impact on international migration:

- The paradox of "marginalisation" and its impact in terms of increase of extra-regional permanent migration and brain drain.
- The paradox of "regionalisation" and its consequences in terms of intra-regional temporary migration.
- The paradox of "securitisation" and its consequences in terms of irregular migration.

All of those paradoxes follow from the structural nature of globalisation and the emergence of a New Global Division of Labour and Power, all notions discussed in the chapter.

[1] See Dicken 2003.

Chapter 3: The paradox of marginalisation: the case of migration from the MENA area to the EU

Starting from a qualitative definition of globalisation, as proposed by transnationalists, this chapter addresses the marginalisation of the MENA area and its impact on migration to the EU. It is important to underline that marginalisation from the global political economy is assessed here in relative terms. In fact, although in absolute terms the countries marginalised from globalisation show some bettering of their main economic or social indicators, relative to global dynamics they fare progressively worse. Therefore, the chapter does not claim that the countries considered did not improve at all their situation, but that they have improved it less than other countries, thus worsening their relative position in the global political economy.

This derives from a definition of globalisation, such as the one adopted by transnationalists, as a dynamic process. In particular, attributing to technological progress a central role in the globalisation process implies a need for societies to constantly update and upgrade their skills in order to keep pace with scientific knowledge (Mittelman 2011; Robinson 2004).

If societies fail to keep up with technological progress, although in absolute terms they may gather some benefits of globalisation, their marginalisation from the global political economy will increase. This chapter demonstrates that the three MENA countries considered, Morocco, Egypt and Tunisia, are a clear case of marginalisation within globalisation. Moreover, the chapter also notices that although no causal relation can be easily identified, this marginalisation from the global political economy impacts on the decision of people to move, facilitating a process of extra-regional migration. The chapter then proceeds by reporting on the dynamics of migration from the three countries of the MENA area to the EU.

Chapter 4: The paradox of regionalisation within globalisation: migration to the EU and migration to the US compared

Chapter 4 discusses the paradox of regionalisation within globalisation and, in particular, how regionalisation (or the lack of it) produces different migratory patterns, with reference to the cases of migration to the EU and to the US.

The paradox of regionalisation within globalisation means that the process of regional integration is happening only in some areas of the world, the so-called triad, while other regions are unable to increase their level of economic and institutional interdependence.[2] This exacerbates the marginalisation of the latter and their progressive detachment from the global core.[3] Consequently, their populations are incentivised to move to more integrated parts of the globe in search of better standards of life. This phenomenon interests both low-skilled workers and workers with high levels of education, thus producing both mass migration and brain drain. Moreover, migratory flows, both regular and irregular, follow regional dynamics, for historical, geographic, social or cultural reasons. Thus, there is an attempt to address them through regional migratory schemes, with varying results, as the cases of both the EU and US immigration policies show.

This chapter compares migration to the EU and the US as they present different dynamics related to the paradox of regionalisation. The case of migration from the MENA region to the EU is an instance of extra-regional permanent migration due to its marginalisation from the global political economy and due to lack of regionalisation. On the contrary, migration to the US from mainly Mexico is a case of intra-regional migration following regionalisation. This discussion includes remittances, brain drain and the impact of these phenomena on sending countries.

Chapter 5: The paradox of securitisation: Is there a common migratory policy in the eu?

The paradox of securitisation within globalisation is particularly evident in the case of the development of a common EU approach to migration. Here the debate around whether or not it is possible to talk about the existence of a common migratory policy in the EU vis-à-vis the migration of third country nationals (TCNs) has led to the idea of the creation of the so-called 'Fortress Europe'. The term 'Fortress Europe' refers to the institutionalisation of an area of internal mobility compounded by the

[2] See, for example, Breslin et al. 2002 and Hettne, Inotai & Sunkel 1999.

[3] On this subject, see, for example, Murshed 2002.

establishment of external barriers vis-à-vis the entry of TCNs.[4] According to the literature, 'Fortress Europe' was born with Council Regulation 1612/68, a piece of legislation distinguishing, for the first time, between the right of free movement of TCNs from that of citizens of a EEC Member State. [5] After the establishment of the freedom of movement for EU citizens, with the Single European Act of 1986 and then with the 'EU citizenship' provisions of the Maastricht Treaty in 1992, this divide has progressively become insurmountable. This has led to the adoption of a very restrictive and security-oriented approach to migration from outside the EU, at both the EU and national level.[6] The consequences of this have been an increase of irregular migration and the progressive 'securitisation' of migration, leading to the paradox of 'securitisation' within globalisation. The notion of 'securitisation' is defined in the literature as the conceptualisation and treatment of migration as a 'security issue', whose competence is attributed to security agencies such as the police or the courts. [7]

The chapter will review the developments leading to the creation of 'Fortress Europe' and will identify the migratory regimes existing within the EU. The chapter will also give some insights about the explanations proposed to the securitisation of migratory policy, and to its externalisation, especially after the refugee crisis.

Chapter 6: Towards a political economy of irregular migration: theory and practice

Chapter 6 attempts to construct a political economy of irregular migration based on the theoretical paradigms discussed in the first section of the book. In particular, both mainstream and critical approaches to IPE allow for a suitable theoretical analysis of irregular migration although this is not always explicit. For example, from the realist standpoint, illegal migration responds to the needs of host countries, as 'the existence and persistence of a black market for immigrants implies that all parties participating in

[4] See Geddes 2003.

[5] See Huysmans 2000; also, Ugur 1995.

[6] For a detailed analysis of some EU member states' migratory policies, see Geddes 2003.

[7] See Huysmans 2000; also, Guiraudon 2000.

these exchanges benefit from these voluntary transactions' (Borjas 1990: 58).

Institutionalists notice how migration acquires regional patterns, due to historical, geographic, social or cultural reasons. Therefore, all responses to migration take the form of regional policies, like US or EU immigration policies, which are becoming stricter and less liberal (Geddes 2000a; 2000b; 2002; Thielemann 2003). Clearly, among the reasons why regional policies are becoming more and more restrictive towards extra-regional migration are the traditional concerns over political unrest, social conflict, cultural clashes or religious struggles (Freeman 1979; Freeman 2008; Freeman et al 2008). However, the chapter notices how the political economy dimension of irregular migration should not be overlooked. This points towards the benefits that particularly employers can obtain from irregular migration in terms of the reduction of the costs of production of both the local and migrant labour force and increase of bargaining power vis-à-vis organised labour (Freeman et al 2008).

Moreover, irregular migrants can only be employed in the underground economy which, therefore, becomes a relevant pull factor for irregular migration (Soysal 2000; Reyneri 1999). However, the existence of an underground sector pre-dates migratory flows and is by no-means produced by them. On the contrary, it is precisely because there is already a thriving informal sector that irregular migrants are attracted to one country instead of another as this guarantees them the possibility to find gainful work despite the lack of proper documents. Indeed, because of the securitisation of migration, most of migrants from the global south are unable to enter and stay in receiving countries legally and remain undocumented throughout most of their time in the hosting country (Reyneri 1999).

Thus, from the political economy point of view, irregular migration favours employers and producers more than the working class as it allows for the flexibilisation of the internal labour market and its precaritisation. On the other hand, legal immigrants should be integrated in the existing labour protection and welfare state provisions system and would be as expensive as the local workforce. As Saskia Sassen writes, 'those who deal with the real political economy of the city already know this: the rhetoric of the law-and-order Republican mayor here in New York City is surprisingly friendly to illegal immigrants' (Sassen 1998: xiii).

Chapter 7: The insertion of migrants into the labour force of receiving countries: competition or complementarity?

In Chapter 7, the insertion of migrants into the labour markets of receiving countries is analysed with the aim of identifying instances of job competition with the local population. Already in the economic literature there is plenty of evidence that job competition is not a problem. For example, the work of Giovanni Peri analyses the long-run impact of immigration on employment, productivity and its skill bias concluding that a no crowding out effect on local population employment can be identified.[8] This is a relevant contribution for the analysis of the insertion of regular migrants in the labour markets of receiving countries. However, realist perspectives still put political considerations over economic ones and advocate for the closure of borders (Borjas 2016, Rodrick 2018).

From a neo-institutionalist viewpoint, it is generally recognised that the liberalisation of migration could have even more economic benefits than free trade. The liberalisation of migration would translate into more jobs also for the local population eliminating at the source the discussion about job competition. Thus regular migrants more than a burden could be considered an asset for hosting societies' welfare states, especially in the context of ageing populations.

From a transnationalist point of view, the issue of whether immigrants represent a competitive threat for domestic workers is addressed in the context of the labour structure of the receiving countries. Some European countries are certainly characterised by high levels of unemployment. However, such unemployment concerns mainly educated young people, living with their families and who can afford waiting for the right job proposal. They are therefore unlikely to compete with migrant workers for low-skilled, often irregular jobs in the underground economy.[9] This phenomenon is actually considered in the literature as a powerful pull factor for irregular migration. This brings scholars to conclude that a demand for migrant workers does exist even in countries where the

[8] See Giovanni Peri, (2013), The economic consequences of the proposed immigration reform, Intereconomics, 2013, Vol. 48(3), pp. 191–192; Peri, Giovanni, (2012), The Effect Of Immigration On Productivity: Evidence From U.S. States, Review of Economics and Statistics, 2012, Vol. 94(1), pp. 348–358; Peri, Giovanni, (2012), Immigration, labor markets, and productivity, Cato journal, Jan 2012, Vol. 32(1), pp. 35–54.

[9] See Reyneri, (1999), pp. 7–14

overall unemployment rate is high. However, the insertion of migrants in the informal sector strongly impacts their ability to contribute to the welfare state, producing instances of hostility towards irregular migrants and migrants in general and further contributing to their marginalisation and even criminalisation.[10]

Chapter 8: Migration and social capital: from islamophobia to social unrest and terrorism

Chapter 8 will discuss the relationship between migration and social capital, to verify how the latter is connected to social unrest, terrorism and Islamophobia. With respect to this, the author subscribes to the niche view that among the many reasons of terrorism there could be poor integration among second and third-generation immigrants. This, in turn, could be referred to the lack of bridging and linking social capital. With respect to the different dimensions of integration, the economic literature seems to agree that there is a Muslim disadvantage in economic integration for second-generation immigrants to Europe.[11] For example, a paper by Claire L. Adida, David D. Laitin and Marie-Anne Valfort demonstrates that in the French labour market, anti-Muslim discrimination exists as a Muslim candidate is 2.5 times less likely to be called for a job interview than their Christian counterparts.[12] However, in the chapter the economic dimension of integration will be compounded by a more holistic approach to the lack of integration of Muslim communities. This will entail tackling the question of Islamophobia[13] and the dynamics of

[10] See Reyneri, (1999), pp. 7–14

[11] Identifying barriers to Muslim integration in France, (2010), Claire L. Adida, David D. Laitin, and Marie-Anne Valfort, PNAS December 28, 2010. 107 (52) 22384–22390; https://doi.org/10.1073/pnas.1015550107.

[12] Identifying barriers to Muslim integration in France, (2010), Claire L. Adida, David D. Laitin, and Marie-Anne Valfort, PNAS December 28, 2010. 107 (52) 22384–22390; https://doi.org/10.1073/pnas.1015550107.

[13] See Body-Gendrot, S., (2007a), Urban "riots" in France: anything new?, In Paul Ponsaert, ed. Local security policy in the Netherlands and Belgium, Den Haag, Netherlands: Boom Juridische Utig, 2007; Body-Gendrot, S., (2007b), Order and disorder in the urban landscape, in Burdett, R. and Sudjik, D., eds, The urban age book, Tashent; Bourdieu, P. (1980) "Le capital social: notes provisoires". Actes de la Recherche in Sciences Sociales, 31, 2–3; Brehm J, Rahn W. (1997) "Individual-Level evidence for the causes and Consequences of Social capital", American Journal of political Science, 41, 3: 999–1023;

integration of especially Muslim migrants into host societies, both before and after the events of September the 11th. Special attention will be paid to the role of social capital in facilitating or hindering integration. [14]

After defining social capital and analysing the issue of how to measure it, the chapter discusses the finding that the emergence of "Muslim voices", instead of increasing the level of trust between Muslim communities and local civil society, may have indeed reduced it. This might enhance diversity and multiply the occasions of confrontation between host communities and communities of migrants. In a few words, it could reduce social capital and facilitate Islamophobia and social unrest, possibly representing a fertile ground for the spread of radical versions of Islam and terrorist tendencies mostly among second and third-generation migrants.[15]

Chapter 9: Populism and migration

This chapter addresses the contemporary debate on the relationship between populism and migration. It assesses whether anti-migrant feelings are inherently part of the populist credo or are the consequences of strategic political considerations.

Coleman J. (1988) "Social Capital in the Creation of Human Capital", American Journal of Sociology, 94: 95–120

[14] See Body-Gendrot, S., (2007a), Urban "riots" in France: anything new?, In Paul Ponsaert, ed. Local security policy in the Netherlands and Belgium, Den Haag, Netherlands: Boom Juridische Utig, 2007; Body-Gendrot, S., (2007b), Order and disorder in the urban landscape, in Burdett, R. and Sudjik, D., eds, The urban age book, Tashent; Bourdieu, P. (1980) "Le capital social: notes provisoires". Actes de la Recherche in Sciences Sociales, 31, 2–3; Brehm J, Rahn W. (1997) "Individual-Level evidence for the causes and Consequences of Social capital", American Journal of political Science, 41, 3: 999–1023; Coleman J. (1988) "Social Capital in the Creation of Human Capital", American Journal of Sociology, 94: 95–120

[15] See Body-Gendrot, S., (2007a), Urban "riots" in France: anything new?, In Paul Ponsaert, ed. Local security policy in the Netherlands and Belgium, Den Haag, Netherlands: Boom Juridische Utig, 2007; Body-Gendrot, S., (2007b), Order and disorder in the urban landscape, in Burdett, R. and Sudjik, D., eds, The urban age book, Tashent; Bourdieu, P. (1980) "Le capital social: notes provisoires". Actes de la Recherche in Sciences Sociales, 31, 2–3; Brehm J, Rahn W. (1997) "Individual-Level evidence for the causes and Consequences of Social capital", American Journal of political Science, 41, 3: 999–1023; Coleman J. (1988) "Social Capital in the Creation of Human Capital", American Journal of Sociology, 94: 95–120

After discussing the literature on the definition of populism and its relationship to migration, the chapter selects a political economy approach to explain recent populist occurrences and their anti-migrant and xenophobic manifestations.

Following Frieden (2017),[16] in this chapter the roots of populist phenomena are identified in political discontent. This, in turn, is related to a crisis of compensation, which refers to the inability of the state to provide sufficient social safety provisions to compensate the losers of economic events. It is also due to a crisis of representation, meaning the inability of traditional parties to represent the interests of the discontents.

In Frieden's conceptualisation the origin of the failure of compensation and the increase of economic and social inequalities is identified in globalisation. However, this chapter posits that, although globalisation did indeed increase discontent and disparities, as widely discussed in this book, the growth of populism is a more recent phenomenon.

Therefore, following a growing literature, this contribution hypotheses that the global economic crisis represented the catalyst for the contradictions of globalisation to manifest themselves. This is predicated around the case of Italy, where the global financial crisis, thanks to the asymmetries of the eurozone, lead to the sovereign debt crisis. Indeed, the Italian populist parties, the Five Star Movement and the Lega, were originally vehemently against the Euro.

Thus, at the origins of the extraordinary success of the Italian populist movements are the economic difficulties experienced especially by the Italian manufacturing sector as a consequence of the establishment of the euro. With the eurozone crisis, ensuing from the global financial crisis, those difficulties brought political discontent. The austerity measures adopted by EU institutions during the eurozone crisis produced a crisis of compensation. This, together with the crisis of representation, ensuing from the unelected governments of Monti, Letta and Renzi, convinced the Italian electorate to accept the populist offers of the Five Star Movement and the Lega.

[16] Frieden, J., (2017), The political economy of the globalization backlash: Sources and implications, Jeffry Frieden Harvard University December 2017

Conclusion: Migration and the Dark Side of Globalisation

Within the background discussed in the first chapters (Chapters 1-2-3), globalisation is defined as a highly uneven process, creating geographical marginalisation and social inequalities. This leads to what the author calls the 'dark side' of globalisation. In the conclusion, it is argued that globalisation inherently has a dark side, which derives from its uneven and unequal nature.

The negative consequences of globalisation in relation to international migration range from the creation of inequalities in the labour markets and modern slavery (Chapter 7) to the marginalisation of some regions which produce mass migration of an irregular nature (Chapter 4 and 5). Such marginalisation is not only characteristic of the global south, but increasingly also of the periphery of Europe (Chapter 6). Moreover, the securitisation paradox, seemingly aimed at limiting global terrorism, might actually conceal instances of discrimination especially against Muslim migrants, paradoxically fuelling terrorist tendencies in second and third-generation migrants (Chapter 8). Finally, the highly technological content of this definition of globalisation impacts also on the modalities of action of organised crime, heavily involved in the irregular migration business and more and more global in nature although keeping its local roots (Chapter 6). This can lead to the growing antagonisation, if not even criminalisation, of international migrants by host societies reflected in the widespread success of right-wing populist parties and populism (Chapter 9).

Contents

List of Figures

List of Tables

CHAPTER 1

IPE and Migration: The Role of the State

This chapter addresses a classical question in the International Political Economy of migration, the question of what is the role of the state in international migration. Although this question is contiguous to the one of whether the state can effectively regulate migration, the so-called "policy gap" (Czaika & de Haas, 2013: Chapter 6), this chapter will only touch upon this latter issue as this will be further developed in Chapter 6. Here the role of the State in the migration debate will be discussed in theoretical terms by making reference to the mainstream literature in IPE, in particular neo-realist and neo-institutionalist authors. To be sure, this is a highly theoretical question as, according to which IPE approach is made reference to, the answer to the question is different. For example, neo-realists, a la Weiner (1980, 1995), but also Freeman (1979, 2008, 2018), Borjas (2014, 2015, 2016) and Rodrik (2017, 2018), strongly emphasise the role of the State in international migration. Indeed, from this perspective, if the State is unable to contain and regulate migration, it is because it lacks the political will to do so. This also means that the state has a vested interest in keeping irregular migration, as seen in Chapter 6.

On the other hand, neo-institutionalists, such as Brettell and Hollifield (2014), posit that nation states have lost their ability to control their borders in the age of globalisation and this is increasingly becoming the responsibility of international institutions. This leads to the debate

L. S. Talani, *The International Political Economy of Migration in the Globalization Era*, International Political Economy Series,
https://doi.org/10.1007/978-3-030-79321-0_1

about global and regional governance schemes of international migration (Chapter 4). At the roots of the increase of international migration in the age of globalisation, there should be, according to neo-institutionalists, the formation of coalitions of interests supporting them in liberal democracies. This should eventually lead to regional and global governance solutions for international migration. Furthermore, the presence of international institutions, also under the form of regulatory regimes able to regulate migratory flows, should make it more likely for liberal states to open their borders, not only to trade, but also to migration.

This chapter will address the theoretical foundations of this discussion after positioning this contribution in the International Political Economy debate.

Positioning the Book in the IPE Debate: The State of the Debate

This book discusses international migration in the era of globalisation as well as confronting various issues in international migration from a distinctive international political economy perspective. It examines migration as part of a global political economy while addressing the theoretical debate relating to the ability of the state to regulate international migration and the so-called 'policy gap' or 'gap hypothesis' between migration policies and their outcomes (Boswell & Geddes, 2011; Czaika & de Haas, 2013; Guiraudon & Lahav, 2006).

This contribution seeks to understand how IPE has approached the issue of international migration. It examines arguments from the literature as well as advancing the author's own thesis that migration is one of the structural components of globalisation, and, as such, it entails a number of paradoxes. But what do we mean by the international political economy of migration?

The definition of international political economy adopted in the book is a strictly political scientists' one. This divide between the economic and political scientists' definition of IPE has been broadly equated in Cohen's intellectual history of IPE with the growing distinction between the so-called 'American' and 'British' schools of IPE (Cohen, 2007, 2008). This characterisation is very controversial and the related controversy is reflected in the special issues of two of the leading Journals in IPE. The Review of International Political Economy (RIPE) reported the views of the so-called 'American School of IPE'[1] while a special issue of New Political Economy (NPE) published the reactions to Cohen's typology of the so-called 'British School'.[2] In the American IPE tradition, Cohen's categorisation has been criticised by scholars such as Peter Katzenstein, Kate McNamara, Henry Farrell and Marty Finnemore. They underlined that proposing a mostly economic oriented view of the American school of IPE might be due to focusing too much on journal outlets and not enough on books.[3] On the other hand, some of the members of the 'British School' of IPE took issue with the softer characterisation of the discipline made by Cohen in the UK and Europe.[4] Whether or not an economic approach to IPE prevails in the US and a political scientists' one is more relevant in Europe, the question remains of how the two approaches differ from each other.

This question is related to the fact that, in the late 1980s, the field of international political economy saw the spread of a new school, called 'open economy politics' (OEP) (Lake, 2009a, 2009b). OEP reverted to the use of public choice and rational choice to address issues in the international political economy, applying methodological individualism and relying on formal modelling. The methodology adopted by OEP is the typical neo-classical economics individualist one, based on ranked preference maximisation by individual actors within institutional constraints and entailing strategic interactions with other actors. The analytical toolkit is the same as formal economics, with formal models and econometric testing of hypotheses. In OEP the focus is on methodological individualism, which in the case of IPE means equating the interests of aggregate actors, as classes, institutions or even states, to the sum of individual preferences. In some occasions, aggregate actors are simply considered as single units. In the words of Lake (2009a: 50):

[1] See Review of International Political Economy (2009), Vol. 16, No. 1, "Special Issue: Not So Quiet on the Western Front: The American School of IPE".

[2] See New Political Economy (2009), "Special Symposium: The 'British School' of International Political Economy", New Political Economy, Vol. 14, No. 3.

[3] See Review of International Political Economy (2009), Vol. 16, No. 1, "Special Issue: Not So Quiet on the Western Front: The American School of IPE".

[4] See New Political Economy (2009), "Special Symposium: The 'British School' of International Political Economy", New Political Economy, Vol. 14, No. 3.

> OEP begins with sets of individuals – firms, sectors, factors of production – that can be reasonably assumed to share (nearly) identical interests. Relevant units of analysis vary by technology, institutions, and other factors. Cohen's (2007: 199) characterization notwithstanding, units are not ontologically given nor are states the primary unit of analysis, although state policy is often the object to be explained.

Clearly then, OEP derives its methodology, main assumptions and units of analysis directly from neo-classical economics. It is indeed considered an application of rational choice and public choice theory (Lake, 2009a, 2009b), disciplines that represent the intellectual bases of the International Political Economy Society, co-founded by Lake and Helen Milner and promoting OEP (Keohane, 2009).

From the perspective of these scholars, the political scientists' interpretation of IPE is lacking in analytical structure, its methodology is at best historical, if not merely descriptive or even anecdotal, and econometric testing is inexistent. In a few words, it is hardly a scientific approach and can, at best, propose problems and contextualise them historically. Therefore, a more formalised economic approach is needed to allow for testable hypotheses and generalisable conclusions (Frey, 1984: 4).

On the other hand, scholars who belong to the tradition of International Relations take issues with a simplistic application of the economic method to the study of IPE, emphasising the centrality of sociopolitical structures in their definition of international political economy (Gilpin, 1975: 43; Keohane, 1984: 1; Keohane, 2009: 37; Murphy & Tooze, 1991: 4). Their understanding of international political economy requires explanation of social change, including the interaction between the political, the economic and the social.

Hence, the political scientists' approach to international political economy entails asking a set of questions about the nature of the relations between the states and the markets, the socio-economic structure and the political structure, by means of different perspectives and different methodologies (Keohane, 2009: 37; Murphy & Tooze, 1991: 4).

This book is firmly set within the political scientists' approach to IPE, and therefore, it is not adopting a formalised, economic characterisation of the international political economy of migration, although, inevitably, some reference to the economic literature will be made. This means that generally the book will adopt a macro-level of analysis to address international migration, hence the focus on the state and on supranational institutions. It therefore leaves micro, individual level of analysis to economists, who are certainly better suited to address it.

The distinction between the macro-level of analysis and the micro-level of analysis in migration theories has been proposed by Brettell and Hollifield (2014), in their discussion of the different social sciences perspectives on migration. To systematise the various disciplinary approaches to the issue, the authors classify theories across disciplines according to their main research questions, levels of analysis, relevant methodologies and main hypotheses.

On the basis of such classification, it is possible to distinguish social sciences that address mostly the micro-dimension of migration, such as history, anthropology and, importantly, economics, from those that focus on the macro-dimension, such as sociology, demography and, importantly, political science. It is already clear from this distinction that the two disciplines of economics and political science are ontologically different in the way in which they address the question of migration as they focus on two different levels of analysis. Differences in the social sciences exist, however, also in relation to the methodologies adopted and to the questions asked. For historians, for example, the basic questions relating to the migration process are, who moves, when do they move, why do they move? Why do some people not move? How do those who do move experience departure, migration and settlement? The answer to these questions is then sought through historical methodology, often relying on archives, and focusing on the single migratory experiences of small groups or individuals due to the nature of their research (Brettel & Hollifield, 2014). Similarly, anthropologists usually adopt an ethnographic methodology to investigate a specific case despite their endeavour to engage in cross-cultural comparisons to allow for broader generalisations. In economics then, the micro-level of analysis is also dictated by a specific methodology, methodological individualism. This refers to the rational decision-making of the single actor as the origin of the decision to migrate. The latter thus becomes the outcome of a cost–benefit analysis normally effected through mathematical maximisation models (Brettel & Hollifield, 2014).

Whereas lawyers are more eclectic in their choices of levels of analysis and methodology, for sociologists the heuristic variables are represented by macro-phenomena, such as social relations and structures. These are summoned to explain both questions relating to the origins of migratory flows and to the dynamics of integration within traditional structuralist theoretical frameworks. Their methodology can vary, implying the use of ad hoc surveys or more aggregate data. For demographers, statistical tools are adopted to understand the nature of population change through the identification of predictive models. Political scientists, instead, focus on aggregate political entities such as institutions and importantly, states to ask questions relating to their ability to regulate migration, the impact of migration on institutions of sovereignty and citizenship and questions relating to the integration of migrants into host societies. Here, again the methodologies used can vary according to the different theoretical frameworks adopted (Brettel & Hollifield, 2014).

It thus seems clear that the approach of political scientists and economists is hugely different in terms of the methodology selected, which in the case of economics is almost invariably methodological individualism and formal modelling of the rational choice process behind the decision to migrate. However, also the focus of analysis, as well as the research questions addressed differ significantly, in that political scientists mostly ask questions relating to the role of macro-political aggregates, and the state in particular, in facilitating, regulating or hindering migratory flows.

This book is concerned with a macro level of analysis and the study of macro-aggregates in immigration policy. Also, it is not addressing questions relating to the rights of migrants or to integration policies, although it may begin to do so to a limited extent in Chapter 8. For these reasons, some of the scholars who are currently active in the field of the economic approach to the political economy of migration are not explicitly included in the discussion.[5]

[5] This is the case, for example, for the work of authors such as Claire Adida on migration in Africa mainly addressing issues relating migrant policy and migrant integration. For example, her article 'Too Close for Comfort? Immigrant Exclusion in Africa'mainly focuses on the question, 'why do some immigrant minorities in the developing world integrate into their host societies whereas others face exclusion and hostility?'. Similarly, Adida et al. (2010) address the question of whether there is a Muslim disadvantage in economic integration for second-generation immigrants to Europe. Also Dancygier (2014) mainly works on the integration of migrants. In her book, Immigration and Conflict in Europe, she presents an in-depth overview of the book's theory of immigrant conflict beginning by providing a conceptualisation of both types of conflict, immigrant–native and immigrant–state conflict, and goes on to describe their indicators. Similarly, her other works address the question of political representation and the integration of migrants in host societies and are, therefore, not tackling the macro-analysis of immigration policy

More closely to the interests of this contribution, Adrian Shin (2017) proposes a comparative political economy of migratory policies in authoritarian regimes. However, the interest of the present title is on migration from the Global South to western liberal democracies. Differently, David Singer (2015)[6] provides a very relevant criticism of the role of migrants in the global political economy,[7] which reveals the mismatch between macro-level theorising and micro-level data analysis thus pointing to the need of macro-level theorising as proposed in this book.

This macro-level perspective allows for an understanding of the motivations behind the increase of mass migration and the brain drain from marginalised regions of the world, whereas more micro-analyses, such as the ones of many economists, only address the question of the integration of migrants in host societies (Singer, 2015).

For example, the work of Giovanni Peri analyses the long-run impact of immigration on employment, productivity and its skill bias concluding that a no-crowding out effect on local population employment can be identified.[8] This is a relevant contribution for the analysis of the insertion of regular migrants in the labour markets of receiving countries, as we will see in Chapter 7, but not really to understand why economic migrants move. On the other hand, Peri's work with Mayda addresses the political

and policy-making which is the subject of most political scientists' IPE contributions to the debate in migration studies as further elaborated below.

Hainmueller and Hangartner (2013) focus on immigrant integration in host societies, with a particular emphasis on discrimination in naturalisation processes. Also in the case of David Leblang (2010), it is unclear how his work could contribute to the understanding of international migration in the age of globalisation. He published an article on portfolio investment, and on dual citizenship in comparative politics, while, in his article with Sarah Blodgett Bermeo (2015) the author intervened on the issue of migration and foreign aid.

[6] See "Migration, labor, and the international political economy." *Annual Review of Political Science, 18*, 283–301. 2015. (with Layna Mosley).

[7] See "Migration, labor, and the international political economy." *Annual Review of Political Science, 18*, 283–301. 2015. (with Layna Mosley).

[8] See Peri, G. (2013). The economic consequences of the proposed immigration reform. *Intereconomics, 48*(3), 191–192; Peri, G. (2012a). The effect of immigration on productivity: Evidence from U.S. states. *Review of Economics and Statistics, 94*(1), 348–358; Peri, G. (2012b). Immigration, labor markets, and productivity. *Cato Journal, 32*(1), 35–54.

impact of immigration on national parties, thus again focusing more on integration, in this case, political integration.[9]

When they do address the question of why migrants leave their countries, economists, as rightly pointed out by Brettel and Hollifield (2014), tend to do so by adopting a rational choice approach based on methodological individualism. For example, the work of Peri and Ortega[10] simply confirms the economic theory of migration that international migration flows are highly responsive to income per capita at destination.

Summing up, this contribution is set firmly within the political sciences' approaches to international migration, in particular within the international relations/international political economy tradition.

So what is the state of the debate in international relations/international political economy with regard to international migration?

Realist Approaches to the IPE of Immigration: The State in the Debate

Myron Weiner[11] is the most prominent representative of the realist IPE approach to international migration.[12] In a very early intervention on the

[9] See Mayda, A. M., Peri, G., & Steingress, W. (2018). The political impact of immigration: Evidence from the United States, NBER Working Paper No. 24510, Issued in April 2018 NBER Program(s):International Trade and Investment, Political Economy.

[10] See Ortega, F., & Peri, G. (2012). *The effect of income and immigration policies on international migration*. NBER Working Paper No. 18322, Issued in August 2012; Ortega, F., & Peri, G. (2009). *The causes and effects of international migrations: Evidence from OECD countries 1980–2005*. NBER Program(s):International Trade and Investment, Labor Studies, NBER Working Paper No. 14833, Issued in April 2009, NBER Program(s):Economic Fluctuations and Growth, International Trade and Investment, Labor Studies.

[11] See The Global Migration Crisis: Challenge to States and to Human Rights (HarperCollins, 1995 ISBN 978-0-06-500232-4); Threatened Peoples, Threatened Borders: World Migration and US Policy (co-editor, W. Norton, 1995); The New Geopolitics of Central Asia and its Borderlands (co-editor, Indiana University Press, 1994); The State and Social Transformation in Afghanistan, Iran and Pakistan (co-editor, Syracuse University Press, 1994); and International Migration and Security (editor, Westview Press, 1993).

[12] In his 1980 article, Weiner contrasts his security/stability framework to international migration with the International political economy/political economy one with the latter meaning mainly what is now considered the globalisation approach to international migration. This is shown in note 2, page 4 of the article where he says: on the political

question of the different approach between economists and political scientists to international migration (Weiner, 1980: 3–4),[13] he notices how the latter "turn our attention from individual decision-making by migrants to the larger social, political, and economic context within which individuals act" (Weiner, 1980: 3). In his understanding, the political science approach pays more attention to "the behavior of states and to the importance of borders" (Weiner, 1980: 3) and not on the individual choice to migrate or the individual determinants of integration.

His seminal work on the security/stability framework of international migration, the so-called "national security thesis", moves from the consideration that the global migration crisis is mostly the consequence of both sending and receiving states' policies (Weiner, 1995: 25). Evidently, there are many incentives for sending countries to allow migration. These range from political and economic advantages, to social or even ethnical-cultural ones. In political terms, forced migration can eliminate political opponents and class enemies or it can be used as a foreign policy measure to dominate ethnic enemies or to force other countries to accept undesired policies. As a macroeconomic policy measure, migration can benefit sending countries by relieving unemployment, reducing the costs of welfare, increasing wealth and foreign currency reserves through remittances and importing technology and innovation from abroad. On the other hand, receiving countries can decide to allow migrants into their territories for a number of reasons, although in his opinion, ease of access is the main determinant of the increase of immigrants in host societies. In fact, Weiner is convinced that there is a limit to the number of migrants

economy of international migration see, for a neo-Marxist perspective, Saskia Sassen, The Mobility of Labor and Capital (Cambridge: Cambridge University Press, 1988), Alejandro Portes and John Walton, Labor. Class. and the International System (New York: Academic Press, 1981) and Stephen Adler, International Migration and Dependence (Westmead: Saxon House, 1977), Stephen Castes and Godula Kosack, Immigrant Workers and Class Structure in Western Europe (London: Oxford University Press, 1973). For other political economy interpretations see Charles P. Kindleberger, Eurone's Postwar Growth: The Role of Labor Supply (Cambridge, Ma: Harvard University Press, 1963), Michael Piore, Birds of Passage: Migrant Labor in Industrial Societies (Cambridge: Cambridge University Press, 1979), Wolf R. Bohning, The Migration of Workers in the United Kingdom and the European Community (London: Oxford University Press, 1972), Wolf R. Bohning, Studies in International Labour Migration (London: Macmillan, 1984).

[13] Weiner, M. (1980). *Security, stability, and international migration*. MIT. https://www.files.ethz.ch/isn/19789/Security_Stability_Migration.pdf. Accessed 5 May 2020.

that any society is able to absorb, and this explains why xenophobic backlashes in Western Europe can be considered as a reaction to the security threat represented by uncontrolled migration. In his words, '[s]tates will not and cannot allow others to decide who will permanently live and work in their own societies' (Weiner, 1995: 22).

It follows that states have the right, even the obligation, to control their borders essentially by any means. In particular, in the age of the globalisation of information, Weiner supports the introduction of ever stringent border controls, including the introduction of identity cards, the use of repatriation schemes, expulsions, border patrols and even the criminalisation of migration used as a deterrent. The author even considers it legitimate for a sovereign state to intervene in conflicts abroad to avoid the occurrence of refugee flows (Weiner, 1980, 1995).

If Weiner conceptualises migration and refugee policy as a question of national security, other scholars within the realist IPE tradition are more keen to allow states to open their borders when this leads to an advantage for them, whether by enhancing their power and stance in the international system or by improving their economic situation relative to other states (Borjas, 2016). This approach is known as the "immigration market thesis" and it posits that migration flows and policies are a consequence of internationally systemic factors. In particular, they are a function of the relative position of states in the international system and its related distribution of political and economic power. Sovereign states can, by allowing refugees and economic migrants in their territories, improve their relative position in the international system and the related balance of power. Here the works of Borjas, although mainly an economist, appear very relevant (Borjas, 2014, 2015, 2016). In the opinion of this scholar, international migration is akin to the regular labour market, in that economic migrants, and also refugees, decide which country to move to on the basis of purely rational considerations relating to which benefits those countries have to offer. On the other hand, states who want to maintain or enhance their stance in the international system can do so by attracting the best and brightest immigrants through adopting suitable policies. One such policy is to reward highly educated migrants by offering advantageous working packages and higher wages. States who fail to attract highly skilled migrants are unlikely to see their power and capabilities increase in the international system (Borjas, 1990: Chapter 13).

Overall, in Borjas opinion, eventually immigration benefits both immigrants and the receiving countries. Quoting the author:

> Receiving countries typically support immigration for a simple reason: they perceive that immigrants generate and overall benefit for natives". (Borjas, 2016: Chapter 8, loc.2119 kindle ed.)

This is true not only for regular migrants, but also for irregular ones as

> the existence and persistence of a black market for immigrants implies that all parties participating in these exchanges benefit from these voluntary transactions'. (Borjas, 2016: 58)

Indeed, Borjas often talks of an "immigration surplus" (Borjas, 2016: Chapter 8, loc. 2171 kindle ed.), which is the economic bonus added to the wealth of host societies by allowing immigrants to enter their territories. This does not mean that all groups in receiving countries win from migration. Looking at the position of the unskilled labour force in receiving countries, it is possible, if not likely, that they will be on the losers' side of migration as their wages are likely to be reduced by the increase of labour supply for unskilled jobs (Chapter 7 in this book).[14] This is not necessarily the case in all countries and for all unskilled jobs, but in Borjas' calculations, this is what happens in the US (Borjas, 2016: Chapter 8). However, the wage losses of unskilled labour are more than compensated by what the rest of society gains, especially producers. Thus immigration is still adding to the wealth of the nation because:

> As with free trade, the laws of supply and demand imply that the dollar gains accruing to the natives who gain must be numerically larger than the dollar losses suffered by the natives who lose. (Borjas, 2016: Chapter 8, loc.2156 kindle ed.)

It does however entail a distributional pain, which is the other side of the economic gain:

> The greater the distributional pain, the greater the economic gain (Borjas, 2016: Chapter 8, loc.2156 kindle ed.)

[14] This assertion is however very contested in the political economy debate over the advantages and disadvantages of economic migration as discussed in Chapter 7.

In Borjas' view, though, opening up the borders to all migrants, as neoliberals would advocate,[15] would not increase the wealth of all nations. In his article on Immigration and Globalisation (Borjas, 2015), the author dispels the myth that by removing all immigration restrictions, the global gains would amount to trillions of dollars annually, the so-called "trillion-dollar bills" lying on the sidewalk (Chapter 7 this book). Similar conclusions are reached also by Collier (2013) and Ruhs (2013). In Collier's analysis, it is emphasised how the short-run impacts of immigration are more positive than what would happen if migration were not restricted by the state:

> Contrary to the prejudices of xenophobes, the evidence does not suggest that migration to date has had significantly adverse effects ... Contrary to self-perceived 'progressives,' the evidence does suggest that without effective controls migration would rapidly accelerate to the point at which additional migration would have adverse effects. (Collier, 2013: 245)

What Borjas (2016) finds interesting here is that if all countries, according to their characteristics, adopt specific types of immigration restrictions, this can only be explained by the fact that unrestricted immigration is not beneficial. Thus, the state is perfectly able to identify the right amount of restrictions necessary to maximise the benefits of immigration and to enforce them effectively.

The "immigration market approach", therefore, is still a realist approach, but it answers the question of whether a state should risk migration much more favourably than the "national security one", as immigration always entails an economic "surplus".

There is however a problem with similar realist interpretations of international migration and it lies in the absolute confidence they have in the ability of the State and of state politics and policies to govern international migration in the age of globalisation. This primacy of national politics over any other consideration is debatable (Brettel & Hollifield, 2014).

A more nuanced, "economic interests" realist approach to understanding immigration policy is the one proposed by Freeman (2018, interview with the author). In the understanding of this scholar, still the nation state is at the centre of the decision-making relating to whether or not to accept migration and what kind of immigrants to allow in.

[15] Representative studies include Clemens (2011).

However, here the state is not conceived of as a unitary actor, but as responding to the interests of particular of socio-economic groups. This is an economic interests approach to the domestic decision-making process (Freeman & Kessler, 2008). As in the case of Borjas, though, there will be groups that win from migration and groups that lose from it. Thus the question remains, to say it with the words of the author, whether: 'those who gain from immigration compensate those who lose to produce a net social gain' (Freeman, 2008: 3).

The starting point of Freeman's analysis is the typical realist one, i.e. how and when to allow migration (Freeman, 2018, interview with the author). Also the answer is a typically realist one: the state should allow more migration when it is in its interest to do so, and stop it when it is not in its interest. However, as mentioned above, in Freeman's analysis the interest of the state is defined by domestic variables and internal dynamics, not by international systemic ones like in the case of Borjas above. In particular, he singles out as the main heuristic variable to explain a successful management of migratory issues, the configuration of national institutions and interest groups. More specifically, states with a high level of centralisation of both policy-making and capital-labour relations are supposed to address the negative dimension of immigration more effectively (Freeman, 2018, interview with the author).

In ascertaining the role of interest groups, Freeman focuses on three: economic, ethnic and issue or ideological ones (Freeman & Jupp, 1992: 83). In his conceptualisation, employers would usually be very active in lobbying the government in favour of more migration, as it is the case in the US. Here, the coalition supporting a relaxed immigration policy is represented by organised business and its affiliated politicians, and liberals, who however have humanitarian and ideological motivations (Freeman & Jupp, 1992: 84). On the other hand, the working class is less keen on increasing immigration, and has a tendency to lobby against it (Freeman, 1979: 255). The final outcome depends on how strong the respective interest organisations are and how permeable state institutions prove to be to lobbying. In particular, the strength of labour organisation is singled out as playing a very relevant role in defining immigration policy. This happens for three reasons. First, stronger trade unions can lobby state institutions more effectively; second, the stronger trade unions, the easier it will be for the government to control the increased labour supply due to immigration; third, strong working-class organisations and parties can tame the reaction of the lower classes stemming from the pressures of a higher intake of immigrants (Freeman & Jupp, 1992: 17).

By using different economic models, it is possible to predict the impact of migration on different interest groups and, therefore, identify the position of different interest coalitions in the debate over immigration policy. However, according to which economic model is adopted, the hypotheses on the societal effect of migration will be different, thus leading to different predictions. For example, the product or output model would predict that the main distinction would be a class one, with the capitalist one in favour of more immigration, and the working class against it due to its depressing effects on wages. However, adopting the Heckscher–Ohlin model, the impact of immigrants on wages would depend on their skills, therefore not all the working class would lose from their increase. Finally, in the specific-factors model, only the socio-economic actors with exactly the same skills as the migrants will lose, whereas the income of the other factors will increase. This will give rise to sectoral cleavages, with mobile factors winning and specific factors losing (see Freeman, 2008: Table 1.1). The range of possible coalitions is then further complicated when fiscal variables are introduced. Indeed, tensions between groups may arise from competition for public goods, but then it depends also on geographical considerations relating to where different groups are located (Freeman, 2008: 4).

There are, in turn, four possibilities identified by Freeman in terms of mobilisation of interest groups for or against a more or less expansionary immigration policy. Adopting a two-dimensional model, with economic costs/benefits on one axis and size of groups on the other, interest mobilisation is a function of whether groups are concentrated or diffuse and whether the costs are large or small. With large economic costs/benefits

Table 1.1 Freeman: Political mobilisation style on immigration policy

	Large costs/benefits	*Small costs/benefits*
Concentrated groups	Client politics	Entrepreneurial politics
Diffused groups	Interest groups politics	Majoritarian politics

Source Elaboration of the author from Freeman and Kessler (2008: 671)

and concentrated groups, the mobilisation style is predicted to be client politics. If socio-economic groups are diffused with large economic benefits/costs, the mobilisation style would be interest group politics. With small economic benefits/costs, entrepreneurial politics is the mode ensuing with concentrated groups, and majoritarian politics with diffuse groups (Freeman & Kessler, 2008: 671).

In general, however, Freeman believes that "migration is economically advantageous for leading economic sectors. Those for which it is not beneficial can be compensated for their losses, or if not, typically wield limited influence in national politics" (Freeman, 2008: 11).

This "interest groups" approach to immigration is thus still based on a cost/benefit analysis to determine whether or not migration should be allowed. However, contrary to the "national security" thesis and the "immigration market" one, here it is mainly economic interest groups that exert pressure on the government to limit or allow immigration. Of course this approach discards issues of identity and ethnicity, which, as the author himself recognises, are relevant in defining the impact of immigration on host societies (Freeman, 2018, interview with the author). Perfectly stable societies can be destabilised by the acquired ethnic diversity linked to immigration producing an erosion of social cohesion and trust as further analysed in Chapter 8.

Overall, however, this political economy approach to identifying the interest coalitions behind a state's immigration policy is worth studying as it allows for a systematic analysis of the issue (Freeman & Kessler, 2008). It does not, though, help solving the question of why, in the age of globalisation, states are prone to allowing more immigrants on their territories, whether regularly or irregularly.

More emphasis on the relation between globalisation and migration from a realist perspective is to be found in Rodrik (2018, Harvard module).

According to Rodrik, hostility towards migration is one of the cultural manifestations, together with a tendency towards authoritarian populism, of the backlash against globalisation, what he calls "the globalization backlash" (Rodrik, 2018). Globalisation, if left unhindered, can manifest itself either under the form of increased migratory and refugee flows, in western developed democracies, or under the form of Foreign Direct Investment and financial flows, in peripheral countries such as Latin America and also South Europe (Rodrik, 2017). The backlash against the economic consequences of globalisation, then, takes the form of anti-migrant populism in western democracies and anti-capitalist populist movements in more peripheral countries. Therefore, as western states cannot allow globalisation to erode the democratic roots of society, they have to implement anti-immigration policies (Rodrik, 2018, Harvard module; 2017; see also Chapter 9). What is relevant here is, again, the high level of confidence this author shares with other realist scholars in the ability of the nation state and of national politics to regulate and control globalisation, in this case, in its manifestation as increase of migratory and refugee flows.

Such a declaration of faith in the nation state, however, is not endorsed by neo-institutionalist interpretations of the IPE of immigration as will be seen in the section below.

The Liberal-State Thesis and the Role of the Judiciary in the Neo-Institutionalist View

Neo-institutionalist accounts of international migration differ substantially not only from realist ones, but also from neoliberal economic ones and from the globalisation thesis (Chapter 2). The difference between liberal institutionalists and neoliberal economists is that the former emphasise the role played by international/supranational institutions, especially regulatory and legal frameworks, to understand why states accept more migrants. On the other hand, neoliberal economists[16] point at the economic advantages brought about by immigrants in terms of increased productivity and GDP (Chapter 7). However, liberal institutionalists also differ substantially from the interpretation to international migration given by globalisation theorists who would stress the structural nature of the current wave of migration (Chapter 2).

The neo-institutionalist approach to the IPE of immigration would propose a number of hypotheses with respect to the question of why has migration increased in the globalisation era. To start with, from this standpoint, an increase in migratory flows should be conceptualised as the consequence of the emergence of interest coalitions in liberal states supporting more migration. These coalitions should include all those interested in more free-trade and in liberalisation policies. However, not everyone in favour of opening up borders to trade is necessarily

[16] Clemens (2011).

supporting opening up the borders also to foreigners as this latter policy is far more contentious from a political, social and even cultural point of view (Brettel & Hollifield, 2014).

Moreover, neoliberal interdependence theorists would suggest that liberal states will be more favourable to allowing more migrants into their territories in the presence of a well-defined international regime on the issue, whether under the form of ad hoc supranational institutions or regulatory systems. The problem being that, whereas for international trade the regulatory framework is well established, in the case of international migration it is hard to talk about the existence of such an international regime. Still, migratory flows have increased exponentially in the age of globalisation.

For these reasons, neo-institutionalist analyses of the current trends in international migration are mostly focusing on regional migratory regimes, as explained in Chapter 4. One of the neo-institutionalist theorists of migration is Hollifield (1992, 1998, 2000, 2004), Hollifield et al., (2014). His approach has been termed the 'liberal-state thesis' (Brettel & Hollifield, 2014), and it is a very well established approach in the IPE of immigration. According to this thesis, the main reason why migrants are attracted to liberal democracies is because of their legal regimes attributing rights to immigrants and foreigners. Thus, as once rights have been extended, it is very difficult if not even impossible to withdraw them, liberal states are constrained by their own laws and institutions when they try to control or limit immigration (Hollifield et al., 2014: 4–5). This partially explains also why migratory flows have increased in the last years, in parallel with the increase in rights-based liberalism (Brettel & Hollifield, 2014). The existence and protection by liberal states' courts of civil, political social, and religious rights to migrants, in some cases even to irregular ones, account for the increase in the number of people that want to enter the territories of liberal democracies. This is, in a few words, a potent pull factor for international migration in the neo-institutionalist view. It is true that this perspective is mainly a domestic one. However, according to this approach, an international migratory regime is emerging, under the form of international law and human rights conventions.

Based on the above, Hollifield explains the "gap hypothesis", i.e. the difficulty states encounter to control immigration, precisely with "the rise of rights-based paradox" (Hollifield et al., 2014: 8). In his words:

> Civil rights– based policies help immigrants not only to get in but also to remain and settle. At the same time, human rights and refugee conventions have underscored the rights of asylum seekers, migrant workers, and their families. (Hollifield et al., 2014: 8)

This is what the author calls a "dilemma or paradox for the liberals" (Hollifield et al., 2014: 9). By extending civil and humanitarian rights to foreigners, the State loses its ability to control immigration. In fact, if those rights are limited only for selected groups of immigrants, under the pressure of right-wing movements with the aim of restricting immigration, then the State cannot really be called "liberal" any longer. Indeed:

> How can a "liberal" society tolerate the presence of individuals in it who are members but not citizens?. (Hollifield et al., 2014: 9)

Liberal states will eventually have to come to terms with the fact that limiting the rights of foreigners entering their countries could undermine their own legitimacy (Hollifield et al., 2014: 28–29). This new "liberal paradox" by Hollifield somehow modifies the ontology of his previous version of the "liberal paradox" given by the contrasting needs of an open economy and the political costs of liberalising migration (Hollifield, 2004). This new version of the liberal paradox puts Hollifield more clearly into the theoretical context of liberal institutionalism.

In her 2016 volume, Gender and Migration (Polity Press), Brettell further contributes to this perspective by analysing how the law and policies of receiving countries have a role in attracting gendered immigration (Brettell, 2016: Chapter 2), thus acting as a pull factor for a very specific kind of foreigner.

From the empirical point of view, it is possible to verify how rights contribute to attracting and retaining immigrants in two ways. First, an analysis of policy changes and their impact on migration should help clarifying how legal regimes modify migratory trends. Second, the role of the judicial system should be studied to ascertain how this has limited de-facto the ability of the state to control migration.

At the international level, the situation is more complicated. Although, according to Hollifield, the rule of law and orderly movements of people could represent an organising principle, its implementation is unlikely to materialise as it would require authority to be centralised in order to overcome collective action problems. This is particularly difficult especially in the absence of a hegemon, or a group of hegemonic countries, promoting it. In the meantime, co-ordination could be achieved, through international issue-linking, log-rolling and moral suasion (Hollifield, 2000: 105).

In sum:

> As states come together to manage this extraordinarily complex phenomenon, it may be possible to construct a truly international migration regime, under the auspices of the United Nations. But we are not sanguine about this possibility because the asymmetry of interests between the North and the South is too great. Even as states become more interdependent in an era of globalization, they are likely to remain trapped in a liberal paradox for decades to come. (Hollifield et al., 2014: 31)

With respect to the creation of an international regulatory regime for migration, Soysal has a more optimistic position. In her opinion, there is already an international institutional system protecting the rights of migrants outside the national level. This configures what the author terms a "post-national notion of citizenship", based on a universalistic definition of personhood, which guarantees migrants' rights of inclusion and cannot be superseded by the State. From this perspective, the notion of national citizenship has already been transcended by a universalistic understanding of this institution (Soysal, 1998, 2000, 2012, 2015). In fact, contemporary developments have underlined two paradoxes of the national definition of citizenship that make it obsolete. On the one hand, there is an increase in the decoupling of rights and identities which together represent the main components of citizenship. By this it is meant that increasingly, in contemporary societies, rights are disconnected from identities. This makes the category of national citizenship not applicable. On the other hand, there is a tendency to recognise as legitimate a universalistic notion of personhood entailing the recognition of particularistic rights in the public domain. Thus, national citizenship is not the only legal category attributing rights to individuals, including migrants, as a universalistic conception of personhood is increasingly legitimised (Soysal, 2000; see also Soysal, 2012).

As the recognition of human rights at the international level is institutionalised, Soysal's approach is well within the liberal institutionalist tradition. As such, it sets itself apart from constructivist approaches attributing heuristic primacy to identities and narratives. Indeed, she is explicit in recognising her theoretical background in complex interdependency (Soysal, 2015). This theoretical affiliation is articulated in two dimensions:

1. Internationalisation, which is considered by the author as:

> the institutionalization of nation-transcending frameworks, models, and standards, and their constitution of actors' orientations and strategies. (Soysal, 2015: 3)

2. Interdependence, which means looking at:

> the world as an interdependent state system, in which nation-states, organizations (public and private), and individuals constitute the major actors. (Soysal, 2015: 3)

Systemic approaches to international migration from a neo-institutionalist standpoint are however not very frequent in the literature as the demand for international co-operation in the field of migration is not yet very pressing. As international migratory regimes do not appear high on the agenda of policy-makers, liberal institutionalists are missing a relevant dependent variable. However, there certainly are pressures for the formation of regional governance schemes to regulate migration and, increasingly, for international co-ordination of border control policies (Ghosh, 1999, 2000).

Nowhere are similar tendencies more developed than in the case of the European Union and the Schengen system (see Chapter 5). For this reason, neo-institutionalist analyses of the process of European migratory policy integration are more abundant. Geddes (2003, 2014: 435), for example, points out, similarly to Soysal, that the 'transnational advocacy' of migrants' rights has found in the high degree of technocratic and bureaucratic insulation of EU bodies a fertile terrain for its institutionalisation. The fact that European Institutions, such as the European Court of Justice, the European Parliament and the European Commission, are not following national electoral logics, makes them more prone to accept pro-migrant lobbying activities (Acosta & Geddes, 2013; Geddes, 2014: 440). The EU framework for migratory and refugee policy is a testament to this phenomenon (Geddes, 2014). In a typical neo-institutionalist fashion, Geddes underlines how, in reality, the EU supranational migratory system is actually not only able to change, but is also already changing the orientation of national policy frameworks towards more convergence (Boswell & Geddes, 2011; Geddes, 2014: 434).

For Thielemann, the necessity to embed the protection of human rights in the EU legal framework is at the basis of the development of a common EU approach to migration and refugee policy. This is coupled by neo-functionalist spillovers and functionalist considerations of efficiency (Thielemann, 2003, 2004). Said that, given the mostly economic nature of the project of European integration, also in the case of migrants' rights, economic ones are privileged with respect to social and political ones (Geddes, 2014). Furthermore, the securitisation[17] agenda of nation states still represents a strong limit to the implementation of a more rights-oriented agenda to migration and refugee policy at the EU level (Geddes, 2000a, 2000b).

The question of securitisation has been investigated more in depth by Guiraudon (2018) from a liberal institutionalist standpoint. Contrary to other liberal institutionalist conceptualisations of the EU common approach to migration and refugee policy, this author does not emphasise its rights oriented nature, but stresses the security-oriented elements. In fact, the competition of domestic security actors to increase their power position with respect to the judiciary has led to a progressive shift of the level of governance of migratory issues from the national to the EU level. This has resulted in the progressive securitisation of the EU's home and justice affairs. Trying to escape the national constraints set by courts and judges, national security agencies, mostly in charge of border control and the criminalisation of migration, have "venue shopped" at the supranational level thus increasingly attributing to EU interventions on migratory and refugee policy a "security" dimension. As such, the trasnationalisation game clearly favoured law and order officials, such as home and justice affair personnel and the police, over more rights-oriented actors in defining the EU common approach to migration (Guiraudon, 2000). This led to the securitisation of the EU common approach to home and justice affairs or even its militarisation after the 2015 refugee

[17] By securitisation is meant the transformation of migration and refugee policy into a security issue and the adoption of security-related tools to deal with it. See Huysmans (2000, 2006: Chapter 7).

crisis (Guiraudon, 2018: 156). Similarly, Huysmans (2000, 2006, 2014) conceptualises the EU approach to migration as a security-oriented one, emphasising the role played in its evolution and implementation by home and justice agencies, although the analysis effected by this author is a more neo-constructivist one. Eventually, in the opinion of this author, the security agenda is destined to undermine the democratic nature of the EU (Huysmans, 2014). Again from a neo-constructivist viewpoint, and following Risse (Borzel et al., 2018), Lavenex (2001, 2015, 2018) understands the contraposition between the 'liberal', human rights approach to migration and the 'security' realist one as reflecting the antagonism between two different ideological propositions and the way in which they are embedded in political institutions.

Conclusion

The present contribution addresses the macro-level of analysis of the political economy of immigration. As such, it differs substantially from the economists' interpretations focusing on the micro-level of analysis and utilising methodological individualism as the main heuristic tool. In the political scientists' understanding of the IPE of migration, the main questions to be answered relate to the role of macro-political entities and socio-economic groups in verifying why migratory flows happen and why they are increasing in size and frequency in the age of globalisation. In particular, the role of the state in facilitating, hindering and regulating migration is often at the centre of the enquiry.

However, depending on the specific IPE approach used, the role of the state is more or less relevant. While realists posit that the State is perfectly able to control and regulate migratory dynamics in the age of globalisation, neo-institutionalists are convinced that liberal democracies have lost their ability to implement migratory policies. This is due to the specific role played by rights-oriented institutions, both at the national, regional and international levels. As a consequence, migrants are attracted to western democracies by the possibility of having their rights recognised and protected by national, regional and international courts and legal systems.

Overall, though, mainstream IPE theories are not able to shed light on precisely how globalisation has impacted international migration. This happens because they attribute too much importance to the political dimension, whether at the national or at the supranational level. In the next chapter, the globalisation thesis will be explained with the aim of answering the hows and whys of the impact of globalisation on migration.

References

Acosta, D., & Geddes, A. (2013). The development, application and implications of an EU rule of law in the area of migration policy. *Journal of Common Market Studies, 51*(2), 179–193.

Adida, C. L., Laitin, D. D., & Valfort, M.-A. (2010, December 28). Identifying barriers to Muslim integration in France. *Proceedings of the National Academy of Sciences, 107*(52), 22384–22390. https://doi.org/10.1073/pnas.1015550107.

Bermeo, B., & Leblang, D. (2015). *Migration and foreign aid*. https://doi.org/10.1017/S0020818315000119. Published online: 20 May 2015. As Accessed on August 4, 2018.

Boswell, C., & Geddes, A. (2011). *Migration and mobility in the European Union*. Palgrave Macmillan.

Borjas, G. J. (1990). *Friends or strangers*. Basic Books.

Borjas, G. J. (2014). *Immigration economics*. Harvard University Press.

Borjas, G. J. (2015). Immigration and globalization: A review article. *Journal of Economic Literature, 53*(4), 961–974. http://dx.doi.org/10.1257/jel.53.4.961. Accessed 13 September 2019.

Borjas, G. J. (2016). *We wanted workers: Unraveling the immigration narrative*. W. W. Norton & Company.

Börzel, T., Draude, A., & Risse, T. (Eds.). (2018). *Oxford handbook of governance in areas of limited statehood*. Oxford University Press.

Brettell, C., & Hollifield, J. (2014). *Migration theory: Talking across disciplines* (3rd ed.). Routledge.

Brettell, C. (2016). *Gender and migration*. Polity Press.

Clemens, M. A. (2011). Economics and emigration: Trillion-dollar bills on the sidewalk? *Journal of Economic Perspectives, 25*(3), 83–106.

Cohen, B. (2007). The transatlantic divide: Why are American and British IPE so different? *Review of International Political Economy, 14*(2), 197–219.

Cohen, B. (2008). *International political economy: An intellectual history*. Princeton University Press.

Collier, P. (2013). *Exodus: How migration is changing our world*. Oxford University Press.

Czaika, M., & de Haas, H. (2013). The effectiveness of immigration policies. *Population and Development Review, 39*(3), 487–508.

Dancygier, R. M. (2014). Electoral rules or electoral leverage?: Explaining muslim representation in England. *World Politics, 66*(2), 229–263.

Freeman, G. (2008). *Migration and the political sustainability of welfare states*. http://cartss.colorado.edu/Freeman.doc. Accessed 23 June 2008.
Freeman, G. (2018). *Interview with the author*. Harvard University.
Freeman, G., & Jupp, J. (1992). *Nations of immigrants*. Oxford University Press.
Freeman, G. P., & Kessler, A. K. (2008). Political economy and migration policy. *Journal of Ethnic and Migration Studies, 34*(4), 655–678.
Freeman, G. P. (1979). *Immigrant labour and racial conflict in industrial societies*. Princeton University Press.
Frey, B. S. (1984). *International political economy*. Basil Blackwell.
Geddes, A. (2000a). *Immigration and European integration: Towards the fortress Europe?* Manchester University Press.
Geddes, A. (2000b). Lobbying for migrant inclusion in the European Union. *JEPP, 7*(4), 632–649.
Geddes, A. (2003). *The politics of migration and immigration in Europe*. Sage.
Geddes, A. (2014). The European Union supranational governance and the remaking of European migration policy and politics. In J. Hollifield, et al. (Eds.), *Controlling immigration: A global perspective* (3rd ed.). Stanford University Press.
Ghosh, B. (1999) *Challenges and opportunities of economic globalisation: Some implications for labour, industry and nation states*. Paper prepared for BundesanstaltfürArbeit workshop, 'Internationalisierung der Arbeitsmarkte', 27–28 April, Frankfurt a/M.
Ghosh, B. (Ed.). (2000). *Managing migration*. Oxford University Press.
Gilpin, R. (1975). *US power and multinational corporation*. Basic Books.
Guiraudon, V. (2000). European integration and migration policy. *JCMS Journal of Common Market Studies, 38*(2), 251–272.
Guiraudon, V. (2018). The 2015 refugee crisis was not a turning point: Explaining policy inertia in EU border control. *European Political Science, 17*(1), 151–160.
Guiraudon, V., & Lahav, G. (2006). *Immigration policy in Europe: The politics of control*. Routledge.
Hainmueller, J., & Hangartner, D. (2013, February). Who gets a Swiss passport? A natural experiment in immigrant discrimination. *American Political Science Review, 107*(1), 159–187. https://www.jstor.org/stable/23357762.
Hollifield, J. F. (1992). *Immigrants, markets and states: The political economy of postwar Europe*. Harvard University Press.
Hollifield, J. F. (1998). Migration, tade and the nation state: The myth of globalization. *UCLA Journal of International Law and Foreign Affairs, 3*(2), 595–636.
Hollifield, J. F. (2000). Migration and the 'New' international order: The missing regime. In B. Ghosh (Ed.), *Managing migration* (pp. 75–109). Oxford University Press.

Hollifield, J. (2004). The emerging migration state. *The International Migration Review, 38*(3), 885–912.

Hollifield, J., Martin, P. L., & Orrenius, P. (Eds.). (2014). *Controlling immigration: A global perspective* (3rd ed.).

Huysmans, J. (2000). The European Union and the securitisation of migration. *Journal of Common Market Studies, 38*(5), 751–777.

Huysmans, J. (2006). *The politics of insecurity: Fear, migration and asylum in the EU*. Routledge, 191 pp.

Huysmans, J. (2014). *Security unbound: Enacting democratic limits*. Routledge.

Keohane, R. O. (1984). *After hegemony: Cooperation and discord in the world political economy*. Princeton University Press.

Keohane, R. O. (2009). The old IPE and the new. *Review of International Political Economy, 16*(1), 34–46.

Lake, D. (2009a). Open economy politics: A critical review. *Review of International Organizations, 4*(1), 219–244.

Lake, D. A. (2009b). TRIPs across the Atlantic: Theory and epistemology in IPE. *Review of International Political Economy, 16*(1), 47–57.

Lavenex, S. (2018). Migration. In T. Börzel, A. Draude, & T. Risse (Eds.), *Oxford handbook of governance in areas of limited statehood*. Oxford University Press.

Lavenex, S. (2015). Justice and home affairs: Institutional change and policy continuity. In H. Wallace, M. Pollack, & A. Young (Eds.), *Policy-making in the European Union* (7th ed., pp. 367–387). Oxford University Press.

Lavenex, S. (2001). Migration and the EU's new eastern border. *JEPP, 8*(1), 24–42.

Leblang, D. (2010, August). Familiarity breeds investment: Diaspora networks and international investment. *American political science review, 104*(3), 584–600. https://www.jstor.org/stable/40863770.

Mayda, A. M., Peri, G., & Steingress, W. (2018). *The political impact of immigration: Evidence from the United States*. NBER Working Paper No. 24510, Issued in April 2018 NBER Program(s): International Trade and Investment, Political Economy.

Murphy, C. N., & Tooze, R. (1991). *The new international political economy*. Lynne Rienner Publishers.

Ortega, F., & Peri, G. (2009). *The causes and effects of international migrations: Evidence from OECD countries 1980–2005*. NBER Program(s): International Trade and Investment, Labor Studies, NBER Working Paper No. 14833, Issued in April 2009, NBER Program(s): Economic Fluctuations and Growth, International Trade and Investment, Labor Studies.

Ortega, F., & Peri, G. (2012). *The effect of income and immigration policies on international migration*. NBER Working Paper No. 18322, Issued in August 2012.

Peri, G. (2012a). The effect of immigration on productivity: Evidence from U.S. states. *Review of Economics and Statistics, 94*(1): 348–358.

Peri, G. (2012b). Immigration, labor markets, and productivity. *Cato Journal, 32*(1), 35–54.

Peri, G. (2013). The economic consequences of the proposed immigration reform. *Intereconomics, 48*(3), 191–192.

Rodrik, D. (2017). *Populism and the economics of globalization*. https://drodrik.scholar.harvard.edu/files/danirodrik/files/populism_and_the_economics_of_globalization.pdf

Rodrik, D. (2018). *The political economy and its future, Harvard module followed by the author*. https://drodrik.scholar.harvard.edu/classes/political-economy-and-its-future.

Ruhs, M. (2013). *The price of rights: Regulating international labor migration*. Princeton University Press.

Sassen, S. (1998). *Globalization and its discontents: Essays on the new mobility of people and money*. The New York Press.

Shin, A. J. (2017). Tyrants and migrants: Authoritarian immigration policy. *Comparative Political Studies, 50*(1), 14–40.

Singer, D., & Mosley, L. (2015). Migration, labor, and the international political economy. *Annual Review of Political Science, 18*, 283–301.

Soysal, Y. (1998). Towards a post-national model of membership. In G. Shafir (Ed.), *The citizenship debates: A reader* (pp. 189–220). University of Minnesota Press.

Soysal, Y. N. (2000). Citizenship and identity: Living in diasporas in post-war Europe? *Ethnic and Racial Studies, 23*(1), 1–15.

Soysal, Y. N. (2012). Citizenship, immigration, and the European social project: Rights and obligations of individuality. *The British Journal of Sociology, 63*(1), 1–21.

Soysal, Y. N. (2015). *Transnational trajectories in East Asia: Nation, citizenship, and region*. Routledge Taylor & Francis. 9781138819351.

Thielemann, E. R. (Ed.). (2003). European burden-sharing and forced migration. *Journal of Refugee Studies, 16*(2) (special issue).

Thielemann, E. R. (2004). Why European policy harmonization undermines refugee burden-sharing. *European Journal of Migration and Law, 6*(1), 43–61.

Weiner, M. (1980). *Security, stability, and international migration*. MIT. https://www.files.ethz.ch/isn/19789/Security_Stability_Migration.pdf. Accessed 5 May 2020.

Weiner, M. (1995). *The global migration crisis: Challenge to states and to human rights*. HarperCollins College.

CHAPTER 2

The Three Paradoxes of Globalisation and Migration

In this chapter, the definitions and dynamics of globalisation will be discussed with the aim of identifying the impact of globalisation on migration. The chapter will proceed by first addressing the recent developments in debates on globalisation, specifically the ones relating to whether globalisation (still) exists. This will be predicated around the identification of different definitions of globalisation, notably a quantitative one and a qualitative one. After discussing, from the different IPE perspectives, the likelihood of de-globalisation and its significance, it will move to the identification of the main consequences of globalisation, with an eye to the ones more strictly relevant to the international movement of people. The focus will be on the New Global Division of Labour and Power and its consequences on migratory processes. Finally, the globalisation approach to migration will be presented by elucidating its origins in the sociological debate and then moving to the transnationalist IPE perspective.

Concluding, by adopting a qualitative definition of globalisation, the globalisation thesis hypothesises that the structural dynamics in place in the global political economy produce the structural need for populations to move both within regions and extra-regionally. This is articulated around three paradoxes that impact on the decision of people to migrate:

L. S. Talani, *The International Political Economy of Migration in the Globalization Era*, International Political Economy Series,
https://doi.org/10.1007/978-3-030-79321-0_2

- The paradox of "marginalisation" and its impact in terms of increase of extra-regional permanent migration and brain drain.
- The paradox of "regionalisation" and its consequences in terms of intra-regional temporary migration.
- The paradox of "Securitisation" and its consequences in terms of irregular migration.

Such paradoxes follow from the structural nature of globalisation and the emergence of a New Global Division of Labour and Power, notions to which we turn in the next sections of this chapter.

Does Globalisation (Still) Exist? Between Sceptics and the De-Globalisation Debate

The global financial crisis of 2007–2008, and following events such as the great recession and euro-zone crisis, gave rise to a flourishing research agenda on de-globalisation and whether globalisation did ever or does still exist. Looking at IPE theorists, it is possible to identify three distinct stances in the globalisation debate (Dicken, 2015: Chapter 1). The first position includes those authors who are unconvinced about the peculiar nature of globalisation and are generally considered 'sceptics' (e.g. Hirst & Thompson, 1999a; Hirst et al., 2009). A second group of IPE scholars, on the contrary, are known as being (hyper)globalists (Dicken, 2015) in that they identify this specific phase of capitalist development as a truly global phenomenon, although they tend to use a quantitative approach to define it (e.g. Garret, 1998; Held & McGrew, 2000, 2007[1]; Holm & Sørensen, 1995). On the other hand, the so-called 'transnationalists' underline the structural nature of globalisation and revert to a qualitative definition (e.g. Dicken, 2015; Hay & Marsh, 2000; Mittelman, 2011).

Sceptical accounts of globalisation notice how, historically, there have been various waves of increased international and cross-border interactions and how, in relative terms, these were even more intense than the current one. As such, globalisation loses any original features and becomes a recurrent historical phenomenon governed by the state. This

[1] Globalisation Theory: Approaches and Controversies, David Held and Anthony McGrew, eds., Oxford: Polity, 2007.

is a typical neo-realist reading of the current phase of capitalist development. Indeed, from this perspective, globalisation is mainly a myth, devised and utilised by the most powerful states in order to consolidate their hegemonic position in the international system. Thus, contrary to widespread belief, globalisation is neither a major problem, like the anti-globalists would maintain, nor a major solution, like neoliberals would say. In sum, the importance of globalisation and its characterisation as a new phenomenon are, according to neo-realists, grossly overstated in the public and academic debate, perhaps on purpose, to allow for this narrative to spread and for hegemonic states to increase their power. In reality, however, nation states are still firmly in charge of the international system which continues to mostly rely on traditional intergovernmentalist solutions (Gilpin, 2000; Hirst et al., 2009). Paradoxically, considering that the United States is currently still at the centre of the international system, rhetoric about globalisation can, from this perspective, acquire characteristics similar to the discourse on westernisation or even imperialism.

Neo-realists rely on a quantitative definition of globalisation as underlined by Dicken (2015: 5). This definition is based on measuring the various elements normally included in the globalisation process, such as migratory and trade flows, social and economic interactions, financial transactions. By effecting a comparison between an ideal notion of the globalised economy and the traditional internationalised economy, neo-realists conclude that the measures obtained do not justify talking about the existence of a truly globalised economy (Hirst & Thompson, 1999a, 1999b, 2009).

Not dissimilarly, in his early work Gilpin concluded that there is not such a reality as a global political economy (2000, 2001). On the contrary, the world economy continues to be firmly based on the nation state which is still in charge of devising the legal and political framework for its national and international activities. Perhaps, these are relying more heavily on regional governance schemes, but they are still aimed at maximising the interests of hegemonic national state(s). Also for Gilpin (2001) the definition of globalisation is a quantitative one, and his conclusion is that this is 'highly uneven, restricted to particular economic sectors and not nearly as extensive as many believe' (Gilpin, 2001: 364). No doubt, for him, other historical periods, such as the gold standard, saw much higher levels of economic integration. In relative terms, even in the late nineteenth century, financial transactions and trade flows were much greater than today (Gilpin, 2001). Concluding, with respect to previous periods, in the current one, more than a tendency towards globalisation, one can identify a clear trend towards regionalisation led by powerful regional states (Chapter 3).

Moreover, according to Gilpin, financial integration is far overstated in the literature as, in reality, this is only referring to speculative capital and short to very short term investment. In fact, there is no evidence that the 'law of one price' is being respected worldwide. Therefore, as this is a very reliable measure of economic integration, it is evident that we are far from achieving a truly global market (Gilpin, 2001: 368).

Other neo-realist scholars, such as Hirst et al., (2009), Hirst and Thompson (1999a, 1999b), insist on the idea of globalisation as a powerful myth. In their conceptualisation, globalisation is not even a process, but simply a rhetorical devise allowing the most powerful states to control others especially in the regional context. This is why there is a tendency towards regionalisation, albeit one which is dominated by regional hegemons. These are the US in the American region, Japan and increasingly China in Asia and Germany and France in Europe. As further discussed in Chapter 3, this tendency is clear when discussing regional governance schemes for migration and has been recognised by both neo-realist and neo-institutionalist scholars. However, whereas the latter believe that regionalisation is a step towards globalisation, neo-realists have no teleological understanding of this process and deny that we are moving towards an integrated global political economy. Hirst et al. (2009) reach this conclusion after contrasting the ideal-type of a globalised economy with that of an internationalised one where 'the principal entities remain national economies, or agents that continue to be primarily located in a definite national territory' (Hirst & Thompson, 1999b: 140).

Still relying on a quantitative definition of globalisation, one of the measures used to verify the extent of economic integration is given by, for example, the stock of foreign-owned productive capital. This measure moved from only 10.1% of the global GDP in 1995, to 22.7% in 2005, still not enough to talk about a truly globalised economy. Moreover, looking at the contribution of inward FDI to domestic fixed capital formation, these do not appear to be particularly relevant (Hirst et al., 2009: 78). Much more revealing, in the opinion of these scholars, is the investment of multinational companies in their own countries of origin. Using such an indicator to assess to what extent multinational companies are still embedded in their nation states, the result is that they remain substantially nationally and regionally centred (Hirst et al., 2009: 83–84). There remains the question of the off-shoring of production and the creation of global value chains, a measure of global integration that can hardly be denied, although Hirst et al. claim that this is often overstated and it is still very much centred in western countries (Hirst et al., 2009: 92: Fig. 3.6).

On the other hand, indication of the continued importance of the nation state for the economy is given by the extent to which even financial markets still have to abide to the legal framework of the countries where they are physically located. Moreover, their assets are mostly raised within national territories, they provide credit to national citizens and are constrained by national or at best regional macroeconomic policies, exchange rate regimes and interest rates (Hirst & Thompson, 1999a). This is far from configuring a situation of uncontrollable financial markets which are deemed to be a feature of globalisation, while, in reality, financial institutions are subject to national political, legal and macroeconomic frameworks. As a consequence, there is no better way to address their governance than to rely on intergovernmental cooperation (Hirst et al., 2009: Chapter 8).

Overall, neo-realists coincide in believing that the notion of globalisation is a myth in the hands of the most powerful nation states devised to perpetrate their hegemony in the regional and international context. Therefore, the only way to address the shortcomings of the world economy is to rely on traditional intergovernmentalist institutions.

To be sure, this belief is the consequence of adopting a characteristic quantitative definition of globalisation which fails to identify the structural nature of the phenomenon and the underlying structural links between its various components. Once the measurement of the elements identified within the globalised economy does not confirm its existence, the logical conclusion is that it is not a reality but a myth.

On the other side of the spectrum, neo-institutionalists totally endorse the globalisation process and are therefore referred to as (hyper) globalists. The sceptical approach of realists is here completely reversed. While the former had considered globalisation a myth to justify the imposition of the hegemony of the most powerful nation states, for the latter the current globalisation phase is a distinct and original manifestation of transnational economic integration (Held & McGrew, 2007).

This conclusion is achieved by adopting a quantitative definition of globalisation, similarly to neo-realists, which however, in the case of neo-institutionalists provides ample evidencc of the actuality of the phenomenon. Data and indicators of globalisation not only corroborate its existence in neo-institutionalist analyses, but also confirm that global economic integration is well advanced and produces 'transformations' at the level of the national, regional and global governance systems. It does, therefore, require new institutional solutions at all of these levels.

Globalisation is again defined in terms of an increase in its various constituent components. Keohane and Nye (2012), for example, talk about the intensification of both international and transnational relations. David Held notices that globalisation 'suggests a growing magnitude or intensity of global flows such that states and societies become increasingly enmeshed in worldwide systems and networks of interaction' (Held & McGrew, 2000: 3). For Phil Cerny globalisation is manifested through both internationalisation, i.e. denser economic relations amongst states, and transnationalisation, which are denser relations cutting across states (1999, 2010).

Such an intensification of both international and transnational interactions is, for neo-institutionalist undeniable on the basis of the evidence available and this makes the current capitalist phase a completely new and distinct one from any previous ones. In the current phase of the global political economy not only are there more interactions between states, but also more activities that transcend the level of states to acquire a truly global dimension. The origins of this current phase are, for Held, to be traced back to the end of the Second World War in 1945. The difference between this wave of internationalisation and the ones historically preceding it can be identified, according to Held, by looking at four measurable and quantifiable spatio-temporal dimensions: extensity, intensity, velocity and reciprocity between local and global development (Coleman & Sajed, 2013: 122–123).

By analysing and measuring these four dimensions it is possible to categorise globalisation in four different ways. When the four dimensions all score very high we have 'Thick Globalisation'. When the first three dimensions are high but reciprocity is low, we have 'Diffused Globalisation'. With only high extensity and intensity there is 'Expansive Globalisation'. 'Thin Globalisation' occurs when the extensity of connections is high, but the other three dimensions score low. We are now, in the opinion of Held, clearly experiencing 'Thick Globalisation' (Coleman & Sajed, 2013: 122–123).

The adoption of a quantitative definition of globalisation, which, as noticed, is typical of both mainstream IPE approaches, brings about the possibility of a decline in its indicators. This would configure a situation of de-globalisation. Concentrating on a numerical definition of globalisation, some scholars in IPE have rekindled the debate on 'the end of globalisation' or 'the end of the world as we know it' (Dicken, 2015: 1). Of course, neo-realists would dispute that globalisation ever existed. As a consequence, it would be inappropriate for them to talk about de-globalisation.

From a quantitative point of view, de-globalisation is, as quoted in Kaminska (2015), evident, '[g]lobalisation is dead (or at least very ill). (...) Capital account globalisation has reversed to levels seen in the early 1990s'.[2]

Barbieri (2017)[3] proposes an interpretation of de-globalisation from a liberal institutionalist perspective. Based on a neo-institutionalist definition of globalisation, he notices that globalisation's first wave took place between 1880 and the early 1930s. He underlines that de-globalisation was what followed the 1929 crash of Wall Street, and the policies adopted by the US and the UK to recover their economies, as analytically discussed by James (2001).[4] The consequences of these policies were, in Barbieri's opinion, first of all economic ones, with a reversal of the increase in trade and financial flows globally and the impoverishment of the world as a whole. However, he also puts the blame on de-globalisation, and its ensuing social discontent, for the rise of Nazism in Germany, Mussolini's fascism in Italy and the demise of the international institutional system of the League of Nations (Barbieri, 2017). In the words of the author, '[o]n the whole, the world was a lot poorer when the United Kingdom and the United States prioritized their own economic recovery without regard for

[2] Kaminska, I. (2015, September 24). *How do you solve a problem like de-globalisation?*. FT Alphaville. https://ftalphaville.ft.com/2015/09/24/2140786/how-do-you-solve-a-problem-like-de-globalisation/.

[3] Barbieri, P. (2017, November 13). The losers of deglobalisation. *Foreign Affairs*. https://www.foreignaffairs.com/articles/2016-11-13/losers-deglobalisation. Accessed 3 October 2019.

[4] *The end of globalisation: Lessons from the Great Depression*. Harold James Harvard University Press, 2001.

the international system' (Barbieri, 2017). A similar analysis is proposed by James (2017).[5]

Thus, Barbieri (2017) identifies a parallel between the Great Depression and the Global Financial Crisis of 2007–2008 with its ensuing austerity policies which produced a wave of political discontent leading to the success of populism in the US with Trump (Inglehart & Norris, 2016; Rodrik, 2017; see also Chapter 9)[6] and in the UK with Brexit (Farrell & Newman, 2017).[7] The causes and consequences of populism, especially on migration, are the subject of the last chapter of this book. Here it suffices to say that it is debatable whether Brexit and Trump can be properly considered populist phenomena. Moreover, as discussed below, a structural definition of globalisation would dispute first that in the period between 1880 and the early 1930s we can actually talk of globalisation and second, that the notion itself of de-globalisation is based on a numerical definition of the event.

In fact, it is precisely the insistence on measurable phenomena which leaves unsolved a number of questions. These relate first of all to the nature of the relationships existing between the economic, political and social dimensions of globalisation. Furthermore, insisting on quantitative elements does not help identifying the causes and consequences of globalisation, making it extremely difficult to relate it to other phenomena, such as social and geographical marginalisation and political upheavals. It is therefore necessary to deepen the perspective by adopting a qualitative definition.

[5] James, H. (2017, September 12). *Deconstructing deglobalisation*. Project Syndicate. https://www.project-syndicate.org/commentary/three-emotions-underlying-deglobalisation-by-harold-james-2017-09.

[6] Inglehart, R. F. & Norris, P. (2016). *Trump, Brexit and the rise of populism: Economic have-nots and cultural backlash*. Faculty Research Working Paper Series, RWP16-026. Cambridge, MA: Harvard Kennedy School of Government.

[7] Farrell, H., & Newman, A. (2017). BREXIT, voice and loyalty: Rethinking electoral politics in an age of interdependence. *Review of International Political Economy, 24*(2), 232–247.

Moving to a qualitative approach to globalisation allows us to overcome the question of relating the existence of the phenomenon to the intensification of its components. In fact, the numerical dimension becomes irrelevant as the process of globalisation is a process of structural transformations. These are ontologically related to each other and their ensemble concurs to identify this historical period as unique and not repeatable (Dicken, 2015; Mittelman, 2011). The different components of the globalisation process, and their relations to each other, prompt some authors to talk about a globalisation 'syndrome' (Mittelman, 2011). There is, in fact, an intrinsic connection between each element of globalisation which does not, however, configure a cause-effect relationship, but more a dialectical one, which still allows for the identification of relevant trends and possible reactions. Indeed, there is a structural link between technological innovation, which is often considered in the literature the engine of the process of globalisation, and the transformations affecting both the productive sphere, through the creation of global value chains, and the financial one, with the 24/7 digital access to financial markets all over the globe. In turn, the geographical restructuring of production brings about a 'New Global Division of Labour and Power' with all that means in terms of geographical and social inequalities and marginalisation, as discussed in the next section. The structural transformations described above, impact on the nature and the role of the institutional system, both national and international, leading to structural global inequalities, commodification, migratory flows and the subordination of the political dimension to the economic one (Overbeek & Léon, 2015; Mittelman, 2011).

In the opinion of the author of this book, at the root of transformations induced by globalisation is technological progress which is the driving force behind the process of the global restructuring of production and its related geographical reallocation (similarly Overbeek & Léon, 2015; contrary Dicken, 2015: 6). It is the change in the technological paradigm which makes it convenient for the business sector to produce abroad creating global commodity chains. It is indeed the first time in history that, thanks to technological progress, it makes economic sense to produce the different components of goods in different coun-

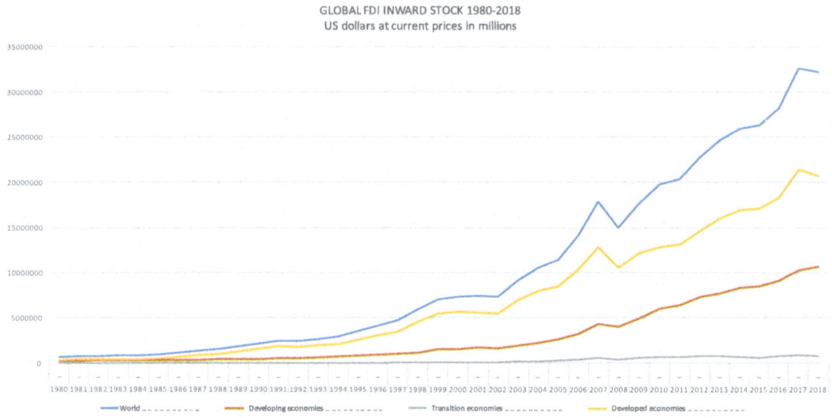

Fig. 2.1 Global stock of FDIs 1980–2018 (Millions of Us $ at current prices) (*Source* UNCTAD web-site: https://unctadstat.unctad.org/wds/TableViewer/tableView.aspx)

tries, exploiting cost-reducing economies of scale, clusters of expertise and lower labour costs. This happens through foreign direct investment (FDIs), mergers and acquisitions (M&As) and the creation of export processing zones (EPZs) (Mittelman, 2011).

Understood this way, it does not seem plausible that the global financial crisis reversed the globalisation process as it did not touch its structural components (Dicken, 2015: 5). To be sure, even a cursory look at the global stock of FDIs confirms that the level is constantly increasing, in both developing and developed countries (see Fig. 2.1). Indeed, in 2018, ten years after the global financial crisis, the stock of world FDIs, representing structural investment in foreign countries, was 80% more than the peak in 2007 and more than double what it was in 2008.

From the qualitative point of view, structural de-globalisation can only happen in cases of a "reversal" of the technological changes that led to the structural modifications of the production and financial spheres of the economy. This scenario is highly unlikely, unless the widespread alarm about the environmental crisis becomes a reality and global warming provokes a disaster of such proportions as to turn back the technological clock significantly (Malm, 2015[8]; Moore, 2016[9]).

Much more likely is that the progress in technology will lead to a new historical phase of capitalism, a sort of paradox of relocalisation within globalisation, where physical movement from one geographical location to another for the purpose of production will become less necessary and cost-effective. This could be the outcome of the increased use of Artificial Intelligence and robots in productive processes, as well as the widespread use of three-dimensional printers, or the already common practice of distance-working. The COVID-19 crisis has most likely hastened these tendencies. However, the analysis of these dynamics exceeds the purpose of this discussion.[10]

The next section will concentrate on the consequences of globalisation to identify its relation to migratory processes.

[8] Malm, A. (2015). *Fossil capital: The rise of steam-power and the roots of global warming*. London: Verso.

[9] Moore, J. (2016). *Anthropocene or capitalocene? Nature, history, and the crisis of capitalism*. Oakland, CA: PM Press.

[10] For a discussion of a similar scenario see *The Economist*. (2020). web-site https://www.economist.com/special-report/2019/07/11/supply-chains-are-undergoing-a-dramatic-transformation?al_applink_data=%7B%22target_url%22%3A%22https%3A%5C%2F%5C%2Fecon.st%5C%2F2P4FQWl%22%2C%22extras%22%3A%7B%22fb_app_id%22%3A237759909591655%7D%2C%22referer_app_link%22%3A%7B%22url%22%3A%22fb-messenger%3A%5C%2F%5C%2F%5C%2F%3Fapp_id%3D237759909591655%22%2C%22app_name%22%3A%22Messenger%22%7D%7D.

The Consequences of Globalisation on Migration and the New Global Division of Labour and Power[11]

The debate in IPE about the consequences of globalisation differs according to the different approaches analysed. On one extreme of the

[11] *The new international division of labour global transformation and uneven development*; Authors: Starosta, Guido; Editors: Charnock, Greig, Starosta, Guido (Eds.); Palgrave, 2016. See also Amiti, W., & Wei, S. J. (2006). Service offshoring and productivity: Evidence from the United States. National Bureau of Economic Research Working Paper No 11926, Cambridge, Mass. Antkiewicz A., & Whalley, J. (2006). *Shifting global economic power to the BRICSAM from the OECD.* Centre for International Governance Innovation (CIGI), Waterloo, Ontario. (mimeo) Bhagwati J., Panagariya, A., & Srinivasan, T. N. (2004). The muddles over outsourcing. *Journal of Economic Perspectives, 18*(4), 93–114. Dreher, A., & Gaston, N. (2005). *Has globalisation really had no effect on unions.* Swiss Institute for Business Cycle Research Working Paper No. 110. Feenstra, R. (1998). Integration of trade and disintegration of production in the global economy. *Journal of Economic Perspectives, 12*, 31–50. Feenstra, R., & Hanson, G. H. (1996). Globalisation, outsourcing and wage inequality. *American Economic Review Papers and Proceedings, 86*, 240–245. Feenstra, R., & Hanson, G.H. (1997). Foreign direct investment and relative wages: Evidence from Mexico's Maquiladoras. *Journal of International Economics, 42*(3/4), 371–393. Feenstra, R., & Hanson, G. H. (1999). The impact of outsourcing and high-technology capital on wages: Estimates for the U.S., 1979–1990. *Quarterly Journal of Economics, 114*(3), 907–940. Gaston, N. (2002). Unions and the decentralization of collective bargaining in a globalized world. *Journal of Economic Integration, 17*, 377–396. Dumont, M., Rayp, G., & Willemé, P. (2006). Does internationalization affect union bargaining power? An empirical study for five EU Countries. *Oxford Economic Papers, 58*(1), 77–102. Grossman, G., & Helpman, E. (2002). Integration versus outsourcing in industry equilibrium. *Quarterly Journal of Economics, 117*(1), 85–120. Mundell, R. (1957). International trade and factor mobility. *American Economic Review, 48*(3), 321–335. OECD. (2006). *The internationalization of production, international outsourcing, and OECD labor markets*. Economics Dept., OECD, Mar 2006. (mimeo) Rodrik D. (1999). Globalisation and labor, or: If globalisation is a bowl of cherries, why are there so many glum faces around the table?. In R. E. Baldwin (Ed.), *Market integration, regionalism and the global economy*. Cambridge U.P. for CEPR. Tomiura, E. (2005). Foreign outsourcing and firm-level characteristics: Evidence from Japanese manufacturers.*Journal of the Japanese and the International Economies, 19*, 225–271. Whalley, J., & Xian, X. (2006). *China's FDI* and Non-FDI economies and the sustainability of future high Chinese growth. The University of Western Ontario. (mimeo) Panel 1 Whalley, J. (2006). *The post MFA performance of developing Asia.* (mimeo) Xiao, G. (2004). People's Republic of China's round-tripping FDI: Scale, causes, and implications. Asia Development Bank Institute Discussion Paper No. 7.

spectrum, neo-realists essentially do not believe that globalisation is a distinct phenomenon that cannot be controlled by national states. Therefore, any discussion about its consequences becomes redundant. This applies also to the consequences of globalisation on migration. Indeed, as already seen in Chapter 1, neo-realist scholars approach the question of the role of the state on migration by emphasising its unchanged ability to control and regulate it, even in the age of globalisation.

In fact, international migratory dynamics themselves are determined by the interests of the most powerful nation states, either in the security dimension, like in Weiner, or in the economic one, as Borjas and Freeman would say (Chapter 1).

On the other hand, the neo-institutionalist perspective asserts that the national level of governance is transcended by the supranational one as a consequence of the process of globalisation, a process whose actuality they firmly confirm. Indeed, the ontology of globalisation from this perspective is precisely the change in the role played by the national, as opposed to the transnational and international political system, and the progressive transformation of their respective institutional settings. In their view, the state does not succumb to globalisation, as the transnationalist would maintain, but is profoundly transformed. For this reason, neo-institutionalists are also often called transformationalists (Cerny, 2010).

In particular, Cerny (1999, 2010) notices how the state is modified by globalisation in the following two aspects:

1. The domestic aspect: the state changes its role in the economy and with respect to the provision of 'public goods'.
2. The international aspect: the state modifies its economic role in the international system which becomes more and more interdependent and transnational.

Domestically, and as a consequence of the creation of a complex system of transnational and international interdependence, the state loses its ability not only to provide public goods, but also to define them. In fact, the notion itself of public goods is modified by the new necessities created by the complex interdependence and emergence of a multi-layered institutional system (Heritier, 2002; Talani, 2003).

We can distinguish, with the literature, the impact of globalisation on the four main categories of public goods: regulatory, productive, distributive and redistributive (Cerny (1999, 2010).

Regulatory public goods are very revealing of the impact of globalisation on the ability of the state to provide them. These goods are represented by the regulatory system imposed by the state on its own institutions and operations. Most evidently, globalisation has profoundly limited the ability of the state to provide regulatory frameworks in many policy realms, from the financial one, to the environmental and migratory ones, to name a few. This regulatory function needs to be provided at a higher level of governance in order to guarantee its applicability and efficacy (Cerny, 1999, 2010: 9). In a nutshell, globalisation has shifted the level of governance from the national to the regional/international level, increasing the limits and constraints that supranational institutions can effectively impose on nation states (Held & McGrew, 2007).

In the case of productive goods, the impact of globalisation is most evident. Productive goods are those goods that the state needs to provide as it is the only entity able to exploit economies of scale. Typical examples are energy goods. However, in the globalisation era productive goods can now be produced and distributed by entities, such as multinational companies, whose dimensions often exceed the size of the nation state and which can therefore provide them in a more efficient way by exploiting even bigger economies of scale.

Other forms of public goods, such as distributive public goods, used to be provided by the State due to the collective nature of the users, like transport for example. Here it is the commodification of consumption ensuing from globalisation which makes the role of the State in their provision redundant (Cerny, 1999, 2010; Heritier, 2002).

Finally, in the case of redistributive public goods, the impact of globalisation is dramatic. These are goods provided by the state on the basis of political considerations, with redistributional aims in mind, typically the welfare state (Cerny, 1999; Heritier, 2002). Here globalisation deeply impacted the ability of the state to provide them by reducing state capacity to control its budgetary policy and expenditure. Having to compete internationally to attract business from abroad, and also as a consequence of the liberalisation of financial markets, the State is unable to allocate funds to spend on redistributive policies. Moreover, given global pressures on the labour markets, due to both geographical relocation of production and mass migration, the working class is increasingly fragmented and precarious which impairs the representational capacity of trade unions and, with it, their ability to intervene effectively in the bargaining process for the provision of the welfare state. In effect, the state's role has changed from that of providing the 'welfare state' to establishing the 'competition state' (Cerny, 1999: 101). The competition state is a state able to identify its dynamic competitive advantage, which substitutes the old notion of the 'comparative advantage' and implies that the state actively pursues policies to achieve it (Cerny, 2010: 32). Similar policies are aimed at providing a new kind of public good, defined as 'immobile factors of capital': infrastructure, human capital, technology and innovation, high educational standards and attractive living standards for the middle class (Cerny, 1999: 101).

In the international realm, the 'competition state' has to contend with the other states to attract business, mostly under the form of foreign direct investment, but also skilled labour and expertise, as well as technology and innovation. This deeply modifies states' ways of interacting with the international system, moving from the security agenda to the political economy one. This occurs not only in developed countries, but also in developing ones. In turn, the competition state, to win the race against other states to attract business, will have to adopt suitable policies by, for example, shifting from the macroeconomic policy-making to the micro-economic one to implement flexibility of the labour markets; cultivating their competitive advantage by adopting a typical neoliberal agenda based on low taxation; adopting liberalisation policies and low inflation with all that means in terms of curtailing expenditure and the welfare state;and promoting an efficiency agenda in both the private and the public sector with appropriate neoliberal labour policies (Cerny, 1999).

In terms of migratory policies, as already seen in Chapter 1, this means that liberal institutionalists believe that the nation state, in the globalisation era, is increasingly constrained by the international legal system in its ability to regulate migration. Moreover, migrants are actually attracted to liberal regimes, such as the EU, precisely because of the rights that they guarantee to them (Chapter 1).

A completely different approach to the definition of globalisation is the one proposed by the so-called transnationalists who stress the structural nature of this process. In their opinion, the kind of transformations brought about by globalisation does not refer only to the institutional dimension, as in the case of neo-institutionalists, but also to the modalities in which economic activities are organised, both in the realm of production and in the financial structure (Dicken, 2015; Mittelman, 2011; Overbeck, & Léon, 2015). To reach these conclusions, trasnationalists rely on a qualitative approach to globalisation, one which does not require to measure the phenomena which are included in it to verify whether it exists or not. The ontology of globalisation is instead represented by the presence of a series of interconnected structural transformations. In Peter Dicken's words:

> The crucial diagnostic characteristic of a 'global economy', therefore, is the qualitative transformation of economic relationships across geographical space, not their mere quantitative geographical spread. This involves 'not a single, unified phenomenon, but a syndrome of processes and activities'. (Dicken, 2015: 6)

Similarly, Mittelman (2000: 6; 2011) refers to the notion of a 'syndrome' to signify the concomitant presence of a number of interrelated structural phenomena characterising the ontology of the current phase of capitalist development.

Critically, the relationship between the different components of globalisation is not casual. They are, instead, connected in an ontological way, meaning that the existence of each one of them brings about modifications in all of the other elements of globalisation in a dialectical and historically determined way, as further elaborated below. Consequently, it is not appropriate to refer to this approach as "vague" as mainstream and economic theorists often do (see Hirst & Thompson, 1999a).

At the origin of this qualitative approach to globalisation are the extraordinary developments in technological progress experienced in the last decades. In the literature there is a debate about whether technological innovation is an exogenous component of globalisation, or whether it is driven by globalisation itself and, in particular, by the investment patterns of big transnational corporations moved by profit maximisation (Hay, 2002[12]; Dicken, 2015: 6). Hay (2002), for example, re-introduces agency in the debate about globalisation, but he mainly refers to the

[12] Hay, C. (2002). Globalisation as a problem of political analysis: Restoring agents to a "Process without a Subject" and politics to a logic of economic compulsion. *Cambridge Review of International Affairs, 15*(3), 379–392. Here, however, globalisation is discussed as exogenous whereas, in the analysis of this book, it is technological progress to be considered an exogenous component of globalisation.

actors behind the project of globalisation itself, not to those determining technological progress. However, here is not the place to address such an essential question, which has already been tentatively discussed by the author elsewhere (Talani, 2014). For the purposes of the present analysis, it is suffice to say that the current level of technological progress has brought about structural transformations in the two most important dimensions of the economy: the productive sphere and the financial one (Mittelman, 2011; Talani, 2012). Those transformations have also modified the modalities of functioning of the social and the political realms.

Starting with the financial structure, it is thanks to technological innovation that the working of financial markets has been completely transformed. It is now possible to access markets 24 h a day, seven days a week, all over the globe at the click of a mouse (or, perhaps, more recently, at the blink of an eye!) (Strange, 1996).

It is because of this structural transformation that capital movements have become much more sensitive to variations in interest rates globally, rendering national macroeconomic policies less effective. For example, national attempts to fix interest rates in order to stimulate growth of GDP and employment are impaired by the tendency of financial capital to move towards high interest rates countries. This also produces changes in the ability of the state to control their exchange rate policy. Overall, the nation state's grip on economic policies is substantially curtailed, enhancing the power position of financial markets. What is important to notice here, however, is that this is the direct consequence of technological changes and that without them financial liberalisation would not make any economic sense (Gill, 1997; Padoa-Schioppa, 1994; Obstfeld & Taylor, 2004).

This is what explains the crisis of the nation state and its subordination to economic forces in both the developed and the developing world. There is no scope to downplay the constraints imposed by globalisation on the ability of the nation state to implement effective macroeconomic policies. The impact of such a situation can risk creating serious political instability leading to unrest and even regime change (Overbeek, 2000; Van der Pijl, 2011). Moreover, different to neo-institutionalists, the structural nature of these phenomena make trasnationalists believe that international organisations will not be able to overcome the power of economic forces through the creation of a global economic governance system, at least not effectively. The ensuing crisis of legitimacy and authority of political institutions, both at the national and at the supranational level, is deemed by transnationalists as an extremely worrying consequence of the subordination of politics to economics due to globalisation and the loss of control over economic policy-making (Overbeek, 1999, 2000, 2015; Van der Pijl, 2011).

Similarly, structural changes in the production process are to be referred to technological progress. Thanks to the digital revolution, the reduction in transport prices and communication, it is now profitable to produce different components of goods in different, even very distant locations. This is behind what is known as the geographical relocation of production and the creation of global value chains. The main modalities allowing this process are foreign direct investment (FDIs), mergers and acquisitions (M&As) and the creation of export processing zones (EPZs). Thanks to technological progress, it is possible for transnational corporations to exploit global production chains in a profit maximising way by relying on bigger economies of scale, lower costs of transport and distant labour control. As in the case of financial globalisation, although in a less dramatic fashion, the power of the nation state, but also of international institutions for that matter, to control and constraint transnational corporation is much curtailed. In the opinion of the author, although, obviously, neoliberal policies aimed at liberalising trade are very important for the functioning of this new system of production, without technological changes they would not be sufficient to produce globally as this would be inconvenient. Equally, it is highly possible, and it is actually already happening, especially after the COVID-19 crisis, that global value chains will be shortened, and the geographical relocation of production reversed in part or even entirely, as a consequence of new production technologies, such as AI and 3D printers, making it more cost effective to produce locally. This would lead to a sort of paradox of relocalisation within globalisation.

To be sure, globalisation is not an even process. The creation of inequalities and marginalisation is again a structural component of globalisation, as is its tendency towards regionalisation (Dicken, 2007: 33; Mittelman, 2000: 41). Regionalisation is a structural phenomenon of geographical relocation of production. For reasons relating to minimisation of costs and maximisation of profits, this tends to happen in the regional context, although there are different patterns of regionalisation (see Chapter 4). Moreover, regionalisation is not happening all over the globe. There are regions that are not integrated within themselves nor in the global political economy, as further discussed in Chapter 3 with the case of the MENA region. This leads to what the author, following the literature, calls the paradox of marginalisation within globalisation (Dicken, 2007; Mittelman, 2000: 56).

The uneven nature of the globalisation process, including regionalisation and marginalisation, leads to what Mittelman defines as the new Global Division of Labour and Power (GDLP) (Mittelman, 2000, 2011: 188).

This notion entails the creation of global commodity chains,[13] alongside regional patterns, allowing for the exploitation of cost reduction opportunities, economies of scale or clusters of expertise, made possible by technological progress. This configures the creation of a global market for labour, which relies also on brain drain and mass migration from regions not included in the globalisation and regionalisation processes (Mittelman, 2011; Overbeek, 2015). The overall effect of this GDLP is a reduction of the cost of unskilled labour in the Global North. Indeed, the dynamics of the Global Division of Labour and Power lead to lower costs of labour in the Global North thanks to both the geographical displacement of production and also the hiring immigrants, especially undocumented ones, in host economies. This is compounded by pressures on the costs of labour exerted by the reduced representativeness and ability to influence the national wage policy-making by Trade Unions, thanks also to constitutionalisation of disciplinary institutional arrangements at the national or at the regional level (Gill, 1997). This conclusion is now rather consensual in the literature and clearly demonstrated by the performance of the wage share of global GDP in the last two decades (ILO & OECD, 2015; Piketty, 2013).[14]

The idea of the GDLP was proposed by Mittelman to supplement the notion of the new international division of labour (NIDL), devised with the aim of explaining the displacement of production from advanced capitalist to developing countries, a phenomenon dating back to the second

[13] As originally defined by Hopkins and Wallerstein (1986: 159), a commodity chain is "a network of labor and production processes whose end result is a finished commodity".

[14] ILO and OECD. (2015); The Labour Share in G20 Economies International Labour Organization Organisation for Economic Co-operation and Development with contributions from International Monetary Fund and World Bank Group Report prepared for the G20 Employment Working Group Antalya, Turkey, 26–27 February 2015.

half of the twentieth century (Mittelman, 2011: 187–188). NIDL theorists (Caporaso, 1987; Fröbel et al., 1980; Lipietz, 1985; Mishra, 1999) explained this geographical shift in manufacturing as a consequence of the fragmentation of production made possible by new technologies, especially in transport and communications. On the other hand, skilled jobs and R&D activity would be retained in developed states (Mittelman, 2011: 191–192).

The combination of the displacement of labour-intensive production processes to low-wage countries and increased specialisation in R&D and innovation in the developed economies, often leading to 'jobless growth' there, produces a specialisation of the Global North in highly skilled labour and of the Global South in unskilled workers. It also leads to an overall structural surplus of unskilled labour (Antonelli & Fassio, 2013; Sassen, 1996; Martin, 2005 provides an excellent and detailed overview of developments in this period; Overbeek & Léon, 2015).

The NIDL idea, however, has been criticised for being excessively economically deterministic and giving too much importance to the exploitation of cheap labour in explaining the decision of capital to invest and produce abroad (Mittelman, 2011). Moreover, it also underestimates the role played by migration in allowing for the import of cheap labour in national economies as well as taming the demands of the local unskilled labour force (Overbeek & Léon, 2015). Finally, it neglects the phenomenon of brain drain (Kelly, 1989 as mentioned in Mittelman, 2011).

The notion of the GDLP, as proposed by Mittelman, on the contrary, ties together all the dynamics above by relating them to the structural changes implied in the current phase of capitalist development, the globalisation era. Indeed, globalisation, by simultaneously restructuring global production and global power relations, both geographically and socially, creates new poles of attraction for both skilled and unskilled labour to move to. Workers from marginalised zones of the globe try to escape to regions more integrated in the global political economy where opportunities to find jobs exist. This reflects the uneven nature of globalisation, creating a hierarchy among regions, countries and social groups (Mittelman: 195).

In the words of the author:

> In brief, the GDLP involves a restructuring among world regions, including their constituent units, notably states, cities, and the networks

> that link them. Another element of reordering is massive transfers of population from the global South, Eastern Europe, and the former Soviet Union to the West, though, as noted, there are also significant migratory flows within the South. Acting as magnets attracting imports of labor, global commodity chains form networks that interlink multiple production processes, as well as buyers and sellers. (Mittelman, 2011: 188)

Another usually neglected dimension of the new global labour dynamics is the "informalization of the labor supply" (Mittelman, 2011: 198), directly related to the structural increase in economic migratory flows as a consequence of marginalisation within globalisation (similarly Overbeek & Léon, 2015). To be sure, one of the contradictions of the current historical phase is that, while trade, finance and services are increasingly liberalised, labour mobility between the Global South and the Global North is increasingly restricted, despite new structural incentives to migrate (Overbeek & Léon, 2015). Indeed, little evidence exists that the liberalisation of migration is high on the international agenda while the tendency towards the adoption of more restrictive national, as well as regional migration policies is well documented (see Chapters 4 and 5). This is the essence of what the author of this book calls "the paradox of securitisation within globalisation". Selected entry of highly skilled labour or migrants with skills in demand by the host country is usually allowed legally, while those seeking to enter without the required set of skills do so through illegal means. It has been noticed that this system produces both the disciplining of labour in the domestic context and a very selective mobility of labour internationally, thus creating hierarchies (Overbeek & Léon, 2015). Moreover, this system incentivises the systemic involvement of smuggling networks and international criminal organisations in the management of the irregular migratory flows. Due to their irregular status, then, illegal migrants are typically employed in the informal economy, thus adding to their marginalisation in host societies (Mittelman, 2011: 198, Chapter 6 this book).

Modifications in the productive sphere bring about changes in both the social and political dimension at all their organisational levels (Dicken, 2011; Mittelman, 2000; Overbeek, 2000). As noticed above, transnationalists underline how globalisation structurally entails the subordination of politics to economics resulting in a crisis of authority of the nation state. This does not mean that the state is substituted by other institutions, such as supranational or transnational institutions as the neo-institutionalists believe. The idea is that, given the new structural organisation of economic forces, the state is increasingly more dependent on the policy choices of transnational capital, like transnational corporations. In the new global mode of production, states are competing against each other to attract foreign direct investment and capital from abroad and market forces acquire a much greater bargaining power (Overbeek, 2000). This is again a structural consequence of globalisation as it makes it possible for production to be effectively spread globally thus diminishing the ability of the national political level to control and constrain transnational capital. In a related way, the state loses the ability to guarantee the prosperity and welfare of its citizens, which brings to political and social tensions as elaborated further in Chapter 9 (Rodrik, 2017). In addition, as technological development increases the empowerment and reach of civil society, globalisation can also foster attempts at direct democracy, especially those utilising digital platforms, which may further de-legitimise the state, even bringing to authoritarian outcomes (Chapter 9; Gerbaudo, 2019a, 2019b; Mudde, 2004; Mittelman, 2000; Norris & Inghlhart, 2018, 2019; Rodrik, 2018).

Apart from the risks of populist forms of direct democracy based on digital platforms, the literature also notes opportunities for democratisation ensuing from the loss of power of the state and the increased role of civil society in the globalisation era (Mittelman, 2000: 27–28).

Given the dwindling power of the nation state and its increasing necessity to acquire a better position in the new GDLP by adopting more and more capital friendly policies, there is a need to identify a different geopolitical fix above the level of the nation state (Mittelman, 2011: 203). Considering the transnational nature of capital, the geopolitical fix is increasingly moved from the national to the regional level of governance, where, as already noticed, the majority of productive activities concentrate. Hence an increasing trend towards regionalism, i.e. the institutionalisation of the regional level of governance, in parallel with the economic regionalisation process (Harvey, 2011: Chapter 4).

The regional level becomes the locus of the changing GDLP as it represents the site of the reorganisation of production and therefore also the catalyst for migratory flows. This makes it important to distinguish between intra-regional and extra-regional movements of people as done in this book. Indeed, whereas intra-regional migratory movements are connected to regional integration processes, extra-regional ones are related to instances of marginalisation and the lack of regional integration. Following Mittelman (2015: 204):

> Quite clearly, globalisation is an uneven process (…) Given the disparities between global regions and marginalized regions, there are different globalisation scenarios. While the former are riding the waves of globalisation, the latter are driven by its currents.

Overall, in the transnationalist view, the process of globalisation is characterised by contradictions and disparities, which generate new challenges and opportunities. Not everyone is due to win from it and redistribution of gains is made difficult by the enhanced role of economic forces in defining national, regional and international politics (Mittelman, 2000, 2011). In the next section, the relationship between globalisation and international migration is discussed with reference to the debate in sociology.

Discussing the Relation Between Globalisation and Migration in Sociology

As far as the relationship between globalisation and migration is concerned, the contribution of the sociological perspective is substantial.[15]

In its most basic formulation, the globalisation thesis of migration posits that the transnationalisation of societies and communities followed a process of globalisation of the economy at the end of last millennium. With economic globalisation and the transnationalisation of social relations came a structural increase in international migration and the related loss of border control which is seriously questioning traditional notions of citizenship and sovereignty. Thus, migration is considered as a structural component of globalisation and the state loses its centrality in the international system to be superseded by transnational dynamics. Hence, the use of the term "transnationalists" to define the scholars proposing this approach.[16]

[15] See, for example, Sassen (1991, 1996, 1999, 2006).

[16] See, for example, Sassen (1991, 1996, 1999, 2006).

The globalisation thesis of migration was originally devised by sociologists, but the notion used in this book differs somehow from their understanding, as analysed above.[17] From the sociological standpoint, Castles' early work focused in particular on policy failure with respect to migration.[18] His article, "Why Migration Policies Fail" originally published in Ethnic and Racial Studies in March 2004, proposed a complex understanding of both the forces driving international migration and the policy-making processes at the national level. In particular, Castles' article pointed to three reasons for policy failure, of which one was explicitly related to globalisation and the North–South divide, in line with the focus of this book.[19]As Castles himself states "[t]he North–South divide generates migration" (Castles, 2004: 210). And, quoting again, "Today, the most crucial borders are no longer those between nation-states, but those between North and South: that is, between the powerful nations of North America, Western Europe, Japan, and Oceania, and the poorer countries of Africa, Asia and Latin America" (Castles, 2004: 211). Therefore, "Migration control is essentially about regulating North–South relations (Castles, 2004: 212). This book retains the emphasis on migration from the Global South to the Global North.

However, Castles' early conceptualisation of the role of globalisation in international migration differed from what is proposed here. Castles underlined that "Globalisation generates the cultural, capital, and technical means needed for migration" (Castles, 2004: 212). In his understanding, "Globalisation also creates strong cultural pressures for mobility" (Castles, 2004: 212). Moreover, as globalisation makes it easier for people to live across borders, new transnational communities are formed. This makes it difficult to tackle a transnational problem by using national policies (Castles, 2004: 213).

Thus, the emphasis is on cultural transformations brought about by globalisation allowing for people to adapt to different cultural environments and to form transnational communities.

[17] See, for example, Sassen (1988); also, Sassen (1981).

[18] See The age of migration: international population movements in the modern world, Stephen Castles | Hein de Haas (1969-; Mark J Miller, 2014 | New York: Guilford Press | Fifth edition.

[19] Castles, S. (2004). Why migration policies fail. *Ethnic and Racial Studies, 27*(2), 205–227. https://doi.org/10.1080/0141987042000177306.

Although, as underlined again by Castles in 2017[20] and by Anderson in 2017,[21] these considerations seem to have stood the test of time, they do not coincide with the understanding of the impact of globalisation on international migration proposed here. As further underlined below, this contribution focuses on the structural consequences of globalisation on migration, specifically the structural changes on the production sphere and on the Global Division of Labour and Power, creating the structural need for the populations of the marginalised areas of the global political economy (the "Global South") to move permanently to the core countries of the Global North. In sum, globalisation does not facilitate international migration—only, or even primarily—because it provides the cultural means to move abroad (as stated by Castles in 2004). Rather, most crucially, it creates the structural need to do so by increasingly marginalising areas of the globe (what is called here the paradox of marginalisation within globalisation).

More recently Castles deepened this understanding of the impact of globalisation on migration by relating it to the notion of social transformation.[22] Although the notion of social transformation, directly deriving from Polanyi's work (Castles et al., 2015: 5), is linked by the author to the current global shifts in dominant economic, political and strategic relationships (Castles, 2010), Castles does not define globalisation nor does he explicitly identify the structural changes in the production and financial sphere creating the structural need for marginalised populations to move. Moreover, his contribution does not seem to include the discussion of the dynamics of regionalisation and how they differ/complement globalisation and impact on migration (the paradox of regionalisation within globalisation). Finally, although the study of irregular migration is mentioned in Castles' article (2010) by making reference to Reyneri (2003), a systematic investigation on how irregular migration is conceptualised in the IPE literature is still lacking (paradox of securitisation

[20] See Castles, S. (2017). Migration policies are problematic—because they are about migration. *Ethnic and Racial Studies, 40*(9), 1538–1543. https://doi.org/10.1080/01419870.2017.1308532.

[21] See Anderson, B. (2017). Towards a new politics of migration? *Ethnic and Racial Studies, 40*(9), 1527–1537. https://doi.org/10.1080/01419870.2017.1300297.

[22] Castles, S. (2010). Understanding global migration: A social transformation perspective. *Journal of Ethnic and Migration Studies, 36*(10), 1565–1586. https://doi.org/10.1080/1369183X.2010.489381.

within globalisation). Castles' analysis remains anchored to his previous cultural understanding of globalisation, as it is implicit in his claim that "neoliberalism has emerged as the ideology and theory of political and economic practices behind the market-driven reforms that have characterized this transformation" (Castles et al., 2015: 7). However, neoliberalism and neoliberal globalisation are not only an ideology, but firstly a mode of production and, because of that, they initially transform economic relations (both in the financial and in the productive sphere) and then, consequently, social and cultural relations (what is considered here as the transnationalist conception of globalisation). Even his idea of counter-hegemony, a Gramscian concept as the author acknowledges, is not a structural one, i.e. grounded on economic relations, but a cultural one and therefore, more than a Gramscian, a Polanyian interpretation (Castles et al., 2015: 9).

Certainly Castles' co-edited book, "Social Transformation and Migration" (Palgrave 2015), does contain a number of case studies, by different authors, grounding his idea of social transformation to the evidence. However, its concluding chapter, "International Migration in an Era of Neoliberal Social Transformation", does not represent a comprehensive and systematic assessment of the structural components of globalisation influencing the various forms of international migration such as: temporary, permanent, intra-regional, extra-regional, regular and irregular. Moreover, it does not address the debate in International Political Economy as this contribution seeks to do. Indeed, what the author defines as "the interconnectedness of the social, the economic and the political" (Castles et al., 2015: 305) is the subject matter of the whole discipline of International Political Economy.

Differently, Saskia Sassen, in her early study of intra-regional migration in developing countries increasingly integrated in the Global Political Economy, like Latin America or South-East Asia, moves sociological analysis more in the direction of political economy (Sassen, 1981, 1988: Chapter 4). The problem with these countries is precisely that more integration in the global economy, through foreign direct investment, export processing zones and mergers and acquisitions, does not seem to have reduced intra-regional migration which is, instead, increasing. In Sassen's conceptualisation this happens because of the social dynamics triggered by rapid export-led industrialisation and its consequences on emigration (Sassen, 1981, 1988: Chapter 4):

From an analytical standpoint, it is important to identify various steps in addressing this issue. First of all, it is necessary to verify whether the countries in question are experiencing a rapid, export led integration in global markets through an increase in their global share of FDIs, M&A and Export processing zones.

These new development patterns have a clear impact on the employment rate which is increased both in the agricultural and manufacturing sectors and also in the related service and logistical sectors. Moreover, there is a concentration of productive activities in specific locations to exploit economies of scale and reduce costs of servicing and transportation.

The way this increased labour demand is fulfilled is by including into the labour markets sectors of society which were previously outside it, in particular women. This leads to what Saskia Sassen calls the feminisation of work.

However, although the labour demand is increased, the author also identifies a substitution effect of the newly employed labour categories, especially women, and the traditional male labour force, which being often less mobile and more expensive, is paradoxically thrown out of the labour market and forced to migrate to other countries within the region. This is further incentivised by cultural globalisation making it more acceptable for people to move across the border to improve their standard of living, especially to the more developed countries of the region such as, for example, the US.

Overall, Saskia Sassen's analysis helps shed light on a very specific and paradoxical form of international migration induced by globalisation, i.e. intra-regional migration from countries more and more integrated in the global economy to more developed ones (Sassen, 1981, 1988: Chapter 4). This happens because:

- The increased need for labour in specific location mobilises new sectors of the labour force, especially women. As a consequence, there is a disruption of the traditional labour organisation and an increase in the pool of emigrants, especially among male workers.
- The feminisation of work is especially disruptive of male work patterns both in the industrial sector and in more traditional works.
- Foreign direct investment makes foreign culture and ideology more familiar to less developed countries facilitating intra-regional migration.

Although Sassen successfully addresses the issue of increased migration from less developed countries to western industrialised ones, her analysis is limited to countries that are increasingly involved in the global political economy. This contribution adds to her analysis by referring to the literature on the transnationalist approach to globalisation and migration. This helps answer questions relating to why emigration has increased in the last decades also in places which are more marginalised from the global economy. This allows the author to develop an overarching theoretical understanding of the impact of globalisation on both intra-regional migration and extra-regional migration through the two paradoxes of regionalisation within globalisation and marginalisation within globalisation. Finally, it also endeavours to define a political economy of irregular migration through the paradox of securitisation as further elaborated below.

To this aim, the next section summarises the debate in IPE about globalisation and its impact on migration.

Conclusion: The Three Paradoxes of Regionalisation, Marginalisation and Securitisation within Globalisation

As already discussed above, the transnationalist, non-mainstream IPE perspective has been particularly successful in proposing convincing hypotheses with respect to the impact of globalisation on migration from the Global South to the Global North.

Based on a qualitative definition of globalisation, transnationalist scholars see the current phase of capitalist development as a structurally distinct one, including the reorganisation of productive and financial activities, the transnationalisation of social and economic forces and the subordination of politics to economics at all levels of governance (Dicken,

2015; Talani, 2018[23]: Chapter 1[24]; see also Mittelman, 2011[25]; 2010[26]). The engine of these structural transformations is technological development and the IT revolution characterising the turn of the millennium (Dicken, 2015; Harvey, 2011[27]; Overbeek, 1995, 2000, 2015; Sassen, 1988, 1991, 1996, 2006, 2014: Introduction[28]; Talani, 2018). Moving away from a purely quantitative definition of globalisation, one which is only focusing on quantifying cross-border relations, transnationalist scholars focus instead on structural changes. These affect all dimensions, from the economic one, involving the modification of the productive and financial structure; the technological dimension, which reverberates on the cultural one by modifying the ontology as well as the perception of space–time relations; the political sphere, with the creation of new power structures and relations, as well as the social context, including mass movements of people and the creation of transnationalist communities. All these structural modifications transcend the national space and attribute a transnational character to this new phase of capitalist development, hence the transnationalist definition of globalisation. In the words of Overbeek and Léon:

> As is the case with globalisation, they take place simultaneously in subnational, national and international arenas. The national thus ceases to be the primary constitutive dimension of social relations. Instead, a network of social relations that encapsulates local, national, regional and international structures, dynamics and realities emerges. In this sense, the divide

[23] Talani, L. S. (2018). Globalisation. In T. Show (Ed.), *Palgrave handbook of contemporary international political economy*. London: Palgrave.

[24] Dicken, P. (2015). *Global shift: Mapping the changing contours of the world economy*. London: Sage.

[25] Mittelman, J. H. (2011). *Contesting global order: Development, global governance, and globalisation*. London: Routledge.

[26] Mittelman, J. H. (2010). *Hyperconflict: Globalisation and insecurity*. Stanford: Stanford Security Studies.

[27] Harvey, D. (2011). *The Enigma of capital and the crises of capitalism*. London: Profile Books.

[28] Sassen, S. (2014). *Expulsions: Brutality and complexity in the global economy*. Cambridge: Harvard University Press.

between domestic and international politics becomes even less useful for the analysis of contemporary politics (Overbeek & Léon, 2015[29]).

Following Stephen Gill (1995)[30] globalisation is inherently related to the spread of market relations and, as such, it entails commodification, i.e. the inclusion in the sphere of the market of activities and relations which were previously outside it (Bayart, 2004).[31]

The main factor determining the current phase of capitalist development is, above all, technological innovation, making possible its structural transformations. However, ultimately, the aim is always to maximise profits by exploiting cheaper labour, cheaper raw materials or semi-finished manufactures, expanding to new markets and escaping the internal dynamics of capital-labour relations (Overbeek & Léon, 2015).

As already noted by Strange (1998) and underlined by Sassen (2006, 2014: Introduction), one of the consequences of a similar approach to globalisation is the disempowerment of the national state's political authority and its subordination to transnational economic forces. When a new geopolitical fix is achieved, mostly at the regional level, it does not retain the political prerogatives and legitimacy previously enjoyed at the national level of governance. This is intimately related to the structural changes entailed by globalisation, transforming the ability of the state or any other political entity to govern its processes, including migratory flows. The mechanisms allowing for a similar disempowerment of the political sphere are initiated by technological innovation and its impact on the productive and financial structures. In particular, the modifications of the productive sphere, through the creation of global value chains and geographical reallocations of production, are what mostly impact on global migratory dynamics (Dicken, 2015; Overbeek & Léon, 1995, 2000, 2015[32]; Sassen 1996; 2006; 2014). While some countries manage

[29] <fresh page><cn>2.<em><ct>Neoliberal globalisation, transnational migration and global governance < au > Alba I. León and Henk Overbeek in Talani and McMahon 2015.

[30] Gill, S. (1995). Globalisation, market civilization and disciplinary neoliberalism. *Millennium: Journal of International Studies*, *24*(3), 299–423.

[31] Bayart, J. F. (2004). *Le gouvernement du monde. Une critique politique de la globalisation*. Paris: Fayard.

[32] Overbeek, H. W., & Léon, A. I. (2015). Neoliberal globalisation, transnational migration and global governance In L. S. Talani, & S. McMahon (Eds.), *Handbook of the*

to become more integrated in the global political economy, by relying on foreign direct investment, export processing zones and mergers and acquisitions (Mittelman, 2000, 2011), others are increasingly left behind. The marginalisation of entire regions of the globe is somehow paradoxical, although ontologically related to the working of globalisation, and one of its effects is increasing the incentives to migrate from marginalised countries and regions to core ones. Migration, in this case, takes the form of both mass migration and brain drain (Mittelman, 2000, 2011; Sassen, 1988, 1991).

This phenomenon of marginalisation is compounded by one of regionalisation as production tends to cluster within specific regions. This happens because regionalisation allows for the exploitation of economies of scales, clusters of expertise and reduction of the costs of production and transportation. Thus, one can identify a new paradox of globalisation, a paradox of regionalisation within globalisation, deriving from the creation of increasingly integrated economic regions.

Here migratory dynamics are not eliminated, but follow the logic described by Saskia Sassen on intra-regional migration and the feminisation of work (Sassen, 1988, 1991, 2014).

Overall, the phenomena described above configure what Mittelman defines as the new Global Division of Labour and Power (Mittelman, 2011), whose main components are precisely the global reallocation of production on a regional basis, and the increased use of cheap labour from third world countries both on-site, in integrated regions, and off-site, through the increase of mass migration from the Global South to the Global North. Instances of brain drain from marginalised regions contribute to further marginalisation thus triggering a vicious circle of marginalisation (Overbeek & Léon, 2015). One of the most visible consequences of this new configuration of structural power is the overall decrease of the cost of labour, which is testified by the decreasing trend of the wages share of global GDP (Piketty, 2013). Such an outcome, indeed, is not only the consequence of the geographical displacement of production, but also, and substantially, of the increased use of immigrant labour in developed countries, both regularly and irregularly. This, inevitably, put pressure on the local labour force, decreasing its bargaining power as well as its ability to organise effectively (Mittelman, 2011).

international political economy of migration (pp. 37–53). Cheltenham: Edward Elgar, 17 p.

What is particularly interesting here is the political economy of irregular migration. If the ability of the State to control migratory flows is indeed limited by the globalisation process, regional responses to migratory pressures are increasingly adopting a securitised and restrictive approach to the issue. This is the case, for example, of both American and European regional responses to international migration, which are no doubt becoming less liberal and more rigorous (Overbeek & Léon, 2015: Chapters 4 and 5).

Many are the motivations proposed in the general, as well as in the academic debate, to justify such a severe approach to migration. These range from concerns over political and social unrest, to fears of cultural clashes and religious conflicts, including terrorist attacks (Freeman, 1979, 2008; Freeman et al. 2008). However, one cannot underestimate the role that irregular migration plays in decreasing the overall costs of production inside developed countries and regions at the same time reducing the ability of organised labour to react (Freeman et al., 2008; Overbeek & Léon, 2015; Mittelman, 2011).

There is also the possibility that the existence of a thriving underground economy, with all that it means in terms of availability of irregular jobs, could represent a relevant pull factor for irregular migration (Fellini et al., 2018; Reyneri et al., 2017[33]). Of course the informal economy precedes irregular migration, as this existed already before migratory flows. However, it is precisely the existence of many job opportunities of an irregular nature that makes it possible for irregular migrants to find gainful positions in the receiving countries, thus attracting more of them. Moreover, the recent tendency to create more and more barriers to the legal entry and stay of third country nationals makes it even more important for them to find job opportunities in the informal sector. Thus,

[33] See also Fellini, I., Guetto, R., & Reyneri, E. (2018). Poor returns to origin-country education for non-western immigrants in Italy: An analysis of occupational status on arrival and mobility. *Social Inclusioin, 6*(3), 34–47. Dettaglio; Reyneri, E., Fellini, I., Quassoli, F., & Coletto, D. (2017). Overview of recruitment methods and individual factors for migration. Country report Italy [Working paper]. Dettaglio; Reyneri, E., Fellini, I., Fullin, G., Quassoli, F., Dimitriadis, I., & Coletto, D. (2017). Overview of contextual and institutional factors for migration. Country report Italy [Working paper]. Dettaglio; Fullin, G., & Reyneri, E. (2015). Mezzo secolo di primi lavori in Italia. Per una storia del mercato del lavoro italiano [Half a century of first jobs for Youth. For a story of the Italian labour market]. *Stato E Mercato, 105*(3), 419–467. Dettaglio; Fullin, G., & Reyneri, E. (2013). Introduzione. Gli immigrati in un mercato del lavoro in crisi: il caso italiano in prospettiva comparata. *Mondi Migranti* (1), 21–34. Dettaglio.

the securitisation discourse adopted by legislators and policy-makers to justify an ever restrictive approach to international migration, instead of increasing security, actually decreases it by attracting more irregular migrants who can only be employed in the irregular, if not even in the illegal, economy (Fullin & Reyneri, 2013, 2015).

Looking at the dichotomy between regular and irregular migration, the second is clearly more favourable to the neoliberal system. This happens as irregular migration allows for an uber-flexibilisation of the labour force, which is stripped of the most basic rights, including the right to self-preservation in some occasions. Moreover, relying heavily on irregular work makes it much more difficult to sustain the welfare state, as tax contributions are substantially reduced, thus adding again to the neoliberal credo in the limitations of state intervention in the economy. In the words of Saskia Sassen:

> (…) those who deal with the real political economy of the city already know this: the rhetoric of the law-and-order Republican mayor here in New York City is surprisingly friendly to illegal immigrants. (Sassen, 1998: xiii)

On the whole, the following can be listed as the political economy consequences of migratory waves deriving from globalisation:

- The increase of the precarisation of the working force in both the Global South and the Global North.
- A shift in the balance of power in favour of transnational companies at the global level as well as at the regional and national ones.
- A loss of ability by political institutions to counter the increased power of transnational economic forces in their move of the level of governance from the national to the supranational one.

At the theoretical level, the following paradoxes appear of relevance:

- The paradox of marginalisation within globalisation and its consequences in terms of the increase of brain drain and mass migration from marginalised countries.
- The paradox of regionalisation within globalisation and its consequences in terms of the increase of intra-regional temporary migration.

- The paradox of securitisation and its consequences in terms of irregular migration.

The analysis of each one of the paradoxes is the focus of this book. The next chapter will address the issue of marginalisation within globalisation with reference to the specific case of the Middle East and Northern Africa region.

References

Al-Shawarby, S., & Selim, H. (2012, October). Are international food price spikes the source of Egypt's high inflation?. *Economic Research Forum*. Working Paper 714, Cairo. http://www.erf.org.eg/CMS/uploads/pdf/714.pdf. Accessed 15 January 2013.

Amiti, W., & Wei, S. J. (2006). *Service offshoring and productivity: Evidence from the United States*. National Bureau of Economic Research Working Paper No 11926, Cambridge, Mass.

Anderson, B. (2017). Towards a new politics of migration? *Ethnic and Racial Studies, 40*(9), 1527–1537.

Antkiewicz, A., & Whalley, J. (2006). *Shifting global economic power to the BRICSAM from the OECD*. Centre for International Governance Innovation (CIGI), Waterloo, Ontario. (mimeo).

Arezki, R., & Bruckner, M. (2011). *Food prices and political instability*. IMF Working Paper 11/62. http://www.imf.org/external/pubs/ft/wp/2011/wp1162.pdf. Accessed 11 January 2013.

Ayubi, N. (1996). *Over-stating the Arab State*. I. B. Tauris.

Barbieri, P. (2017, November 13). The losers of deglobalisation. *Foreign Affairs*. https://www.foreignaffairs.com/articles/2016-11-13/losers-deglobalisation. Accessed 3 October 2019.

Bayart, J. F. (2004). *Le gouvernement du monde. Une critique politique de la globalisation*. Fayard.

Bellemare, M. F. (2011, June 28). Rising food prices, food price volatility, and political unrest. *SSRN*. http://ssrn.com/abstract=1874101. Accessed 11 January 2013.

Bhagwati, J., Panagariya, A., & Srinivasan, T. N. (2004). The muddles over outsourcing. *Journal of Economic Perspectives, 18*(4), 93–114.

Breslin, S., Hughes, C. Phillips, N., & Rosamond, B. (2002). *New regionalisms in the global political economy*. Routledge.

Brinkman, H. J., & Hendrix, C. S. (2011). *Food insecurity and conflict: Applying the WDR framework*. World Development Report 2011. August 2, 2010. http://wdr2011.worldbank.org/food.

Buelens, F. (1999). *Globalisation and the nation state*. Edward Elgar.

Busch, A. (2008). *Banking regulation and globalisation*. Oxford University Press.

Caporaso, J. A. (Ed.). (1987). *A changing international division of labor-international political economy yearbook series* (2nd Vol.). Boulder: Lynne Rienner.

Castles, S. (2010). Understanding global migration: A social transformation perspective. *Journal of Ethnic and Migration Studies, 36*(10), 1565–1586.

Castles, S. (2004). Why migration policies fail. *Ethnic and Racial Studies, 27*(2), 205–227.

Castles, S. (2017). Migration policies are problematic—Because they are about migration. *Ethnic and Racial Studies, 40*(9), 1538–1543. https://doi.org/10.1080/01419870.2017.1308532

Castles, S., de Haas, H., & Miller, M. J. (2014). *The age of migration: International population movements in the modern world* (5th ed.). Guilford Press.

Castles, S., Ozkul, D., & Cubas, M. (Eds.). (2015). *Social transformation and migration*. London: Palgrave.

Cerny, P. (1999). Reconstructing the political in a globalising world. In F. Buelens (Ed.), *Globalisation and the nation state*. Edward Elgar.

Cerny, P. (2010). *Rethinking world politics: A theory of transnational neopluralism*. OUP.

Charnock, G., & Starosta, G. (Eds.). (2016). *The new international division of labour global transformation and uneven development*. Authors: Starosta, Guido; Editors: Palgrave, 2016.

Coleman, W. D., & Sajed, A. (2013). *Fifty key thinkers on globalisation*. Routledge.

Cox, R. W. (1981). Social forces, states, and world orders: Beyond international relations theory. *Millenium, 10*, 127–155.

Cox, R. W. (1996). *Approaches to world order*. Cambridge University Press.

Dicken, P. (2015). *Global shift: Mapping the changing contours of the world economy*. Sage.

Dicken, P. (1998). *Global shift, transforming the world economy* (3rd ed.). Paul Chapman Publishing.

Dicken, P. (2003). *Global shift: Reshaping the global economic map in the 21st century* (4th ed.). Sage.

Dicken, P. (2007). *Global shift: Mapping the changing contours of the world economy* (5th ed.). Sage.

Dicken, P. (2011). *Global shift: Mapping the changing contours of the world economy* (6th ed.). Guilford.

Dreher, A., & Gaston, N. (2005). *Has globalisation really had no effect on unions*. Swiss Institute for Business Cycle Research Working Paper No. 110.

Dumont, M., Rayp, G., & Willemé, P. (2006). Does internationalization affect union bargaining power? An empirical study for five EU countries. *Oxford Economic Papers, 58*(1), 77–102.

FAO. (2011, March). *North Africa brief*. FAO/GIEWS Global Watch. http://www.fao.org/giews/english/shortnews/nafria110311.pdf. Accessed 28 July 2011.

Farrell, H., & Newman, A. (2017). BREXIT, voice and loyalty: Rethinking electoral politics in an age of interdependence. *Review of International Political Economy, 24*(2), 232–247.

Feenstra, R., & Hanson, G. H. (1996). Globalisation outsourcing and wage inequality. *American Economic Review Papers and Proceedings, 86*, 240–245.

Feenstra, R., & Hanson, G. H. (1997). Foreign direct investment and relative wages: Evidence from Mexico's maquiladoras. *Journal of International Economics, 42*(3/4), 371–393.

Feenstra, R., & Hanson, G. H. (1999). The impact of outsourcing and high-technology capital on wages: Estimates for the U.S., 1979–1990. *Quarterly Journal of Economics, 114*(3): 907–940.

Feenstra, R. (1998). Integration of trade and disintegration of production in the global economy. *Journal of Economic Perspectives, 12*, 31–50.

Fellini, I., Guetto, R., & Reyneri, E. (2018). Poor returns to origin-country education for non-western immigrants in Italy: An analysis of occupational status on arrival and mobility. *Social Inclusion, 6*(3), 34–47.

Fraser, E., & Rimas, A. (2011, January 30). The psychology of food riots. *Foreign Affairs*. http://www.foreignaffairs.com/articles/67338/evan-fraser-and-andrew-rimas/the-psychology-of-food-riots. Accessed 11 January 2013.

Freeman, G. P. (1979). *Immigrant labour and racial conflict in industrial societies*. Princeton: Princeton University Press.

Freeman, G. (2008). *Migration and the political sustainability of welfare states*. Available at: http://cartss.colorado.edu/Freeman.doc. Accessed 23 June 2008.

Freeman, G. P., & Kessler, A. K. (2008). Political economy and migration policy. *Journal of Ethnic and Migration Studies, 34*(4), 655–678.

Fröbel, F, Heinrichs, J, & Kreye, O. (1980). *The new international division of labour*. Cambridge: Cambridge University Press.

Fullin, G., & Reyneri, E. (2013). Introduzione. Gli immigrati in un mercato del lavoro in crisi: il caso italiano in prospettiva comparata. *Mondi Migranti* (1), 21–34.

Fullin, G., & Reyneri, E. (2015). Mezzo secolo di primi lavori in Italia. Per una storia del mercato del lavoro italiano [Half a century of first jobs for Youth. For a story of the Italian labour market]. *Stato e Mercato, 105*(3), 419–467.

Garret, G. (1998). *Partisan politics in the global economy*. Cambridge University Press.

Gaston, N. (2002). Unions and the decentralization of collective bargaining in a globalized world. *Journal of Economic Integration, 17*, 377–396.

Giddens, A. (1999). *Runaway world: How globalisation is reshaping our lives*. Routledge.

Gill, S. (1995). Globalisation, market civilization and disciplinary neoliberalism. *Millennium: Journal of International Studies, 24*(3), 299–423.

Gill, S. (1997). An EMU or an Ostrich?: EMU and neo-liberal globalisation: Limits and alternatives. In P. Minkkinen, & H. Potomaki (Eds.), *The politics of economic and monetary union*. Kluwer.

Gill, S. R. (1992). The emerging world order and European change. In R. Miliband, & L. Panitch (Eds.), *Communist regimes: The aftermath. The socialist register 1992*. Merlin Press.

Gilpin, R. (2000). *The challenges of global capitalism: The world economy in the 21st century*. Princeton University Press.

Gilpin, R. (2001). *Global political economy: Understanding the global economic order*. Princeton University Press.

Goldstein, J., & Keohane, R. O. (1993). *Ideas in foreign policy*. Cornell University Press.

Gosh, B. (1999). *Challenges and opportunities of economic globalisation: Some implications for labour, industry and nation states*. Paper prepared for the Bundesanstalt fur Arbeit workshop on "Internationalisierung der Arbeitsmarkte", 27–28 April, Frankfurt a/M.

Gramsci, A. (1975). *Quaderni dal carcere*. Einaudi.

Gramsci, A. (1999). *Selections from the prisons' notebooks of Antonio Gramsci*. Elecbook, Electronic Books.

Grossman, G., & Helpman, E. (2002). Integration versus outsourcing in industry equilibrium. *Quarterly Journal of Economics, 117*(1), 85–120.

Harvey, D. (2011). *The Enigma of capital and the crises of capitalism*. Profile Books.

Harvey, D. (2008, September–October). The right to the city. *New Left Review, 53*.

Harvey, D. (2010). *The Enigma of capital and the crises of capitalism*. Profile Books.

Hay, C., & Marsh, D. (2000). *Demystifying globalisation*. MacMillan.

Hay, C. (2002). Globalisation as a problem of political analysis: Restoring agents to a "process without a subject" and politics to a logic of economic compulsion. *Cambridge Review of International Affairs, 15*(3), 379–392.

Held, D. (2002). Law of states, law of people. *Legal Theory, 8*(1), 1–44.

Held, D. (1995). *Democracy and the global order: From the modern state to cosmopolitan governance*. Stanford University Press.

Held, D., & McGrew, A. (2000). *The global transformations reader*. Polity Press.

Held, D. (2010). *Cosmopolitanism: Ideas and realities*. Polity Press.

Held, D., & McGrew, A. (Eds.). (2007). *Oxford globalisation theory: Approaches and controversies*. Polity.
Held, D., McGrew, A., Goldblatt, D., & Perraton, J. (1999). *Global transformation*. Polity Press.
Heritier, A. (Ed.). (2002). *Common goods: Reinventing European and international governance*. Rowman and Littlefield Publishers.
Hettne, B., Inotai, A., & Sunkel, O. (1999). *Globalism and the new regionalism*. MacMillan Press Ltd.
Hirst, P., & Thompson, G. (1999a). *Globalisation in question* (2nd ed.). Polity.
Hirst, P., & Thompson, G. (1999b). The tyranny of globalisation: Myth or reality?. In F. Buelens (Ed.), *Globalisation and the nation state*. Edward Elgar.
Hirst, P., Thompson, G., & Bromely, S. (2009). *Globalisation in question: The international economy and the possibilities of governance* (3rd ed.). Polity Press.
Holm, H. H., & Sørensen, G. (1995). *Whose world order?: Uneven globalisation and the end of the Cold War*. Westview Press.
Hopkins. (Ed.). (2006). *Global history: Interactions between the universal and the local*. Palgrave.
ILO and OECD. (2015, February 26–27). The labour share in G20 economies international labour organization organisation for economic co-operation and development with contributions from international monetary fund and World Bank Group Report prepared for the G20 Employment Working Group Antalya, Turkey.
Inglehart, R. F., & Norris, P. (2016). *Trump, Brexit and the rise of populism: Economic have-nots and cultural backlash*. Faculty Research Working Paper Series, RWP16–026. Harvard Kennedy School of Government.
Jacks, D. S., O'Rourke, K. H., & Williamson, J. G. (2011). Commodity price volatility and world market integration since 1700. *The Review of Economics and Statistics, MIT Press, 93*(3), 800–813.
James, H. (2017, September 12). *Deconstructing deglobalisation*. Project Syndicate. https://www.project-syndicate.org/commentary/three-emotions-underlying-deglobalisation-by-harold-james-2017-09.
James, H. (2001). *The end of globalisation: Lessons from the great depression*. Harold James Harvard University Press.
Kaminska, I. (2015, September 24). *How do you solve a problem like de-globalisation?*. FT Alphaville. https://ftalphaville.ft.com/2015/09/24/2140786/how-do-you-solve-a-problem-like-de-globalisation/.
Keohane, R. O., & Nye, Jr., J. S. (2012). *Harvard University power & interdependence* (4th ed.). Pearson.
Keohane, R. O. (2000). Sovereignty in international society. In D. Held, & A. McGrew (Eds.), *The global transformations reader*. Polity Press, Ch.9.
Keohane, R. O., & Nye, J. S. (1977). *Power and interdependence*. Little Brown.

Lagi, M. (2012). *Economics of food prices and crises*. Publisher: necsi.edu. http://www.academia.edu/1399884/Economics_of_Food_Prices_and_Cri ses. Accessed 11 January 2013.

Lagi, M., Bertrand, K. Z., & Bar-Yam, Y. (2011). *Food crises and political instability in North Africa and the Middle East*. New England complex system Institute. http://necsi.edu/research/social/food_crises.pdf. Accessed 10 January 2013.

Lipietz, A. (1985). The world crisis: The globalisation of the general crisis of Fordism. *IDS Bulletin*, *16*, 6–11. https://doi.org/10.1111/j.1759-5436.1985.mp16002002.x.

Lilley, P. (2000). *Dirty dealing: The untold truth about global money laundering*. Kogan Page.

Lynch, D. (2008, April 30). Tension in Egypt shows potency of food crisis. *USA Today*. http://www.usatoday.com/money/world/2008-04-29-egypt-food-shortage_N.htm. Accessed 11 January 2013.

MacKenzie, D. (2011). Can complexity theory explain Egypt's crisis?. *New Scientist*. http://www.newscientist.com/article/dn20082-can-complexity-the ory-explain-egypts-crisis.html. Accessed 11 January 2013.

Malm, A. (2015). *Fossil capital: The rise of steam-power and the roots of global warming*. Verso.

Maloney, S. (2011). The economic dimension: The price of freedom. In K. Pollack (Ed.), *The Arab awakening* (pp. 66–76). The Brookings Institution.

Mishra, R. (1999). *Globalization and the welfare state civil service systems in comparative perspective series*. London: E. Elgar.

Mittelman, J. H. (2000). *The globalisation syndrome: Transformation and resistance*. Princeton University Press.

Mittelman, J. H. (2010). *Hyperconflict: Globalisation and insecurity*. Stanford Security Studies.

Mittelman, J. H. (2011). *Contesting global order: Development, global governance, and globalisation*. Routledge.

Moore, J. (2016). *Anthropocene or capitalocene? Nature, history, and the crisis of capitalism*. PM Press.

Mundell, R. (1957). International trade and factor mobility. *American Economic Review*, *48*(3), 321–335.

Murshed, M. S. (2002). *Globalisation, marginalisation and development*. Routledge.

Noueihed, L., & Warren, A. (2012). *The battle for the Arab Spring*. Yale University Press.

Obstfeld, M., & Taylor, A. M. (2004). *Global capital markets: Integration, crisis and growth*. Cambridge University Press.

OECD. (2006, March). *The internationalization of production, international outsourcing, and OECD labor markets*. Economics Dept., OECD. (mimeo).

Orlowski, L. T. (2008). *Stages of the 2007/2008 global financial crisis: Is there a wandering asset-price bubble?*. Economics Discussion Papers, No 2008-43. http://www.economics-ejournal.org/economics/discussionpapers/2008-43. Accessed 18 May 2009.

Overbeek, H. (1999, February 16–20). *The Budapest process: Internationalization of migration control*. Paper presented to the 40th annual conference of the international studies association. Mimeo.

Overbeek, H. (2000). Globalisation, sovereignty and transnational regulation: Reshaping the governance of international migration. In B. Gosh (Ed.), *Managing migration: Time for a new international regime*. Oxford University Press.

Overbeek, H. W. (1995). Globalisation and the restructuring of the European labor markets: The role of migration. In M. Simai (Ed.), *Global employment: An international investigation into the future of work*. Zed Books and United Nations University Press.

Overbeek, H. W., & Léon, A. I. (2015). Neoliberal globalisation, transnational migration and global governance. In L. S. Talani, & S. McMahon (Eds.), *Handbook of the international political economy of migration* (pp. 37–53). Edward Elgar. https://doi.org/10.4337/9781782549901.

Padoa-Schioppa, T. (1994). *The road to monetary union in Europe: The emperor, the Kings and the Genies*. Clarendon Press.

Palan, R. (2003). *The offshore world*. Cornell University Press.

Pellerin, H., & Overbeek, H. W. (2001). Neoliberal regionalism and the management of people's mobility. In A. Bieler, & A. D. Morton (Eds.), *Social forces in the making of the 'New Europe'*. Palgrave.

Piketty, T. (2013). *Capital in the twenty-first century*. Cambridge: The Belknap Press of Harvard University Press.

Reyneri, E., Fellini, I., Fullin, G., Quassoli, F., Dimitriadis, I., & Coletto, D. (2017). *Overview of contextual and institutional factors for migration*. Country report Italy [Working paper]. Dettaglio.

Reyneri, E., Fellini, I., Quassoli, F., & Coletto, D. (2017). *Overview of recruitment methods and individual factors for migration*. Country report Italy [Working paper]. Dettaglio.

Roccu, R. (2012). Of middle classes, economic reforms and popular revolts: Why democratization theory failed, again. In S. Panebianco (Ed.), *Winds of democratic change in the mediterranean* (pp. 61–82). Rubbettino.

Rodrik, D. (1999) Globalisation and labor, or: If globalisation is a bowl of cherries, why are there so many glum faces around the table?. In R. E. Baldwin (Ed.), *Market integration, regionalism and the global economy*. Cambridge U.P. for CEPR.

Rodrik, D. (2017). *Populism and the economics of globalization*. https://drodrik.scholar.harvard.edu/files/dani-rodrik/files/populism_and_the_economics_of_globalization.pdf.

Rosenau, J. (2002). Governance in a new global order. In A. McGrew, & D. Held (Eds.), *Governing globalisation: Power, authority and global governance*. Polity Press.

Rosenau, J. (2003). *Distant proximities: Dynamics beyond globalisation*. Princeton University Press.

Rosenau, J. (2008). *People count! Networked individuals in global politics*. Co

Ruggie, J. G. (1993). Territoriality and beyond: Problematizing modernity in international relations. *International Organization, 47*(1), 139–174.

Sasse, S. (2014). *Expulsions: Brutality and complexity in the global economy*. Harvard University Press.

Sassen, S. (1981). *Exporting capital and importing labor: The role of Caribbean migration to New York City*. NYU.

Sassen, S. (1988). *The mobility of labour and capital*. Cambridge University Press.

Sassen, S. (1991). *The global city: New York*. Princeton University Press.

Sassen, S. (1996). *Losing control? Sovereignty in the age of globalisation*. Columbia University Press.

Sassen, S. (1999). *Guests and Aliens*. The new Press.

Sassen, S. (2006). *Territory, authority, rights: From medieval to global assemblages*. Princeton University Press.

Sassen, S. (1998). *Globalisation and its discontents: Essays on the new mobility of people and money*. The New York Press.

Scholte, J. (2011). *Building global democracy? Civil society and accountable global governance*. Cambridge.

Scholte, J. (2005). *Globalisation: A critical introduction* (2nd ed.). Palgrave.

Shafik, N. (1999). Technology: A disintegrative factor in the Arab World. In C. M. Hudson (Ed.), *The Middle East dilemma*. Columbia University Press.

Smith, A. F. (2011). *Starving the south*. St. Martin's Press.

Strange, S. (1996). *The retreat of the state*. Cambridge University Press.

Strange, S. (1998). *Mad money*. Manchester University Press.

Talani, L. S. (2003, March). Avoiding the "G" word in reinventing European and international governance. *International Studies Review*.

Talani, L. S. (2006). *Interview with an anonymous functionnaire of the Egyptian Ministry for Migration and Manpower*. Unpublished.

Talani, L. S. (2009). *From Egypt to Europe*. I. B. Tauris.

Talani. L. S. (2012). *Globalisation, hegemony and the future of the city of London*. London: Palgrave.

Talani, L. S. (2013). *European political economy: An IPE perspective*. Routledge.

Talani, L. S. (2014). *The Arab spring in the global political economy*. London: Palgrave.

Talani, L. S. (2018). Globalisation. In T. Show (Ed.), (2018). *Palgrave handbook of contemporary international political economy*. Palgrave.
The Economist. (2020). https://www.economist.com/special-report/2019/07/11/supply-chains-are-undergoing-a-dramatic-transformation?al_applink_data=%7B%22target_url%22%3A%22https%3A%5C%2F%5C%2Fecon.st%5C%2F2P4FQWl%22%2C%22extras%22%3A%7B%22fb_app_id%22%3A237759909591655%7D%2C%22referer_app_link%22%3A%7B%22url%22%3A%22fb-messenger%3A%5C%2F%5C%2F%5C%2F%3Fapp_id%3D237759909591655%22%2C%22app_name%22%3A%22Messenger%22%7D%7D.
Tomiura, E. (2005). Foreign outsourcing and firm-level characteristics: Evidence from Japanese manufacturers. *Journal of the Japanese and the International Economies, 19*, 225–271.
Van der Pijl, K. (2011, July–August). Arab revolts and nation-state crisis. *NLR, 70*.
Weiner, M. (1995). *The global migration crisis: Challenge to states and human rights*. HarperCollins College Publishers.
Whalley, J. (2006). *The post MFA performance of developing Asia*. (mimeo).
Whalley, J., & Xian, X. (2006). *China's FDI and non-FDI economies and the sustainability of future high Chinese growth*. The University of Western Ontario. (mimeo) Panel 1 =.
Xiao, G. (2004). *People's Republic of China's round-tripping FDI: Scale, causes, and implications*. Asia Development Bank Institute Discussion Paper No. 7.

CHAPTER 3

The Paradox of Marginalisation: The Case of Migration from the Mena Area to the EU

Introduction

According to the qualitative, transnationalist definition of globalisation adopted in this book, the globalisation process is an eminently uneven one, both geographically and socially exclusive, and entailing a new redistribution of power at both the geographical and social level. At the core of this process is the paradox of regionalisation within globalisation. This implies that those regions experiencing a high degree of integration into the global economy also tend to increase their level of economic and, to a certain extent, political regional integration (Mittelman, 2011). There are, however, regions, such as the Middle East and the North African region (MENA), which are progressively less and less integrated in the global political economy, as well as less and less integrated, both politically and economically, within themselves (Talani, 2012). The MENA area is to be analysed as a paradigmatic case of marginalisation within globalisation, and as a clear example of the lack of regionalisation, as the relevant literature repeatedly claimed[1] (see, e.g., Dodge & Higgot, 2002; Valbjorn, 2016; also Hakimian & Moshaver, 2001; Hudson, 1999).

[1] Valbjorn, M. (2016). North Africa and the Middle East. In T. A. Börzel, & T. Risse (Eds.), *The Oxford handbook of comparative regionalism*. Oxford: Oxford University Press; Hakimian, H. (2018). *De las rentas del petróleo al crecimiento inclusivo: Lecciones aprendidas de la región MENA* (pp. 125–148). Awraq 15; Hakimian, H., & Abdulaal, A.

L. S. Talani, *The International Political Economy of Migration in the Globalization Era*, International Political Economy Series,
https://doi.org/10.1007/978-3-030-79321-0_3

The theoretical analysis discussed in the previous chapter of this volume allows us to identify in the economic marginalisation of the MENA area, and its lack of political and economic integration, a relevant political economy factor for recent waves of mass migration and brain drain. As already pointed out, from the theoretical point of view, the following issues appear particularly relevant:

1. The paradox of "marginalisation" and its impact in terms of the increase of extra-regional permanent migration and brain drain.
2. The paradox of "regionalisation" and its consequences in terms of intra-regional temporary migration.
3. The paradox of "securitisation" and its consequences in terms of irregular migration.

(2015). *GCC economic integration: Fiction or reality?* (pp. 1–8). Khamsoon; Hakimian, H. (2011). The economic prospects of the 'Arab Spring': A bumpy road ahead. *CDPR Development ViewPoint, 63*, 1–2; Richards, A. & Waterbury, J. (1996). *A political economy of the Middle East.* Boulder: Westview Press, Chapter Fifteen; Halliday, F. (2002). The Middle East and the politics of differential integration. In T. Dodge, & Higgott, R. (Eds.), *Globalization and the Middle East* (pp. 36–56). London: The Royal Institute of International Affairs; Dresch, P. (Ed.). (2005). *Monarchies and nations: Globalisation and identity in the Arab states and the gulf.* London, New York: I. B. Tauris. DS247.A134 M66 2005; Schaebler, B., & Stenberg, L. (Eds.). (2004). *Globalization and the Muslim world: Culture, religion, and modernity.* Syracuse, N.Y.: Syracuse University Press. DS35.62.G58 2004; Dodge, T. (2002). *Bringing the bourgeoisie back in: Globalisation and the birth of liberal authoritarianism in the Middle East.* In R. Higgot, & T. Dodge (Eds.), *Globalization and the Middle East: Economy, society and politics* (pp. 169–187). London, UK: Brookings Press. ISBN 9781862031340; Higgot, R., & Dodge, T. (2002). *Globalisation and its discontents: The theory and practice of changein the Middle East.* In R. Higgot, & T. Dodge (Eds.), *Globalization and the Middle East: Economy, society and politics.* London, UK: Brookings Press. ISBN 9781862031340.

Kitchen, N., Dodge, T., Lawson, G., El Issawi, F., Stein, E., Ulrichsen, K. C., Alaaldin, R., Phillips, C., Thiel, T., Rafati, N., & Voller, Y. (2012). *After the Arab Spring: Power shift in the Middle East?.* IDEAS reports—special reports, Kitchen, N. (Ed.). (SR011). *LSE IDEAS, London School of Economics and Political Science.* London, UK; Dodge, T. (2012). *After the Arab Spring: Power shift in the Middle East?: Conclusion: The Middle East after the Arab Spring.* IDEAS reports—special reports, Kitchen, N. (Ed.). (SR011). *LSE IDEAS, London School of Economics and Political Science.* London, UK.

Dodge, T. (2012). *After the Arab Spring: Power shift in the Middle East?: From the 'Arab Awakening' to the Arab Spring; The post-colonial state in the Middle East.* IDEAS reports—special reports, Kitchen, N. (Ed.). (SR011). *LSE IDEAS, London School of Economics and Political Science.* London, UK.

The MENA area is therefore paradigmatic of some of the structural developments of the process of globalisation and deserves further attention.

This chapter assesses the state of marginalisation of the MENA area from the global political economy and its lack of regionalisation with a view to its migratory dynamics. For the sake of this analysis, it is important to clarify the geographical definition of the area considered. According to the International Monetary Fund (IMF), the MENA region is composed of a number of countries with close geographical, cultural and historical ties, whose economies are also facing similar issues. These are Algeria, Bahrain, Djibouti, Egypt, Iran, Iraq, Jordan, Kuwait, Lebanon, Libya, Mauritania, Morocco, Oman, Qatar, Saudi Arabia, Sudan, Syria, Tunisia, the United Arab Emirates and Yemen.[2] The main economic disparity within the region is represented by whether a country is oil-exporting or non-oil-exporting (also known as resource-poor country). The latter are Egypt, Jordan, Lebanon, Morocco, Syria, Tunisia and Yemen. The formers are divided into the members of the Gulf Cooperation Council (GCC) (Bahrain, Kuwait, Oman, Qatar, Saudi Arabia and United Arab Emirates) and other MENA oil exporters (Algeria, Iran and Libya). The classification into oil-exporting or non-oil-exporting country is based on the World Economic Outlook (IMF, 2003) convention. This establishes that a country is oil-exporting if its oil earnings constitute more than 50% of total export earnings. Sudan, which was previously an oil-producing country, is now classified as a resource-poor country as its oil earnings are now less than 50% of its export earnings, although it remains within the MENA area.[3]

In this chapter, the economic dimension of integration will be assessed by looking at the three main indicators of integration into the global political economy. These are the share of global foreign direct investment (FDI) stock and flows, the amount of mergers and acquisitions (M&A) and the degree of trade openness. The objective is to ascertain to what extent the MENA area has been integrated into the global political economy and is enjoying regional integration. Then, the migratory flows from the MENA area are discussed to verify their dynamics in the

[2] See IMF World Economic Outlook. http://www.imf.org/external/pubs/ft/weo/2012/02/weodata/weoselagr.aspx#a406. Accessed 8 April 2013.

[3] See IMF website: http://www.imf.org/external/pubs/ft/weo/data/assump.htm. Accessed 8 April 2013.

context of marginalisation. The chapter will focus, in particular, on three countries, Egypt, Tunisia and Morocco, which are particularly relevant in terms of migratory dynamics across the Mediterranean and that have been clearly marginalised from the process of globalisation. Information about the motivations for Moroccan migrations at the point of departure, which are used in this chapter, have been gathered by the author through a survey taking place in Rabat in 2006.

The Marginalisation of the MENA Area from the Global Political Economy

Looking at the three main indicators of integration into the global political economy (Hirst et al., 2009; Mittleman, 2000, 2011), namely, the share of global foreign direct investment stock; the level of mergers and acquisitions; and production for exports; it is possible to verify the extent to which the MENA area has been marginalised in the globalisation process.[4]

The paradox of marginalisation within globalisation emerges clearly when looking at the patterns of world stock foreign direct investments (FDIs). Starting from the early 1980s, as before 1980, the world FDI stock was irrelevant, its impressive increase in the following years mainly interests developed nations (which are still holding the biggest share, around 65%) (Fig. 3.1). As far as developing countries are concerned, Asia and Oceania are the region with the highest stock of FDI (Fig. 3.2).

As already noticed above, the biggest share of world FDI stock is in developed countries, with Northern America and Europe holding around 63%. Eastern and South Eastern Asia, including China, held only 20.5% of total FDI stock in 2018, while Latin America and the Caribbean recorded the second-biggest share of FDI stock amongst developing countries with a mere 6.6% in 2018. Instead, the North Africa and West Asian Region, was clearly lagging behind, with only 3.8% of world FDI stock (Fig. 3.3). It is interesting to note that, compared with 1988, North Africa is the only developing region actually decreasing its share of World FDIs from 2% in 1988 to a tiny 1.4% in 2018.

[4] Others prefer to use absolute numbers which are not provided here for the theoretical reasons so far discussed. See Henry and Springborg (2010).

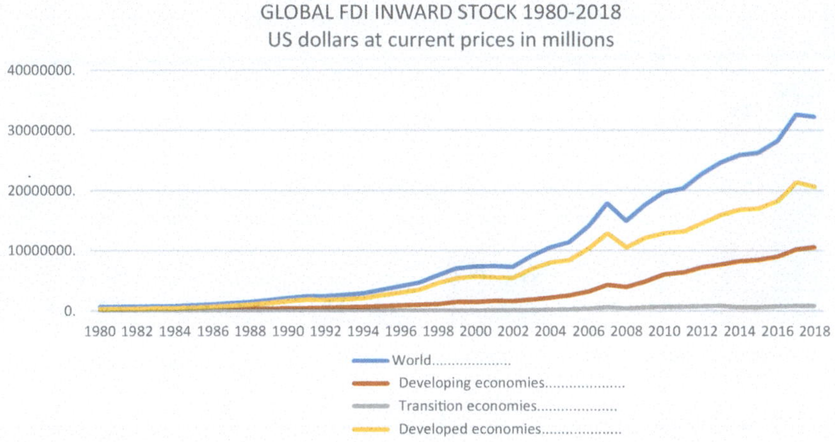

Fig. 3.1 Global FDI inward stock 1980–2018 (US$ ml at current prices) (*Source* UNCTAD, elaboration of the author)

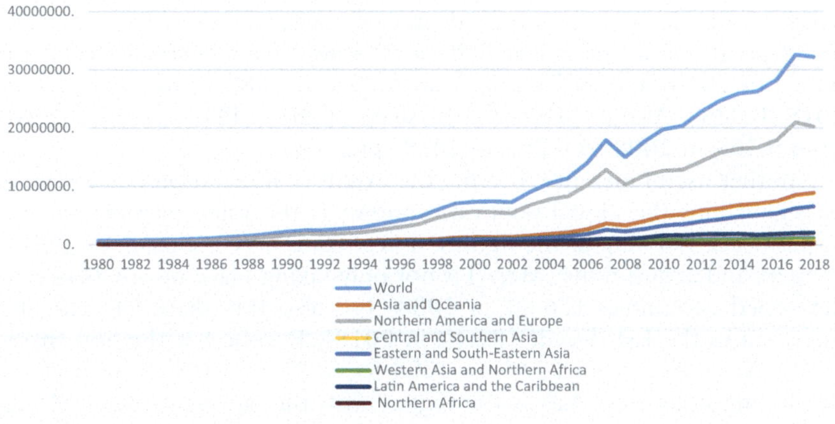

Fig. 3.2 Stock of FDI by regions 1980–2018 (US$ ml at current prices) (*Source* UNCTAD, elaboration of the author)

It is revealing that in the three countries considered in this chapter, the percentage share of total world FDI stock has been decreasing in the last decade. This starts from already incredibly low shares in the 1980s,

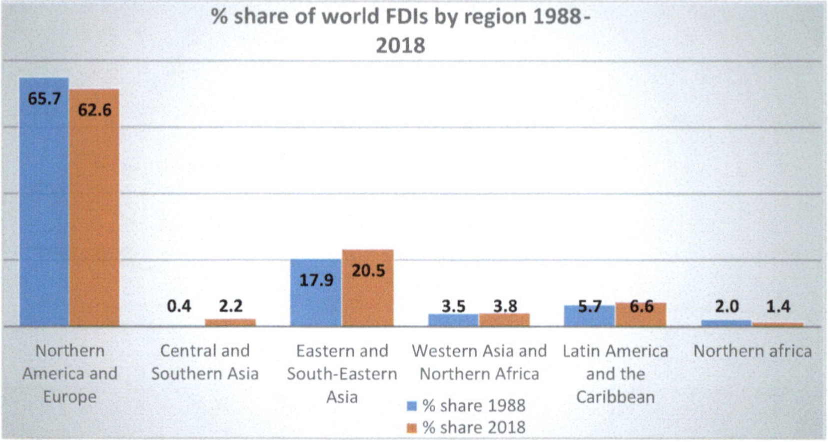

Fig. 3.3 Percentage share of total FDI stock 1988–2018 by region (*Source* UNCTAD, elaboration of the author)

which clearly demonstrates their degree of marginalisation. In particular, in Egypt, the share of global FDI stock moved from 0.59% in 1988 to 0.36% in 2018, and in Tunisia from 0.39% in 1988 to almost zero in 2018 (0.08%). Also in Morocco, the share of global FDI stock decreased from 0.26% in 2008 to 0.20% in 2018 (Fig. 3.4).

Another useful measure to verify the extent of integration of a country or region into the global political economy is the value of mergers and acquisitions. Looking at the figures, it is striking how the share of global mergers and acquisitions (M&A) is not only taking place mostly between developed economies (76.8% in 2008) but also this share has actually increased in the last decade (84.4% in 2018). Europe has the lion share of M&A, moving from 27.9% in 2008 to 46.3% in 2018. In respect to developing countries, Asia is the region with the largest share of M&A sales, although this decreased from 13.7% in 2008 to 10.3% in 2018. Of this, most took place in South East and East Asia (8.8% in 2008 and 4.8% in 2018). The MENA area, on the contrary, is clearly lagging behind with North Africa reporting a decrease in the share from 3.2% in 2008 to basically 0 in 2018 (0.1%). West Asia is faring slightly better as it moved from a share of 2.8% in 2008 to 1.2% in 2018. In particular, the share of

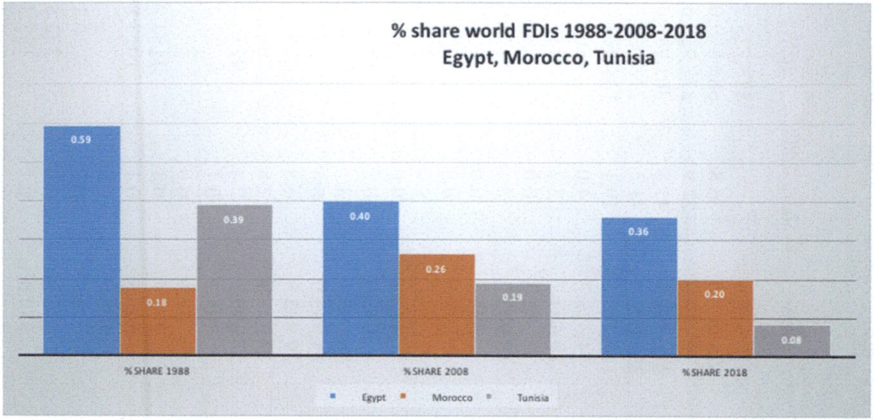

Fig. 3.4 Egypt, Tunisia, Morocco: percentage share of world total FDI stock, 1988, 2008 and 2018 (*Source* UNCTAD)

the countries considered here, notably Egypt, Tunisia and Morocco, has always been more or less around 0% (Table 3.1).

Table 3.1 Mergers and acquisitions (sales) by country and region, 2008–2018

Value of net cross-border M&As by region/economy of seller, 2008–2018

(Millions of dollars)

Region/economy	*2008*	*% Share 2008*	*2018*	*% Share 2018*
World	617,648.6		815,725.7	
Developed economies	474,067.0	76.8	688,859.0	84.4
Europe	172,448.0	27.9	377,958.6	46.3
European Union	258,391.1	41.8	361,943.0	44.4
North America	257,007.0	41.6	223,734.0	27.4
Developing economies	117,713.3	19.1	124 264.5	15.2
Africa	24,539.7	4.0	1569.5	0.2
North Africa	19,494.7	3.2	1143.1	0.1
Algeria	82.3	0.0	–	–
Egypt	18,902.7	3.1	122.9	0.0
Libya	307.4	0.0	–	–
Morocco	79.8	0.0	1020.2	0.1
Sudan	–	–	–	–
Tunisia	122.4	0.0	–	–
Other Africa	5045.0	0.8	426.4	0.1
Asia	84,683.1	13.7	83,769.1	10.3
East and South-East Asia	54,553.0	8.8	39,214.5	4.8
East Asia	29,933.5	4.8	21,903.0	2.7
China	17,475.4	2.8	7645.9	0.9
West Asia	17,598.3	2.8	9713.9	1.2
Bahrain	335.2	0.1	77.2	0.0
Iraq	34.4	0.0	–	–
Jordan	876.9	0.1	1154.9	0.1
Kuwait	505.5	0.1	1024.8	0.1
Lebanon	108.0	0.0	–	–
Oman	9.7	0.0	0.1	0.0
Qatar	123.7	0.0	−272.6	−0.0
Saudi Arabia	330.5	0.1	157.3	0.0
State of Palestine	–	–	−4.4	−0.0
Syrian Arab Republic	–	–	–	–
Turkey	13,982.5	2.3	1362.5	0.2
United Arab Emirates	1291.9	0.2	6214.1	0.8
Yemen	–	–	–	–

(continued)

Table 3.1 (continued)

Value of net cross-border M&As by region/economy of seller, 2008–2018				
(Millions of dollars)				
Region/economy	*2008*	*% Share 2008*	*2018*	*% Share 2018*
Latin America and the Caribbean	9232.8	1.5	39,147.9	4.8
Oceania	−742.2	−0.1	−222.0	−0.0
Transition economies	25,868.2	4.2	2602.2	0.3

Source UNCTAD cross-border M&A database (www.unctad.org/fdistatistics)
Note All values and numbers referring to cross-border M&As in the World Investment Report 2019 (WIR19) and its web annex tables are presented on a net basis. Net cross-border M&As are calculated considering sales of companies in a host economy to foreign MNEs. It excludes sales of foreign affiliates (already owned by foreign MNEs) to other foreign MNEs. Divestments (sales of foreign affiliates to domestic firms) are subtracted from the value (number). Totals exclude the financial centres in the Caribbean. https://unctad.org/en/Pages/DIAE/World%20Investment%20Report/Annex-Tables.aspx

The picture of a marginalised MENA region emerges quite clearly also by looking at the share of MENA region's exports over total exports, another indicator of integration in the globalisation process. The share of exports from other developing regions as clearly increased from 1980 to 2018, with Asia and Oceania almost reaching the share of North America and Europe (42.6% and 49.3%, respectively, in 2018). The case of Northern Africa shows a different dynamic, with a decreasing trend and, in any case, one dangerously close to zero (moving from a mere 1.3% in 1980 to an even more meagre 0.7% in 2018). Even sub-Saharan Africa has a higher share of the world export with 1.8% in 2018 (Fig. 3.5).

Figure 3.6 shows that the share of exports from Egypt, Tunisia and Morocco remained basically stable (when it did not decline) and pretty low (always below 0.2%) over the last four decades.

The regionalisation process in the MENA area has witnessed a series of attempts to promote integration, from free trade areas to customs unions and even plans to establish common markets (Talani, 2014). However, none of them resulted in any significant increase of intra-regional trade. According to data from the IMF Directions of trade (DOTs), the total imports and exports of the MENA area, including Afghanistan and Pakistan, with the rest of the world in 2018 were low (IMF, 2018). Looking at the percentage share of intra-regional exports in the MENA

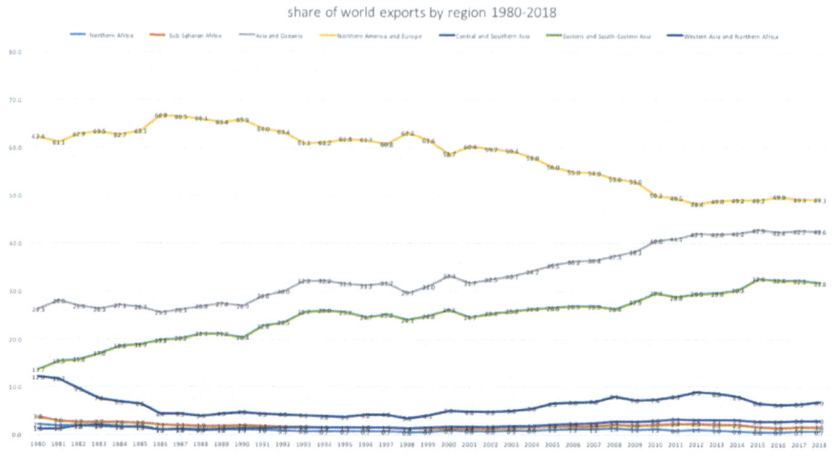

Fig. 3.5 Share of world exports by region 1980–2018 (*Source* UNCTAD database, elaboration of the author)

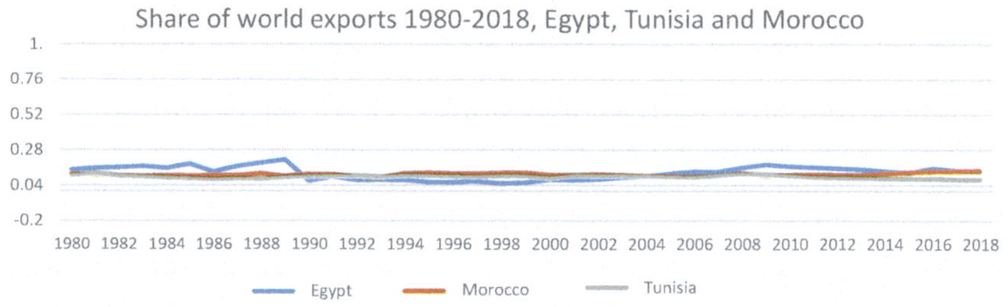

Fig. 3.6 Share of exports to rest of the world from Egypt, Tunisia and Morocco 1980–2018 (*Source* UNCTAD database, elaboration of the author)

area, this appears to be only a fraction of the percentages of regional trade with other parts of the world. IMF figures from 2008 to 2018 show that the MENA area, including Afghanistan and Pakistan, is exporting mainly to advanced economies. On the other hand, the share of intra-regional trade is just around one-tenth of total regional trade (see Fig. 3.7). Even more interestingly, the intra-regional share of total exports and imports of the MENA area has been pretty stable in the last thirty years remaining around the 10% figure (Fig. 3.8).

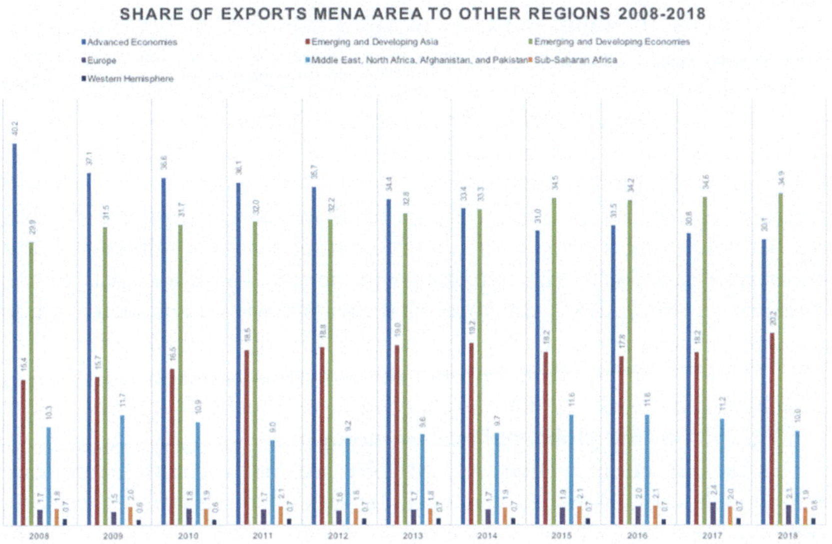

Fig. 3.7 Share of exports from the MENA to other areas, 2008–2018 (*Source* IMF DOTS2019 [https://data.imf.org/regular.aspx?key=61013712])

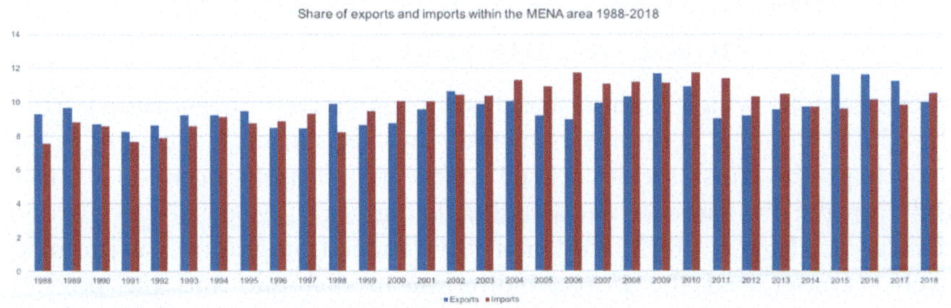

Fig. 3.8 Share of exports and Imports from the MENA to the MENA area, 1988–2018. *MENA area plus Afghanistan and Pakistan (*Source* IMF DOTS2019 [https://data.imf.org/regular.aspx?key=61013712])

Concluding, it seems fair to claim that the MENA area, from the start of the globalisation process, failed to catch up with the rest of the world and is becoming marginalised from the global political economy. Overall, looking at the customary indicators of integration, the percentage of global FDIs, the value and share of M&A sales and the performance of exports and trade openness, the area is undergoing a process of progressive peripherisation. In addition, trade within the region has been very stable and very limited during the last three decades.

Hence, the idea of the regionalisation of the MENA area and its economic and political integration are far from true. As the relevant literature repeatedly pointed out, more than integrating, the MENA area is undergoing a process of (dis)integration from the global political economy. Contrary to the dynamics experienced by other areas of the world, the Arab region is still somewhat divided, with its de facto and de jure integration very much lagging behind. On the other hand, the so-called 'Triad', composed of Europe, the Americas and South-East Asia, has achieved increasingly high levels of intra-regional integration, as well as integration in the global political economy.

This would confirm that the MENA area is a paradigmatic case of marginalisation from globalisation. Its lack of integration, therefore, confirms the existence of a paradox of marginalisation within globalisation. The next sections will assess the marginalisation and migratory dynamics in the three countries considered.

Marginalisation and Extra-Regional Migration: The Case of Tunisia

The extent to which Tunisia has been progressively marginalised from the global political economy is further revealed through analysing what is perhaps the most sensitive issue in the relationship between Tunisia and Europe: the issue of extra-regional migration, especially to Europe (Dodge & Higgot, 2002; Zahlan, 1999). Indeed, migration has progressively monopolised the Euro-Mediterranean dialogue and the policies implemented by the European Union vis-à-vis its Southern Mediterranean neighbours.[5] The history of Tunisian migration to Europe is a long-standing one. Since the mid-1900s, different waves of Tunisian migration were primarily directed to Europe. Mensard (2004) identifies four waves of Tunisian migration, considering the dimensions of the period of emigration, the type of movement and the destination country.

[5] For an overview of the different phases of migration from Tunisia see Mesnard (2004) and Pouessels (2017). Mensard, A. (2004). Temporary migration and self-employment. Evidence from Tunisia. Brussels Economic Review, 47(1).

In the first wave, from the 1950s to the 1960s, the reconstruction process in Western Europe attracted mainly low-skilled Tunisian migrants. They were mostly heading to France and undergoing a less regulated, prevalently male, circular migration process to enter the French labour market through private arrangements (Mensard, 2004; Pouessels, 2017).

In the second period, going from the beginning of the 1960s to the mid-1970s, the migration process of Tunisian migrants became more regulated and the numbers increased substantially. Bilateral agreements were signed between Tunisia and France in 1963, Germany in 1965 and Belgium in 1969. Still, those migratory flows were mainly composed of low-skilled male migrants. At the same time, Libya became a destination country for Tunisians because of its thriving oil industry and the opportunities for irregular work (Mensard, 2004; Pouessels, 2017).

A watershed in Tunisian migration came with the third wave, between the mid-1970s and the early 1980s. This was an oil crisis period for Western European countries, which led to the end of guest-workers' agreements and the introduction of visa for third country nationals. It was also the period during which Tunisian immigrant workers in Europe turned permanent, due to the difficulties to re-enter the host country and the introduction of family reunification facilities by some receiving countries (Mensard, 2004; Pouessels, 2017). On the other hand, because of tensions with Libya, many Tunisian migrants returned to Tunisia. Such changes in circumstances resulted in new destination countries for Tunisians in Europe, such as Spain, Italy and Greece, and in the MENA region, especially towards oil-producing Gulf countries. However, these new migrations were mostly composed of by vulnerable and irregular migrants (Mensard, 2004; Pouessels, 2017).

Finally, a new wave of migration, one displaying the characteristics of a globalisation induced migratory process, started in the early 1980s. This saw the increase of permanent migration to developed countries, especially Europe as it will be shown below. Moreover, a new population of skilled migrants joined previous cohorts of unskilled ones. Indeed, the marginalisation of the country from globalisation processes manifested itself also with an increased shortage of work opportunities for skilled and, especially, highly skilled Tunisians, who, as a consequence, increasingly decided to migrate, further exacerbating the marginalisation of Tunisia (Mensard, 2004; Pouessels, 2017). On the other hand, migrants heading towards the Maghreb are mainly temporary, low-skilled ones and those going to Gulf States mostly doing so through temporary labour migration programmes (Pouessels, 2017: 207, 208).

It is indeed undeniable that the number of Tunisians abroad has been constantly on the rise in the last decades, with an increase of the total number of emigrants registered by UN DESA from 465,676 in 1990 to 813,213 in 2019 (Fig. 3.9).

It is worth noting that in all years considered, extra-regional migration to more developed countries, especially European ones, has prevailed over migration in North Africa and the Middle East (Fargues, 2007; Fargues & Fandrich, 2012). In all years considered, around 90% or above of Tunisian born migrants emigrated to developed countries, mostly to Europe. This is in contrast to the decreasing percentage of the population born in Tunisia emigrating to the Northern Africa and Western Asian region, which moved from 10.3% in 1990 to only 5% in 2019 (Table 3.2). The figures reported below, of course, underestimate the real number of Tunisians abroad as they are estimates of regular migrants, whereas the phenomenon of irregular migration is a substantial one (Pouessels, 2017; Table 3.3).

In 2019 the country with the highest number of migrants born in Tunisia was France with 427,897 Tunisian migrants, 56% of the total number of Tunisians residing in Europe or North America. The second country of residence of Tunisians in Europe was Italy, with a total of

Fig. 3.9 Number of Tunisian migrants 1990–2019 (*Source* UN DESA https://www.un.org/en/development/desa/population/migration/data/estimates2/estimates19.asp)

Table 3.2 Population born in Tunisia by broad regions of emigration 1990–2019

Year	*Major area, region, country or area of destination*	*Tunisia*	*% Over tot. World*
1990	**World**	465,576	
1990	*Geographic regions*		
1990	Africa	4073	0.9
1990	Asia	44,331	9.5
1990	Europe	409,534	88.0
1990	Latin America and the Caribbean	142	3.3
1990	Northern America	6539	1.4
1990	Oceania	957	0.2
1990	**Northern Africa and Western Asia**	47,986	10.3
1995	**World**	476,954	
1995	*Geographic regions*		
1995	Africa	4355	0.9
1995	Asia	43,779	9.2
1995	Europe	418,418	87.7
1995	Latin America and the Caribbean	133	0.0
1995	Northern America	9167	1.9
1995	Oceania	1102	0.2
1995	**Northern Africa and Western Asia**	47,714	10.0
2000	**World**	486,980	
2000	*Geographic regions*		
2000	Africa	4953	1.0
2000	Asia	41,541	8.5
2000	Europe	427,231	87.7
2000	Latin America and the Caribbean	127	0.0
2000	Northern America	11,908	2.4
2000	Oceania	1220	0.3
2000	**Northern Africa and Western Asia**	45,845	9.4
2005	**World**	579,219	
2005	*Geographic regions*		
2005	Africa	5364	0.9
2005	Asia	35,443	6.1
2005	Europe	521,301	90.0
2005	Latin America and the Caribbean	109	0.0
2005	Northern America	15,695	2.7
2005	Oceania	1307	0.2
2005	**Northern Africa and Western Asia**	40,178	6.9

(continued)

Table 3.2 (continued)

Year	*Major area, region, country or area of destination*	*Tunisia*	*% Over tot. World*
2010	**World**	616,386	
2010	*Geographic regions*		
2010	Africa	5918	1.0
2010	Asia	32,972	5.3
2010	Europe	556,442	90.3
2010	Latin America and the Caribbean	174	0.0
2010	Northern America	19,496	3.2
2010	Oceania	1384	0.2
2010	**Northern Africa and Western Asia**	37,839	6.1
2015	**World**	759,817	
2015	*Geographic regions*		
2015	Africa	7337	1.0
2015	Asia	34,447	4.5
2015	Europe	595,520	78.4
2015	Latin America and the Caribbean	207	0.0
2015	Northern America	121,021	15.9
2015	Oceania	1285	0.2
2015	**Northern Africa and Western Asia**	40,056	5.3
2019	**World**	813,213	
2019	*Geographic regions*		
2019	Africa	7946	1.0
2019	Asia	34,563	4.3
2019	Europe	638,385	78.5
2019	Latin America and the Caribbean	233	0.0
2019	Northern America	130,688	16.1
2019	Oceania	1398	0.2
2019	**Northern Africa and Western Asia**	40,743	5.0

Source UN DESA https://www.un.org/en/development/desa/population/migration/data/estimates2/estimates19.asp

109,387 people, 14.2% of the total of Europe and North America (Table 3.4).

The extent to which migration is vital for the Tunisian economy is demonstrated by the level, composition and evolution of its remittances over recent years. Tunisia is heavily reliant on migrants' remittances, as an estimated 10% of Tunisians (first and second generation) live

Table 3.3 Population born in Tunisia emigrated to Northern Africa and Western Asia 2019

Northern Africa and Western Asia	*40,743*
Northern Africa	6765
Algeria	
Egypt	775
Libya	3272
Morocco	2718
Sudan	
Tunisia	
Western Sahara	
Western Asia	33,978
Armenia	
Azerbaijan	
Bahrain	492
Cyprus	36
Georgia	
Iraq	319
Israel	22,894
Jordan	269
Kuwait	1889
Lebanon	
Oman	
Qatar	997
Saudi Arabia	
State of Palestine	595
Syrian Arab Republic	
Turkey	699
United Arab Emirates	5788
Yemen	

Sources UN DESA https://www.un.org/en/development/desa/population/migration/data/estimates2/estimates19.asp

abroad (Pouessel, 2017: 205).[6] Indeed, remittances constituted 10.4% of Tunisian total trade in 2018, up from an already substantial 9.6% in 2005 (Fig. 3.10). In 2019, migrants' remittances to Tunisia as a share of

[6] Pouessel, S. (2017). Tunisia and its diaspora: Between protection and control. In A. Weinar (Ed.), Emigration and diaspora policies in the age of mobility (pp. 205–220). Springer; Brand, L. (2006). Citizens abroad: State and emigration in the Middle East and North Africa. Cambridge: Cambridge University Press; Natteur, K. (2015). Revolution and political transition in Tunisia: A migration game changer?, Migration Information Source, the Online Journal of the Migration Policy Centre, May 28th, http://www.migrationpolicy.org/article/revolution-and-political-transition-tunisia-migration-game-changer.

Table 3.4 Population born in Tunisia emigrated to Europe and North America 2019

EuUROPE AND NORTHERN AMERICA	*769,073*	*% Over tot. Europe and North America*
Europe	638,385	83
Eastern Europe	4221	0.55
Belarus		
Bulgaria	151	
Czechia	925	
Hungary	416	
Poland	433	
Republic of Moldova		
Romania	1784	
Russian Federation	334	
Slovakia	178	
Ukraine		
Northern Europe	15,719	2.05
Channel Islands		
Denmark	1023	
Estonia	21	
Faroe Islands	15	
Finland	983	
Iceland	37	
Ireland	109	
Isle of Man		
Latvia	4	
Lithuania		
Norway	1144	
Sweden	5881	
United Kingdom	6502	
Southern Europe	113,805	14.8
Albania		
Andorra		
Bosnia and Herzegovina		
Croatia		
Gibraltar		
Greece	619	
Holy See		
Italy	109,387	14.22
Malta	719	

(continued)

Table 3.4 (continued)

EuUROPE AND NORTHERN AMERICA	*769,073*	*% Over tot. Europe and North America*
Montenegro		
North Macedonia		
Portugal	139	
San Marino		
Serbia	40	
Slovenia	69	
Spain	2832	
Western Europe	504,640	65.62
Austria	4356	
Belgium	15,682	
France	427,897	55.64
Germany	38,220	
Liechtenstein	23	
Luxembourg	1004	
Monaco		
Netherlands	5001	
Switzerland	12,457	
Northern America	130,688	16.99
Bermuda		
Canada	18,568	
Greenland		
Saint Pierre and Miquelon		
United States of America	130,541	16.97

Source UN DESA https://www.un.org/en/development/desa/population/migration/data/estimates2/estimates19.asp.

the GDP were 5.3%.

Moreover, in 2014, total transfers from emigrants abroad peaked at 2346 million USD dollars, configuring remittances as an extremely important source of foreign currency in Tunisia (Fig. 3.11).

As one would expect given the analysis effected so far, almost the totality of remittances to Tunisia comes from outside the MENA region. European countries provide the highest amount of remittances inflows to Tunisia. In 2017, the transfers of Tunisians residing in European countries amounted to 1692 million dollars, representing 89% of total remittances transfers to Tunisia in that year. France alone provided for a staggering 60% of the total transfer for 2017. The second country contributing the most to Tunisian inflows was Italy with 17% of the total, with Germany

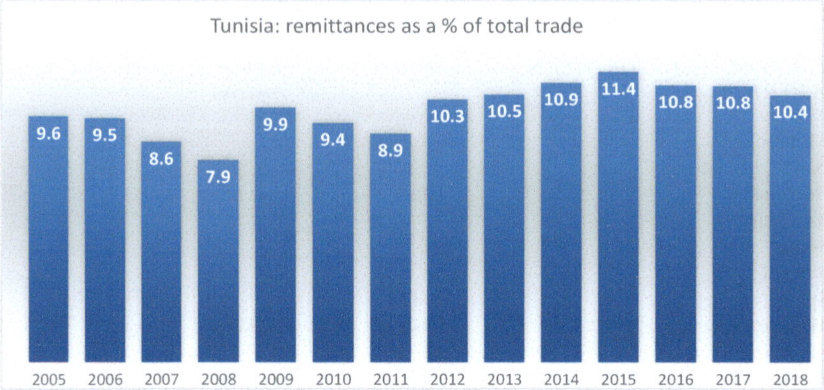

Fig. 3.10 Tunisia: remittances as a percentage of Total trade 2005–2018 (*Source* UNCTAD https://unctadstat.unctad.org/wds/TableViwer/tableView.aspx)

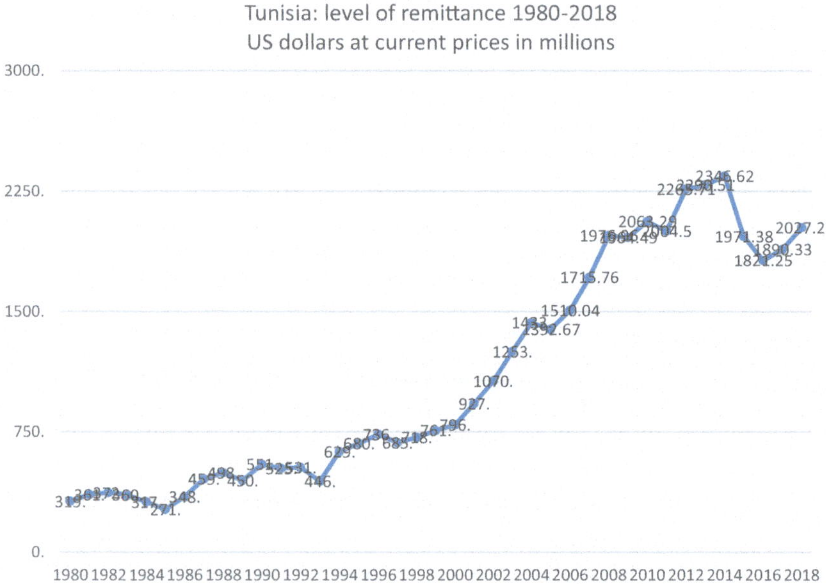

Fig. 3.11 Evolution of remittances in Tunisia 1980–2018 (*Source* UNCTAD https://unctadstat.unctad.org/wds/TableViewer/tableView.aspx)

coming in third with around 5% of the total. These three European countries alone represent 81% of total transfers to Tunisia. This can be compared with around 74 million dollars coming from the rest of the MENA area (excluding Israel), representing a meagre 3.8% of total transfers. Whereas Canada and the US together provided only 53 million dollars in 2017, around 2.8% of total remittances to Tunisia (see Table 3.5). This is a trend which dates back to at least the decade between 1995 and 2006, when the percentage of remittances coming from European countries remained stable at around 90% of the total (Fourati, 2009: 2).

As noted by Pouessel (2017: 206): "The Tunisian State currently considers emigration to be one of its main social and economic issues, but it has come a long way in recognising the need to engage with its emigrants". Experts point to the existence of two axes in Tunisian migratory policy in the frame of the Economic and Social Plan 2010–2014.[7] On the one hand, the Tunisian government tried to promote regular migration by signing agreements with both European and non-European countries (such as Canada, Australia and so on) outside the MENA region (Poussels, 2017: 206). On the other hand, there was an attempt to establish stronger links with the Tunisian diaspora abroad and to involve successful emigrants in investment and development projects at home (Bel Haj Zekri, 2010; Cassarino, 2014[8]; Poussels, 2017: 206). These two strategies did not prove to be particularly successful (Hibou et al., 2011). In fact, despite efforts at effectively regulating emigration with the Organic Law 2004–6 of 3 February 2004 (amending Law 75–40 of 14 March 1975 on passports and travel documents), there was no stopping of irregular migration from Tunisia.[9] Indeed, according to data from Frontex, between January and September 2019, of the 9231 total number of illegal border crossings in the central Mediterranean route, the top country of origin was Tunisia' with 2168 irregular migrants, followed by Sudan with 1313, Pakistan with 962, Côte d'Ivoire with 936 and

[7] Lixi, L. (2019). The ambivalent drivers of migration governance relations between the EU and Tunisia. In A. Geddes, M. V. Espinoza, L. H. Abdou (Eds.), Dynamics of regional migration governance (pp. 91–108). Leiza Brumat Publisher: Edward Elgar Publishing.

[8] Cassarino, J.-P. (2014). Channelled policy transfers: EU–Tunisia interactions on migration matters. *European Journal of Migration and Law, 16*(1), 97–123.

[9] See for example Hibou (2003), Bayart (2007) and Meddeb (2008).

Table 3.5 Bilateral Remittance Estimates for 2017 using Migrant Stocks, Host Country Incomes, and Origin Country Incomes (millions of US$) (April 2018 version)

Remittance-receiving country (across) *Remittance-sending country (down)*	*Tunisia*	*Share of total*
Algeria	2	0.09
Australia	2	0.10
Austria	11	0.55
Bahrain	1	5.71
Belgium	24	1.26
Canada	26	1.37
Czech Republic	4	0.19
Denmark	3	0.14
Egypt, Arab Rep	3	0.16
Finland	2	0.10
France	1119	58.78
Germany	103	5.41
Greece	4	0.19
Hungary	1	0.03
Israel	66	3.49
Italy	322	16.91
Kenya	2	0.10
Kuwait	4	0.23
Libya	7	0.39
Malta	1	0.06
Netherlands	13	0.69
New Caledonia	2	0.13
Norway	3	0.16
Philippines	1	0.07
Poland	1	0.06
Qatar	4	0.21
Russian Federation	1	2.30
Saudi Arabia	36	1.91
Spain	7	0.38
Sweden	14	0.72
Switzerland	34	1.77
Turkey	2	0.1
Ukraine	1	0.04
United Arab Emirates	17	0.88
United Kingdom	26	1.34
United States	27	1.43

(continued)

Table 3.5 (continued)

Remittance-receiving country (across) *Remittance-sending country (down)*	*Tunisia*	*Share of total*
World	1903	

Notes These estimates are based on the methodology developed by Ratha and Shaw, 2007, "South-South Migration and Remittances," World Bank (www.worldbank.org/prospects/migrationandremittances). The remittance data is for 2017, disaggregated using host country and origin country incomes, and estimated migrant stocks from 2017

These are analytical estimates based on logical assumptions and derived from a global estimation of bilateral remittance flows worldwide. They are not actual officially reported data. The caveats attached to these estimates are (a) the data on migrants in various destination countries are incomplete; (b) the incomes of migrants abroad and the costs of living are both proxied by per capita incomes in PPP terms, which is only a rough proxy and (c) there is no way to capture remittances flowing through informal, unrecorded channels

These estimates are based on the Bilateral Migration Matrix 2017

Source World Bank: https://www.worldbank.org/en/topic/migrationremittancesdiasporaissues/brief/migration-remittances-data

Algeria with 776.[10] It has been estimated that already in 2008, Tunisians represented two-thirds of North African nationals (that is, from the Maghreb and Egypt) who irregularly entered Italy (Boubakri, 2013). This happened despite the number of rights that Tunisia granted to regular migrants after the revolution of 2011, such as political and civil rights (voting, multiple nationality, military duty); socio-economic rights (agreements on labour migration, customs/import incentives); social rights (family, social security); and cultural rights (languages, school, media). Moreover, many efforts were made by the new Tunisian regime to reconfigure the relationship between Tunisia and its diaspora, involving non-state actors present abroad in its emigration and diaspora policy (Lexi, 2019: 98; Poussels, 2017: 206).

Despite this, however, there has been no stemming of the level of legal and irregular extra-regional migration in the years following the Arab Spring. In fact, the rate of growth of the stock of Tunisian migrants in the decade between 2000 and 2010 and 2010 and 2019 has remained constant at around 32%. This is a dynamic that has already been established for decades as a consequence of the growing marginalisation of Tunisia from the global political economy (Bel Haj Zekri, 2010;

[10] https://frontex.europa.eu/along-eu-borders/migratory-map/.

Boubakri, 2011: 17; Fourati, 2008). Further proof of the marginalisation Tunisia is the increasing rate of the phenomenon of 'brain drain', meaning the number of highly skilled Tunisians deciding to leave the country (Boubakri, 2011, 2013).

The figures available illustrate an increase by around 38% in the rate of 'brain drain' from Tunisia in 2007–2008 with respect to the period between 2001 and 2007 (Fourati, 2010). Brain drain is mainly directed towards the EU, with France, Germany and Italy as the main destination countries (see Tables 3.6 and 3.7). Looking at the first decade of the 2000s, brain drain was almost completely extra-regional, with a share of 80% over the total (Table 3.6).

The marginalisation of Tunisia from the global political economy plays a crucial role in the dynamics of brain drain. There is indeed a dialectic relationship between marginalisation and brain drain. On one side, Tunisia's lack of technological skills makes it more difficult for the country to join the global value chains established in other regions of the globe (Talani, 2014). This makes it more difficult for skilled personnel to be employed in the economy of the country, thus giving them incentives to emigrate to the Global North (Bel Haj Zekri, 2010). This is clearly shown in the data relating to the occupation of skilled workers in Tunisia in the last three decades. Unemployment for workers with higher education increased almost ten times between 1984 and 2009 from 2.3% to 21.6% (Table 3.7). By contrast, those with no or little education saw their unemployment rate decrease from 15.2% to 5.7% in the first case and from 22.4% to 12.3% in the case of workers with only primary education.

The figures above give account of the de-skilling of the Tunisian labour force which takes place precisely in the period in which an up-skilling of the labour force would have been necessary for Tunisia to join the global political economy. This, inevitably, prompted unemployed Tunisian skilled workers to move to more developed countries where their skills could be gainfully employed thus adding to 'brain drain' (Bel Haj Zekri, 2010: 6).

This dynamic does not seem to have been modified in the aftermath of the so-called Arab Spring (Talani, 2014). The wave of patriotism following the successful outcome of the uprising of 2011 in Tunisia did not convince many Tunisian expatriates either to repatriate or to invest

Table 3.6 Tunisia: evolution of brain drain according to country of destination, 2001–2008

Destination country	*May 2001–May 2002*	*May 2001–May 2002 (%)*	*May 2005–May 2006*	*May 2005–May 2006 (%)*	*May 2006–May 2007*	*May 2006–May 2007 (%)*	*May 2007–May 2008*	*May 2007–May 2008 (%)*
France	2930	50.3	2620	45.6	3070	54.2	3650	45.6
Germany	490	8.4	540	9.4	970	17.1	620	7.8
Italy	320	5.5	330	5.7	410	7.2	830	10.4
Other European Countries	720	12.3	630	11.0	260	4.6	500	6.3
Gulf countries	530	9.1	630	11.0	280	4.9	140	1.8
Other Arab countries	500	8.6	320	5.6	490	8.7	1440	18.0
Other countries	340	5.8	680	11.8	180	3.2	820	10.3
Total	5830	100	5750	100	5660	100	8000	100

Source Based on data by Fourati (2010: 12)

Table 3.7 Tunisia: Evolution of the unemployment rate by level of education (%)

	1984	*1994*	*1999*	*2004*	*2005*	*2006*	*2007*	*2009*
Illiterate	15.2	16.2	11.9	12.7	7.8	8.0	5.9	5.7
Primary	22.4	19.2	18.9	15.7	15.7	15.2	13.5	12.3
Secondary	11.0	13.0	16.4	14.7	14.9	14.3	15.4	15.3
Higher	2.3	3.8	8.6	10.2	14.8	17.5	19.0	21.6
Total	16.4	15.8	16.0	14.2	14.8	14.3	14.1	14.2

Source Bel Haj Zekri (2010: 6)

in the future of the country to reverse its path towards marginalisation (Abdelfatta, 2011).[11]

In fact, the profile of the Tunisian migrant after the Arab Spring rather resembles the profile of the Tunisian migrants before the events of 2011. According to a survey of the Observatoire National de la Jeunesse implemented in 2005, 75.9% of Tunisians aged between 15 and 29 years of age were entertaining the idea of emigrating regardless of their socio-economic position, gender or level of education. This was up from 25% in 1996. They even declared their willingness to engage in an irregular migratory process in 15% of cases.

The profile of the Tunisian migrant after the Arab Spring confirmed the picture of 2005 (Bel Haj Zekri, 2010). Indeed, around 25,800 of the Tunisian migrants arriving to Italian shores between January and June 2011 were young people, their age varying between 20 and 30, with some minors of 13 to 17 years of age. They were coming from different socio-economic milieus, including both from the poorest towns of Tunisia in the centre west, such as Sidi Bouzid, Thala, Kasserine, Gafsa, and the south, like Gabes, El Hamma, Zarzis, Tataouine. But they were also from the worst quarters of the big cities of Tunis and Sousse. In a way, their profile was similar to that of Mohammed Bouazizi, the young man who initiated the Tunisian uprising by immolating himself. They belonged to big families, with four to eight components, sustained by only one bread-winner, with low wages and no social security (Bel Haj Zekri, 2010). Their motivations to leave Tunisia were therefore clearly unaffected by the new political climate and, in a way, the new political freedom did

[11] For a thorough analysis of return migration and investment see Cassarino (2000).

not seem to have reduced their anxiety about the future. When asked, they seemed to still be caught between immolating themselves as Bouazizi did or taking the risk of an irregular migratory journey to Italy, costing as much as between €500 and €1000. Sometimes the expenses of the journey would be funded with money coming from the compensation for the wounds experienced in the course of the 2011 uprising which amounted to around €1500 (Ben Achour & Ban Jemia, 2011).

It seems fair therefore to conclude that liberation from political dictatorship did not change the prospects of this generation of young Tunisians who, as discussed above, are suffering the economic consequences of the marginalisation of their country from the global political economy. Moreover, the continuation of unhindered emigration from Tunisia after the revolution of 2011 illustrates the extent to which discontent was not related to the political dimension, but to economic marginalisation. In a very paradoxical way, young Tunisians decided to use their renewed liberty to leave the country, a phenomenon which is not likely to stop soon.

Egypt and Marginalisation: Extra-Regional Migration and Brain Drain

Looking at the history of Egyptian migration, it emerges that traditionally, Egyptians were reluctant to move abroad. Famously, a study from the 1930s stated that Egyptians preferred their own country and rarely undertook migration processes always coming back to Egypt.[12]

On the other hand, Egypt was a country of transition and immigration and, to a large extent, still is.[13] Even in the course of the hardest moments in Egyptian history, migratory trends were negligible. In 1927, sources report that the number of Egyptian abroad was a mere 21,565, of which 18,383 lived in neighbouring Sudan (Cleeland, 1936: 36). This is despite the fact that Egypt, up until the 1952, was an extremely poor

[12] Cleland, W. (1936). *The population problem in Egypt: a study of population trends and conditions in modern Egypt* (p. 36, 52). Lancaster, Pa.: Science Press Printing Company.

[13] Choucri, N. (1978). *Migration processes among developing countries: The Middle East* Cambridge: MIT. (There are occasional comments on the contrary, see Bohning (1984: 67) and Birks, J. S., & Sinclair, C. A. (1979). *Egypt: A frustrated labour exporter?. Middle East Journal*, *33*(3), 288–303 and Shell, F.F., 1987, p. 27, note 10).

country.[14] The reluctance of Egyptians to move, continued even after the revolution and, before 1974, very few Egyptians emigrated for jobs abroad, and those who did emigrate were mainly were highly skilled or politically determined professionals.

The debate on the motivations of migration from Egypt is a long-standing one. Recently, the debate has been enriched by contributions, stressing in particular the political motivations behind Egyptian emigration. Tsourapas (2019)[15] looks at the relation between authoritarianism and emigration in Egypt. The analysis is divided chronologically into three phases, each one relating to the three successive authoritarian regimes of Nasser, Sadat and Mubarak. From a conceptual point of view, the author distinguishes between the effects of migration at the domestic, regional and international level. The theoretical discussion engages two different debates in political science, one about authoritarianism and one about migration.

Implicitly adopting a realist approach to migration, Tsourapas's contribution stresses how intra-regional migration of Egyptians in the Arab world has been used by successive Egyptian governments to allow for regime survival. The analysis is predicated on the definition of Egypt, from the Nasser government until the Mubarak one, as a non-democratic regime using labour emigration policy to ensure its survival. However, whereas during the Nasser period, emigration policy was very restrictive, with Sadat and then Mubarak, Egypt started using much more permissive policies in the context of the Arab region to achieve regime stability. The approach adopted by Tsourapas is heavily reliant on realist assumptions, a la Myron Weiner, in attributing to the State the ability to control migration. This is an interesting assumption which should, however, be contextualised within existing theoretical debates.

Indeed, political motivations relating to the history of Egyptian migration, as opposed to economic ones, should be discussed in relation to the dynamics of intra-regional and extra-regional migration. In fact, it is difficult to overestimate the role that economic factors played in both intra-regional and, more recently, extra-regional migration from Egypt.

[14] Amin, G. (2000). *Whatever happened to Egyptians* (p. 95). Cairo: The American University in Cairo Press.

[15] Tsourapas, G. (2019). *The politics of migration in modern Egypt—Strategies for regime survival in autocracies*. Cambridge University Press.

With the above considerations in mind, it is appropriate to distinguish between different phases of Egypt's migratory policies, not only with respect to the political dimension, but also the economic situation of the country, the region and the globe. Finally, it is necessary to understand the role that globalisation played in changing Egyptian migratory practices and policies.

In the literature, Egyptian migratory trends after the revolution of 1952 have been divided into five different phases (Amin, 2000; Appleyard, 1999; Dessouki, 1982; Feiler, 2003; Zohry & Harrell-Bond, 2003) (Table 3.8).

During the first phase, mostly students and professionals emigrated for reasons relating to very precise socio-economic and political conditions. They almost inevitably belonged to the upper class, were highly educated and left the country due to nationalisation and socialist policies implemented by Nasser. Thus, this wave of emigration was the direct result of the new political regime imposed by Nasser and the related disruption of the previous socio-economic order. Those leaving the country in this occasion were the privileged groups of the previous political order which felt endangered and marginalised in the post-revolution climate. Their destination countries were mostly Canada and the US where they felt politically secure. Concluding, as far as emigration from Egypt in the course of the 1960s is concerned, its existence was very limited in terms of number and characteristics.

Table 3.8 Egyptian migration phases

Phases:	*Migratory dynamics:*
Phase 1: Before 1974 The early phase of migration	Very limited migration, mostly highly skilled or politically determined
Phase 2: 1974–84 The expansion phase	Expansion of temporary migration in the Arab oil-producing countries
Phase 3: 1984–88 The contraction phase	Contraction of migration to the Arab world
Phase 4: 1988–92 The deterioration phase	Significant flow of return migrants from the Gulf area to Egypt
Phase 5: 1992–nowadays The recent phase	Significant increase of permanent migration to the more developed world

In the literature, it is reported that migration patterns changed in the course of the 1970s, with 1973 as a turning point (Amin, 2000; Appleyard, 1999; Dessouki, 1982; Feiler, 2003; Zohry & Harrell-Bond, 2003). The war with Israel had an impact on migratory dynamics from Egypt, producing a displacement of people to and from the canal area and the mobilisation of the military. Furthermore, Sadat's open door policy eliminated previous restrictions to the movement of Egyptians and facilitated intra-regional migration. However, what impacted intra-regional migration the most was the first oil shock, attracting Egyptians looking for jobs to the booming members of the OPEC. These migratory flows were of a temporary nature and were treated as such by the Egyptian government who actively supported the return migration of its citizens (Talani, 2010).

In the third phase of the evolution of Egyptian migration, from 1984 to 1988, there was a contraction in the number of Egyptians migrating within the MENA region. This was due to geo-politically determined economic reasons, and, in particular, the start of the Iran–Iraq war curtailing job-opportunities in the area.

A massive decline in the number of Egyptian migrants occurred during the second half of the 1980s, with the Iran–Iraq war, the fall of oil prices, the reduction in job opportunities for construction workers in Arab Gulf States and a new policy by Arab countries to employ their own nationals (Zohry & Harrell-Bond, 2003[16]). Between 1988 and 1992, the flows from the Gulf area to Egypt increased further.

Migration in the area was substantially impacted by the first Gulf War. Around two million people, including more than a million foreigners and two-thirds of Kuwait's citizens, had to leave Kuwait, Iraq and Saudi Arabia. With the end of the war, all Egyptians were forced back to their country or to other countries of the MENA area.

Finally, in the last phase of Egyptian migration, the impact of globalisation is fully felt. The country has been increasingly marginalised from the global political economy. As such, one can witnesses a new trend of a substantial increase in extra-regional migration to the global North. In 2000, the number of Egyptians in Europe and North America was 326,706, up by around 43% from 1990 when it was 298,591.

This trend towards an increase in permanent migration was confirmed in the following decades of 2000 to 2019 (Fig. 3.12). Data from

[16] See Zohry, A., & Harrell-Bond, B. (2003). *Contemporary Egyptian migration: An overview of voluntary and forced migration*.

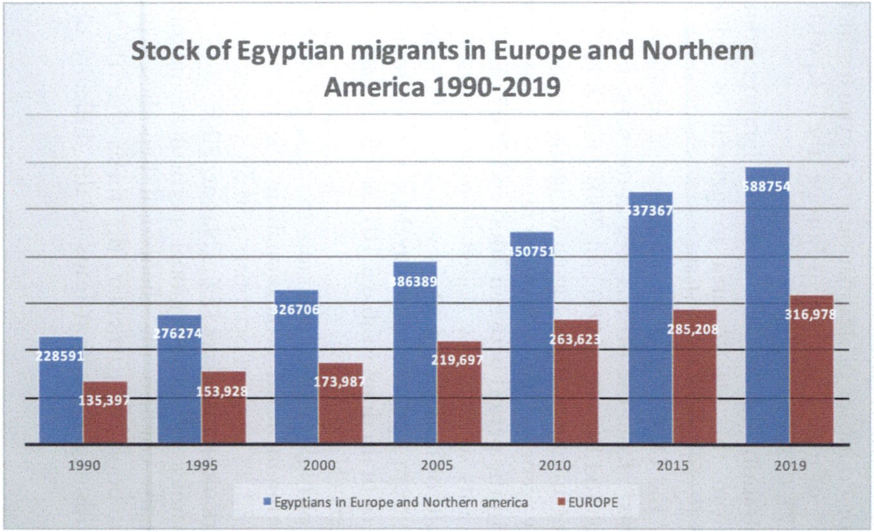

Fig. 3.12 Stock of Egyptian migrants in Europe and Northern America 1990–2019 (*Source* UN DESA https://www.un.org/en/development/desa/population/migration/data/estimates2/estimates19.asp)

UNDESA shows a marked increase in the total number of Egyptians registered as permanent migrants in Europe and Northern America, from 326,706 in 2000 to 588,754 in 2019 (an increase of more than 80%). Of those more than 50% were in Europe. Of course, these figures grossly underestimate the real number of Egyptians abroad because they do not include estimates for irregular migration.

In 2019, the major countries of destination for Egyptians in Europe were Italy, with 38.5% of total Egyptians in Europe, the UK, with 12.5% and France, with 10.7% (Table 3.9).

A further demonstration that this new wave of migration from Egypt in particular, and from the MENA area in general, is induced by marginalisation is the dynamics of brain drain.

Data produced by the OECD reveals an increase of 6% in the rate of over-qualified migrants[17] from the MENA area in the period 2010–2011 as compared to the period between 2000 and 2001 (Fig. 3.13).

Indeed, permanent migrants from Egypt to more developed countries have consistently recorded high education rates, showing a tendency towards brain drain. In the period between 2000 and 2007, 53% of Egyptian permanent migrants were highly skilled,[18] with university degrees or over, while the rate of unskilled workers was only 4.11%. Moreover, permanent migration from Egypt to more developed countries has always represented a source of brain drain (Table 3.10), as it has mainly involved more educated workers (ILO, 2008).

The qualification of highly skilled migrants from Egypt at the beginning of the millennium included professionals in the medical sciences (18.3%); engineers (32.2%); social scientists (36.5%); workers in basic sciences (8%) and agricultural workers (5%) (Nassar, 2005: 8).

Overall, the dynamics of extra-regional permanent migration of highly skilled Egyptians confirm the theoretical hypotheses proposed in Chapter 1 of this book. In particular, the thesis that the paradox of marginalisation within globalisation leads to the difficulty of locally

[17] Over-qualification is defined as the share of tertiary-educated migrants who work in low- and medium-skilled occupations. Data on over-qualification cover 32 OECD destinations (data on Turkey and Korea are not available). The population refers to persons aged 15 to 64.

[18] Highly skilled workers are defined as those studying or having studied towards a university degree or possessing equivalent experience in a given academic field (IOM, 2005).

Table 3.9 Stock of Egyptian migrants in Europe 2019

Europe	*316,978*	*% Over total*
Eastern Europe	5115	1.6
Belarus		
Bulgaria	317	0.1
Czechia	907	0.3
Hungary	1690	0.5
Poland	478	0.2
Republic of Moldova		
Romania	436	0.1
Russian Federation	1059	0.3
Slovakia	228	0.1
Ukraine		
Northern Europe	54,339	17.1
Channel Islands		
Denmark	2022	0.6
Estonia	104	0.0
Faroe Islands		
Finland	1767	0.6
Iceland	41	0.0
Ireland	584	0.2
Isle of Man		
Latvia	53	0.0
Lithuania	51	0.0
Norway	1278	0.4
Sweden	8846	2.8
United Kingdom	39,593	12.5
Southern Europe	157,732	49.8
Albania		
Andorra		
Bosnia and Herzegovina		
Croatia	228	0.1
Gibraltar		
Greece	27,549	8.7
Holy See		
Italy	122,040	38.5
Malta	1045	0.3
Montenegro		
North Macedonia	1970	0.6
Portugal	427	0.1
San Marino		
Serbia	78	0.0

(continued)

Table 3.9 (continued)

Europe	*316,978*	*% Over total*
Slovenia	118	0.0
Spain	4277	1.3
Western Europe	99,792	31.5
Austria	17,578	5.5
Belgium	4895	1.5
France	34,064	10.7
Germany	21,591	6.8
Liechtenstein		
Luxembourg	288	0.1
Monaco		
Netherlands	14,735	4.6
Switzerland	6641	2.1

Source UN DESA https://www.un.org/en/development/desa/population/migration/data/estimates2/estimates19.asp

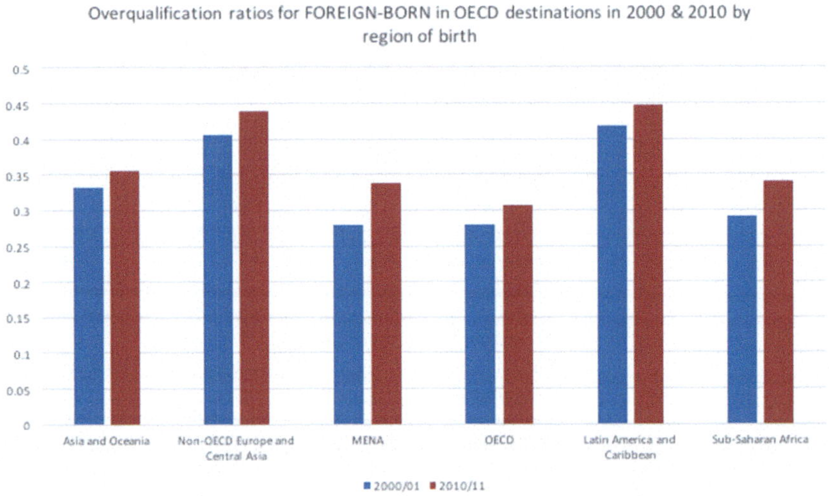

Fig. 3.13 Over-qualification ratios of foreign born in OECD destinations in 2000–2010 by region of birth (*Source* Database on Immigrants in OECD Countries (DIOC) 2000/01 and 2010/11)

Table 3.10 Egypt: Permanent migrants flows by level of education, 2000–2007

Permanent migrants flows by level of education in the period 2000–2007

	Ph.D.	*Masters*	*High diploma*	*University*	*Intermediate*	*No qualifications*	*Total*
2000	8	12	8	243	233	32	536
2001	5	14	12	322	334	31	718
2002	10	10	11	274	323	24	652
2003	4	2	2	167	104	14	293
2004	3	16	9	200	115	14	357
2005	5	15	9	196	180	20	425
2006	4	12	7	210	124	11	368
2007	3	10	10	198	145	7	373
Total	42	91	68	1810	1558	153	3722

Source CAPMAS (2007)

qualified workforces to find employment in the country. This further contributes to the marginalisation of the country triggering a vicious circle (IOM, 2005).

Marginalisation is also proven by the comparison between FDI flows and remittances. Data shows how remittances represent a very important source of foreign reserves, while FDIs score far worse in this respect. In 2018, remittances were almost four times higher than FDIs and in 2019, they represented a staggering 8.8% of the GDP. It is hard to overestimate their relevance for the Egyptian economy (Fig. 3.14).

Turning to the impact of the Arab Spring on Egyptian migratory flows, it seems that, on the basis of the studies conducted so far, the 2011 uprising did not change already established trends. The main characteristics of extra-regional migration after 2011 are indeed comparable to the period before the revolts (Abdelfattah, 2011; Fargues, 2007; Fargues & Fandrich, 2012; Hafez & Ghaly, 2012).

In 2012, the Center of Migration and Refugee Studies of the American University in Cairo assessed the perception of Egyptian migration in the aftermath of the Arab Spring. The pool of people questioned about future migratory patterns in Egypt included professors, political activists, political party founding members, analysts and migration experts. From the analysis, it transpired that migration, especially brain drain, would continue unhindered after the revolution. As one of the participants to the study put it:

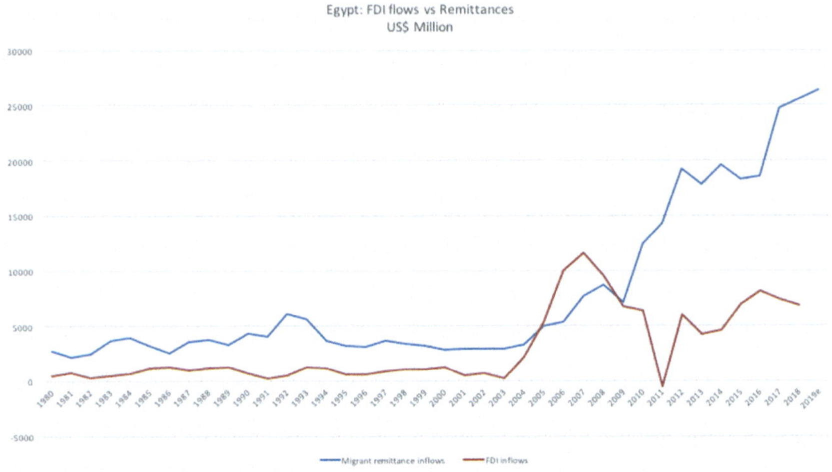

Fig. 3.14 Egypt: FDI flows vs Remittance, US Million, 1980–2019 (*Source* World Bank and UNCTAD. Elaboration of the author)

> push factors exist and I think increased after the revolution. (Hafez & Ghaly, 2012: 11)

The reason why this is the case is related to the theoretical background so far analysed. Indeed, marginalisation was also at the roots of the discontent sweeping the country in the wake of the Arab Spring. As the political change of regime did not manage to resolve the question of marginalisation from the global political economy, the outflow of migrants has no reason to stop (Abdelfattah, 2011: 8). This is confirmed by the survey results conducted by the International Organization for Migration on the motivations for migration after the uprising of 2011 (IOM, 2011). The survey included interviews to 750 Egyptians. Of these, 79% quoted unemployment as the main reason for migration; the low level of wages and salaries were cited by 43% of respondents. Other major motivations to leave were also corruption, security, education and health. On the other hand, the political dimension was not considered a problem, and more than 50% of respondents had positive expectation for the political future of the country. Finally, the revolution did not seem to have changed the intentions of respondents to leave the country, as many of the respondents had already taken this decision before the revolution and had not changed idea after it happened. Thus, the main push factors for migration were the same before and after the events of 2011: a lack of job opportunities and low living standards (IOM, 2011).

If we effect a comparison between the IOM survey after the Arab Spring and the Survey of Young People in Egypt (SYPE) of 2009, implemented by the Population Council, the outcomes are very similar. Of the 15,000 Egyptians between the ages of 10 and 29 interviewed, the percentage of those who were willing to migrate resembled the one of the later survey. The main motivation for departure, the lack of employment, was the same in both surveys. Furthermore, the results replicated the ones of a push and pull factors survey implemented in various MENA countries, including Egypt, in 2000. All in all, academics are consensual in identifying the same underlying factors between migration from Egypt before and after the Arab Spring. Only for very small groups, such as the Copts, the Christians and certain segments of the elite, factors such as political instability and insecurity have influenced their decision to migrate (Abdelfattah, 2011: 8). In conclusion, the Arab Spring did not seem to have modified migratory patterns from Egypt, which followed the same dynamics as before (Fargues & Fandrich, 2012).[19]

The Profile of the Moroccan Migrant in the Globalisation Era

Data on the stock of Moroccan migrants abroad shows an increasing propensity of Moroccans to emigrate. UN DESA reports that between 1990 and 2019, the number of Moroccans abroad doubled, from 763,564 to 1,489,163 with an average annual growth rate of 12%, as opposed to a population growth rate of 1.3% from 1990 to 2017.[20]

Looking at the dynamics of intra-regional and extra-regional migration, in 2019, 93.1% of Moroccans abroad had emigrated in Europe or North America, up from 87.9% in 1990, and only 6.4% in the North Africa and West Asia region, down from 11.5% in 1990. This data clearly shows a substantial increase in extra-regional migration in the last 30 years

[19] On the Egyptian route for refugee see Migration, Mobilities and the Arab Spring Spaces of Refugee Flight in the Eastern Mediterranean Edited by Natalia Ribas-Mateos, Universitat Autònoma de Barcelona, Spain, Edward Elgar, 2016.

[20] UN DESA http://data.un.org/Data.aspx?d=POP&f=tableCode%3a1.

in line with the paradox of marginalisation within globalisation and the increase in migration to western developed countries (Fig. 3.15).

The largest majority of Moroccans, by far, emigrated to Europe. In 2019, around 88% of Moroccans abroad were residing in Europe. 35% of total Moroccan emigrants were in Southern Europe, 21.6% in Spain and 13.5% in Italy, while more than 50% were in Western Europe, with the highest number recorded in France with 33.5% (Table 3.11).

The data above, however, grossly underestimates the real number of Moroccans abroad, given the flourishing of irregular migratory flows from the country. The Moroccan Consulate's data, for example, gives a much higher number for Moroccans residing abroad, namely 3.3 million in 2007. To be sure, since 1981, about 445,000 Moroccans underwent regularisation in France, Belgium, Italy and Spain, clearly pointing to

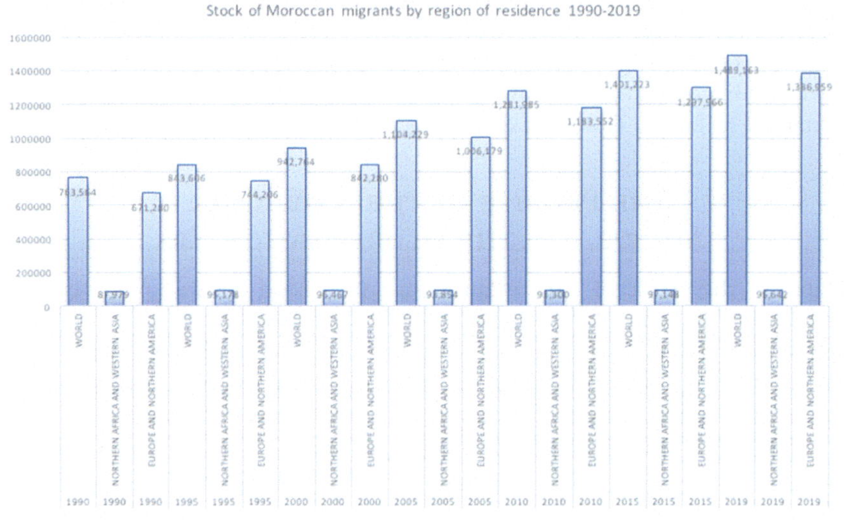

Fig. 3.15 Stock of Moroccan migrants by region of residence 1990–2019 (*Source* UN DESA https://www.un.org/en/development/desa/population/migration/data/estimates2/estimates19.asp)

Table 3.11 Stock of Moroccan emigrants in Europe over world total by regions and countries

	Stock of migrants	*% Over world total*
World	1,489,163	
Africa	10,148	0.68
Asia	89,457	6.01
Europe	1,307,023	87.77
Latin America and the Caribbean	1215	0.08
Northern America	79,936	5.37
Oceania	1384	0.09
Northern Africa and Western Asia	95,642	6.42
Europe and Northern America	1,386,959	93.14
Europe	1,307,023	87.77
Eastern Europe	1416	0.10
Belarus		
Bulgaria	63	0.00
Czechia	67	0.00
Hungary	105	0.01
Poland	52	0.00
Republic of Moldova		
Romania	288	0.02
Russian Federation	339	0.02
Slovakia	10	2.95
Ukraine	492	0.03
Northern Europe	24,362	1.64
Channel Islands		
Denmark	3036	0.20
Estonia	3	0.00
Faroe Islands		
Finland	1006	0.07
Iceland	77	0.01
Ireland	93	0.01
Isle of Man		
Latvia	0	
Lithuania		
Norway	2876	0.19
Sweden	5769	0.39
United Kingdom	11,502	0.77
Southern Europe	525,998	35.32

(continued)

Table 3.11 (continued)

	Stock of migrants	*% Over world total*
Albania		
Andorra	230	0.02
Bosnia and Herzegovina		
Croatia	16	0.00
Gibraltar	374	0.03
Greece	566	0.04
Holy See		
Italy	201,935	13.56
Malta	339	0.02
Montenegro		
North Macedonia		
Portugal	676	0.05
San Marino		
Serbia	21	0.00
Slovenia	9	0.00
Spain	321,832	21.61
Western Europe	755,247	50.72
Austria	1017	0.07
Belgium	109,492	7.35
France	499,379	33.53
Germany	46,283	3.11
Liechtenstein		
Luxembourg	1160	0.08
Monaco		
Netherlands	87,316	5.86
Switzerland	10,600	0.71

Source UN DESA https://www.un.org/en/development/desa/population/migration/data/estimates2/estimates19.asp

the magnitude of the phenomenon of irregular migration.[21] Indeed, the number of illegal border crossing on the Western Mediterranean route, the route between Spain and Morocco by sea and land, was 57,034 in 2018, much more than double the number in 2017 when it was 23,063. The number of illegal border crossings in 2019, between January and

[21] CARIM—Migration Profile Morocco The Demographic-Economic Framework of Migration The Legal Framework of Migration The Socio-Political Framework of Migration Report written by ANNA DI BARTOLOMEO, TAMIRACE FAKHOURY and DELPHINE PERRIN on the basis of CARIM database and publications November 2009.

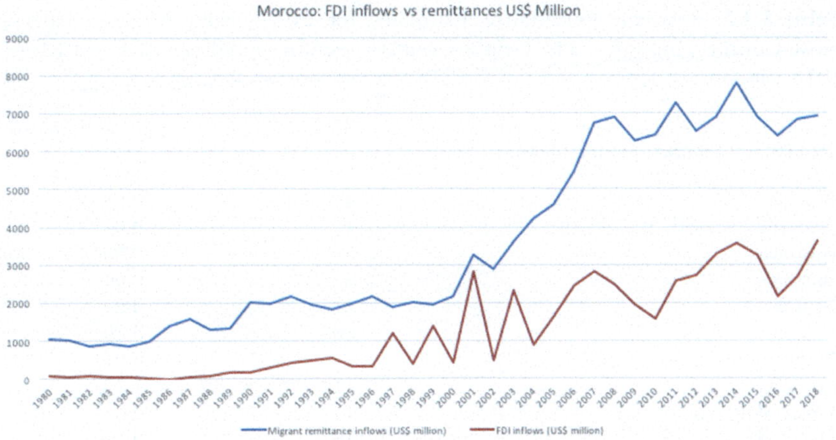

Fig. 3.16 Morocco: FDI inflows and remittances 1980–2018, US$ Million (*Source* World Bank and UNCTAD, elaboration of the author)

September, was 17,706 of which the top 5 migrants' countries of origin were Unknown, 9127; Morocco, 5260; Algeria, 1857; Guinea, 360; Mali, 280 (Frontex, 2019).[22]

Given the relevance of emigration for Morocco, remittances represent a very valuable source of foreign reserves for the country. In 2019, remittances represented a respectable 5.8% of Moroccan GDP.

The importance of remittances for Morocco appears particularly evident when compared with FDI flows, whose performance has never been higher than remittances and in the last two decades, FDI flows have moved towards representing regularly half of remittances or less (Fig. 3.16).

As it might be expected, considering the dynamics of Moroccan extra-regional migration, almost all remittances come from European countries (around 90% of world total), with France first (30.7%), and then Spain (24.75%) and Italy (13.75%) (Table 3.12).

The author underwent a research project gathering data on the characteristics of Moroccan migrants through a field survey in Rabat, in March 2006. 73 structured interviews were implemented to people wanting to

[22] https://frontex.europa.eu/along-eu-borders/migratory-routes/western-mediterranean-route/.

Table 3.12 Bilateral Remittance Estimates for 2017 using Migrant Stocks, Host Country Incomes, and Origin Country Incomes (millions of US$) (April 2018 version)

Remittance-receiving country (across) *Remittance-sending country (down)*	*Morocco*	*% of World tot*
Algeria	2	0.03
Andorra	1	0.02
Australia	5	0.06
Austria	5	0.06
Bahrain	2	0.03
Belgium	508	6.80
Brazil	1	0.01
Canada	114	1.53
Czech Republic	1	0.01
Denmark	14	0.19
Egypt, Arab Rep	4	0.05
Finland	5	0.07
France	2292	30.69
Germany	317	4.24
Greece	2	0.03
Iran, Islamic Rep	0	–
Iraq	1	0.01
Ireland	2	0.02
Israel	399	5.34
Italy	1027	13.75
Jordan	1	0.01
Kuwait	7	0.09
Libya	10	0.13
Luxembourg	1	0.01
Malaysia	1	0.01
Mauritania	1	0.02
Mexico	1	0.01
Monaco	1	0.01
Netherlands	446	5.98
New Caledonia	2	0.02
Norway	13	0.18
Poland	1	0.01
Portugal	5	0.06
Qatar	6	0.08
Russian Federation	2	0.03

(continued)

Table 3.12 (continued)

Remittance-receiving country (across) *Remittance-sending country (down)*	*Morocco*	*% of World tot*
Saudi Arabia	53	0.71
Senegal	6	0.08
South Africa	2	0.03
Spain	1848	24.75
Sweden	22	0.30
Switzerland	47	0.63
Tunisia	12	0.16
Turkey	2	0.02
Ukraine	1	0.01
United Arab Emirates	25	0.34
United Kingdom	57	0.76
United States	184	2.47
Venezuela, RB	2	0.02
World	7467	100

Notes These estimates are based on the methodology developed by Ratha and Shaw, 2007, "South-South Migration and Remittances," World Bank (www.worldbank.org/prospects/migrationandremittances). The remittance data is for 2017, disaggregated using host country and origin country incomes and estimated migrant stocks from 2017
These are analytical estimates based on logical assumptions and derived from a global estimation of bilateral remittance flows worldwide. They are not actual, officially reported data. The caveats attached to these estimates are (a) the data on migrants in various destination countries is incomplete; (b) the incomes of migrants abroad and the costs of living are both proxied by per capita incomes in PPP terms, which is only a rough proxy and (c) there is no way to capture remittances flowing through informal, unrecorded channels
These estimates are based on the Bilateral Migration Matrix 2017

migrate. The questions were devised to identify the motivation for migration at the point of departure and the extent to which people were responding to increasing economic pressures on the country due to its marginalisation from the global political economy. In the survey, the author also tried to verify the dynamics of irregular migration and the willingness of migrants to undertake their migratory process without the necessary documents. The author recognises the limits of the small sample and the reduced significance of the survey. However, the results of this exercise can provide useful insights. All people interviewed, apart from one, declared their intention to emigrate. This makes the sample relevant in order to assess potential migrants' motivations to migrate. Of those interviewed, 31.5% had already worked abroad.

Table 3.13 Motivations for migration: the case of Morocco

	Male %	*Female %*	*Total %*
Lack of employment opportunities in Morocco	47.3	50	47.9
Lower wages in Morocco	50.9	33.3	46.6
Bad standards of life	60	55.6	58.9
To escape family oppression	5.5	5.6	5.6
To help my family	41.8	11.1	34.2
Family reunification	20	33.3	23.3
To increase knowledge	20	5.6	16.4
To escape political problems	10.9	0	8.2
Other	0	11.7	2.9

Source Talani (2006)

In terms of the characteristics of the sample, 75.3% were males; their age spanned between 18 and 44 years with 35.6% between 18 and 24; 35.6% between 25 and 34 and 28.8% between 35 and 44. They were mostly unmarried, both males and females. Only 27.3% of males were married and 33.3% of females. Confirming our hypothesis that globalisation induced migration includes highly skilled workers, and as underlined in the literature,[23] 20.5% of the respondents had a university education, with only 6.8% with no education.[24]

The responses relating to the motivation for migration were overwhelmingly economic. For 47.9% of the respondents, the lack of employment opportunities in Morocco was the main reason to migrate, while 46.6% cited lower wages than in Europe. 58.9% of the sample referenced the difference in the standards of living as the main reason to migrate, whereas 34.2% of respondents cited their desire help their families as a reason to move to Europe. Both for men and women, economic motivations prevailed over any other reason (Table 3.13).

Only men quoted political motivations for migration, but only for a very limited number of respondents (8.2% of the sample). Family unification was a reason to move for both men and women. This is in line

[23] CARIM—Migration Profile Morocco The Demographic-Economic Framework of Migration The Legal Framework of Migration The Socio-Political Framework of Migration Report written by ANNA DI BARTOLOMEO, TAMIRACE FAKHOURY and DELPHINE PERRIN on the basis of CARIM database and publications November 2009.

[24] See Talani (2006).

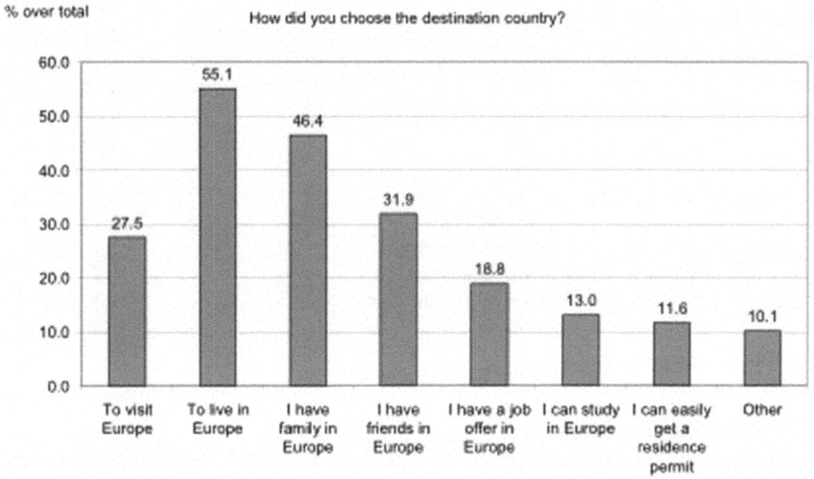

Fig. 3.17 How did you choose the destination country? (multiple answers) *Source* Talani (2006)

with the relevant literature,[25] demonstrating that Moroccan migration is well established and families are formed in the receiving country. Indeed, the great majority of Moroccan migrants gathered information on their destination country from friends and families (87% of the overall sample).

Furthermore, a staggering 81% of respondents already had friends and family working in Europe. Finally, the presence of family in Europe was cited by 46.4% of the sample as the reason for the choice to move there, while 31.9% said that they chose Europe as they already had friends living there (Fig. 3.17).

As we have seen above, the three European countries with the highest number of Moroccans in Europe are France, Spain and Italy. These are also the countries where most of the respondents, 55.9%, claimed to have family and friends.[26] Also for the survey respondents, the main countries of destination in Europe were France, Spain and Italy. The overall

[25] CARIM—Migration Profile Morocco The Demographic-Economic Framework of Migration The Legal Framework of Migration The Socio-Political Framework of Migration Report written by ANNA DI BARTOLOMEO, TAMIRACE FAKHOURY and DELPHINE PERRIN on the basis of CARIM database and publications November 2009.

[26] See Talani (2006).

Table 3.14 Morocco: Where do you think it is easier to migrate? (in a regular or irregular way)

Countries of Destination	*% of Answers*
Spain	57.5
France	6.8
Italy	16.4
Spain/Italy	12.3
Spain/Belgium	1.4
France/Italy	1.4
Belgium/Holland	1.4
I do not know	1.4
Everywhere in Europe	1.4
Total	100

Source Talani (2006)

percentage of interviewees that declared to head to the three countries was 65%, a figure which resembles very closely the overall number of Moroccans already residing there, which is 69.2% as reported above.

It does not come as a surprise that Spain was perceived as the easiest country to move to, given the proximity between the two countries and the existence of very well established routes. Very interestingly, however, potential migrants were fully prepared to move to Europe irregularly. Table 3.14 shows that around 58% of the respondents knew that Spain was the easiest way to reach Europe both regularly and irregularly. This clearly shows that Moroccan migrants are well aware of the existence of irregular routes of entry into Spain.

To gather information about irregular migration, the survey asked both direct and indirect questions. The indirect way to obtain information on the regular or irregular status of migrants' entry and stay in an EU country was to ask whether they have a formal job contract. Indeed, without a job contract, European countries do not release a regular working permit which in turn is a prerequisite for a residence permit and a long term visa.[27] Given the fact that this is an indirect way to ask about the regular status of the migration process, it is more likely to obtain genuine answers and therefore obtain more reliable data on the issue of illegal entry and permanence. There were, however, other questions in the survey to this aim.

[27] See Geddes (2017).

A first indication of the awareness of irregular entry routes into Europe came from the fact that only 18.8% of respondents actually had a job offer. In addition, even those who had already received an invitation to move to Europe, 49 people of 73, had a proper job contract only in 20% of cases, while 40.8% relied on generic job promises by friends or family. A remaining 35% of people just had generic invitations to go to Europe. If one cross-checks with the data of those who had already been in Europe, only 3% of migrants had a job contract while 42.5% declared to have entered with no valid documents.[28]

However, when asked directly whether they would undergo the migratory process irregularly, the answer was yes only for 2% of respondents. Moreover, most of the sample were aware that the migration process would be costly, which is consistent with the idea that they were prepared to undergo it irregularly. In fact, 80% of the respondents estimated that migrating would require between 15,000 and 200,000Dhr (at the time that was between £1,000 and £12,000). Cross-checking with Moroccans in the sample who had already gone to Europe, it emerges that an incredible 73.4% of respondents had paid more than 8000Dhr (around £500), 56.7% more than 15,000Dhr (around £1s000) and 20% more than 35,000Dhr which was around £2200.[29]

The marginalised nature of Moroccan migration to Europe is confirmed by the fact that only 21.2% of those who found a job in the receiving country were employed at their skill and educational level. The biggest number of Moroccans, 47%, were employed in the construction and agricultural sector in the receiving country. This contrasts with the fact that only 2% of them were occupied in building and construction in Morocco, 15% in agriculture, while 67% were in trade and services.

Conclusion

Overall, the analysis above shows that the three countries considered are clear examples of marginalisation from globalisation, while the MENA area does not seem to be in the process of integrating at the regional level neither de jure nor de facto.

[28] See Talani (2006).

[29] See Talani (2006).

In the case of Tunisia, there is a widespread rhetoric about the Tunisian economic miracle before the Arab Spring, and its democratisation after 2011. International organisations, such as the World Bank and, the International Monetary Fund and the European Union, have spoken enthusiastically about the Tunisian economic miracle, considering it a 'Mediterranean dragon'. This was due to its performance in terms of GDP growth and implementation of macroeconomic programmes aimed at achieving stability through structural reforms, liberalisation and privatisations (Talani, 2014).

It is true that Tunisia had achieved some positive macroeconomic results after the restructuring implemented by the IMF at the end of the 1980s. However, this did not obtain the result of facilitating the integration of the country into the global political economy. Quite on the contrary, the marginalisation of the country, as noticed in this chapter, increased in the following decades. This situation exacerbated the socio-economic cleavages and produced a discontent which led to the change of Ben-Ali's regime in 2011 (Talani, 2014).

The same marginalisation dynamics are at the root of the increase in both regular and mostly irregular migratory flows from the country. This phenomenon did not only include low-skilled workers but also highly skilled personnel, confirming that the country was being progressively more and more marginalised from the global political economy. Furthermore, the political upheaval and the establishment of a more democratic regime in the country after the events of 2011 did not seem to have stopped this haemorrhage. Instead, migratory dynamics after the revolution follow the same patterns as before, confirming the hypothesis that Tunisian migration is mainly economic. Indeed, looking at the characteristics of Tunisian migration after the Arab spring, we see that these are the same as before.

Egypt is also clearly following a path towards marginalisation as shown by the performance of its share of global FDI stock in the last decades. Notwithstanding the liberalisation and privatisation policies implemented by Sadat, starting with his 1974 Infitah, Egypt did not become attractive to foreign investors. To the contrary, the country continued to lose shares of FDI stock and flows. Similarly, marginalisation is evident by looking at its share of global mergers and acquisitions, and its trade openness which has remained very low in the last thirty years.

This happened despite the successful implementation of the IMF and World Bank restructuring and stabilisation programmes. One might say that Egypt moved directly from liberalisation to marginalisation, proving that neoliberal policies do not necessarily lead to integration into the global political economy (Talani, 2014).

In fact, despite various policies of liberalisation, privatisation and economic restructuring, one of the most relevant sources of foreign currency for Egypt are remittances coming increasingly from the global north.

There is no denying the fact that extra-regional permanent migration from Egypt to the US and Europe is on the rise. As this includes not only low skilled but also highly educated migrants, it is a sign of the extent to which Egypt is marginalised from the global political economy. This vicious cycle does not seem to have been affected by the various regime changes following the Arab Spring. As in the case of Tunisia, pre and post Arab Spring migratory dynamics seem to have maintained the same characteristics.

Very revealing in this context is also the case of Morocco, a country that did not experience the political unrest of Egypt and Tunisia. Notwithstanding this, its migratory dynamics closely resemble those of the other two MENA countries, with an increasing trend of extra-regional migration towards the global north. Looking at the socio-economic characteristics of Moroccan migration, gathered by the author through field research, it is possible to conclude that they are strongly influenced by social networks, especially family and friends. However, the migratory process is still very economically motivated. Indeed, the majority of migrants are young males, moved by economic considerations and seemingly willing to migrate both regularly and irregularly.

In fact, the majority of the migrants interviewed by the author were not only aware of the possibility of moving abroad without proper documents, but also prepared to do so. Also in the case of Morocco is their insertion into the labour markets of receiving countries below their skill level.

Concluding, the picture coming from the study of the MENA area in general, and of Tunisia, Egypt and Morocco in particular, confirms the paradox of marginalisation within globalisation and its impact on extra-regional permanent migration to the global North.

In the next chapter, we will move to the analysis of the paradox of regionalisation within globalisation.

References

Abdelfatta, D. (2011). *Impact of Arab revolts on migration. CARIM analytic and synthetic notes 2011/68*. EUI.

Al-Atreash, H. M., & Yousef, T. (1999). *Intra-Arab trade—Is it too little?* IMF Working Paper, WP/00/10.

Amin, G. (2000). *Whatever happened to Egyptians?*. American University in Cairo Press.

Appleyard, R. (Ed.). (1999). *Emigration dynamics in developing countries, Volume IV: The Arab Region*. Ashgate.

Bayart, J.-F. (2007). Global subjects: A political critique of globalisation (trans., A. Brown). Polity Press.

Bel Haj Zekri, A., (2010). *Le Cadre socio-politique de la migration hautement qualifiee en Tunisie*. CARIM AS 2010/38, CARIM-EUI 2008.

Ben Achour, S., & Ben Jemia, M. (2011). *Revolution Tunisienne et Migration Clandestine Vers Europe: Reactionnes Europeennes et Tunisiennes*. CARIM AS 2011/65, CARIM-EUI.

Binder, L. (1978). *In a moment of enthusiasm: Political power and the second stratum in Egypt*. University of Chicago Press.

Birks, J. S., & Sinclair, C. A. (1979). Egypt: A frustrated labour exporter?. *Middle East Journal, 33*(3): 288–303 and Shell, F.F., 1987, p. 27 note 10.

Bohning, W. R. (1984). *Studies in international labour migration*. Macmillan Press.

Boubakri, H. (2013). *Revolution and international migration in Tunisia*. Eui Mpc Research Report 2013/04, Badia Fiesolana, San Domenico Di Fiesole (Fi).

Boubakri, H. (2011). *Migrations et development. Reflexion et analyse a partir du cas de la Tunisie*. Carim AS 2011/55, CARIM-EUI.

Brand, L. (2011). *Citizens abroad: State and emigration in the Middle East and North Africa*. Cambridge University Press.

CAPMAS [Central Agency for Public Mobilization and Statistics]. (1991). *External migration survey in Egypt, 1987*. CAPMAS.

CARIM. (2008). *Euro-mediterranean consortium for applied research on international migration*. European University Institute, RSCAS, San Domenico in Fiesole, Florence. http://www.carim.org/. Accessed 3 September 2008.

Cassarino, J.-P. (2014). Channelled policy transfers: EU–Tunisia interactions on migration matters. *European Journal of Migration and Law, 16*(1), 97–123.

Cassarino, J. P. (2000). *Tunisian new entrepreneurs and their past experiences of migration in Europe*. Ashgate Publishers.

Castells, M. (1975). Immigrant workers and class struggle in advanced capitalism: The Western European experience. *Politics and Society, 5*, 33–66.

Castells, M. (2000). *The rise of the network society*. Blackwell.

Choucri, N. (1978). *Migration processes among developing countries: The Middle East*. MIT.

Cleland, W. (1936). *The population problem in Egypt: A study of population trends and conditions in modern Egypt* (p. 36, 52). Science Press Printing Company.

Dessouki, A. E. H. (1982). The shift in Egypt's migration policy 1952–1977. *Middle Eastern Studies, 18*(1), 53–68.

Di Bartolomeo, A., Fakhoury, T., & Perrin, D. (2009, November). *CARIM – Migration Profile Morocco*. The demographic-economic framework of migration. The legal framework of migration. The socio-political framework of migration report written on the basis of CARIM database and publications.

Dodge, T., & Higgot, R. (2002). *Globalisation and the Middle East*. Royal Institute of International Affairs.

Dodge, T. (2002). Bringing the bourgeoisie back in: Globalisation and the birth of liberal authoritarianism in the Middle East. In R. Higgot, & T. Dodge (Eds.), *Globalization and the Middle East: Economy, society and politics* (pp. 169–187). Brookings Press. ISBN 9781862031340.

Dodge, T. (2012). *After the Arab Spring: Power shift in the Middle East?: Conclusion: The Middle East after the Arab Spring*. IDEAS reports—special reports. Kitchen, N. (Ed.). (SR011). *LSE IDEAS*. London School of Economics and Political Science.

Dresch, P. (Ed.). (2005). *Monarchies and nations: Globalisation and identity in the Arab states and the gulf*. I. B. Tauris. DS247.A134 M66 2005.

Egyptian Ministry for Manpower and Emigration. (2006). *Attitudes of Egyptian youth toward migration to Europe*. http://www.emigration.gov.eg/idomsite/Upload/Researches/72/Attitudes%20of%20Egyptian%20youth%20towards%20Migration%20to%20Europe.pdf. Accessed 8 November 2006.

Eurostat. (2000). *Push and pull factors of international migration*. Country report: Egypt, population and social conditions, 3/2000/E/n.7.

Fargues, P. (Ed.). (2007). *Mediterranean migration, 2006–2007*. European University Institute, RSCAS, San Domenico in Fiesole, Florence.

Fargues, P., & Fandrich, C. (2012). *Migration after the Arab Spring*. MPC Research Report 2012/09, EUI, Florence.

Fazio, G. (2009). EMU and the Euro-mediterranean dialogue: Trade interdependence between mediterranean and Euro-area countries. In L. S. Talani (Ed.), *The future of EMU*. Palgrave.

Feiler, G. (2003). *Economic relations between Egypt and the Gulf Oil States, 1967–2000: Petro wealth and patterns of influence*. Sussex Academic Press.

Fourati, H. (2008). *Consultation de la jeunesse et desire d'emigration chez le jeunes en Tunisie 1996–2005*. CARIM AS 2008/47, CARIM-EUI, Florence.

Fourati, H. (2009). *Les Transferts des Tunisiens a l'etranger et leurs impacts sur le neveau de vie des ménages directement beneficiaries 1995–2005*. Carim notes d'analyse de synthese 2009/27, CARIM-EUI, Florence.

Fourati, H. (2010). *La migration hautement qualifiee depuis et vers la Tunisie*. CARIM AS 2010/18, CARIM-EUI, Florence

Geddes, 2017.

Geddes, A. (2000). *Immigration and European integration: Towards the fortress Europe?* Manchester University Press.

Hafez, H., & Ghaly, A. (2012). *The Arab Spring and migration in Egypt: One year on: Impacts, perceptions and attitudes*. MPC Research Report 2012/05, Florence EUI.

Hakimian, H., & Moshaver, Z. (2001). *The state and global change: The political economy of transition in the Middle East and North Africa*. Curzon Press.

Hakimian, H. (2011). The economic prospects of the 'Arab Spring': A bumpy road ahead. *CDPR Development ViewPoint, 63*, 1–2.

Hakimian, H. (2018). De las rentas del petróleo al crecimiento inclusivo: Lecciones aprendidas de la región MENA. *Awraq, 15*, 125–148.

Hakimian, H., & Abdulaal, A. (2015). *GCC economic integration: Fiction or reality?* (pp. 1–8). Khamsoon.

Halliday, F. (2002). The Middle East and the politics of differential integration. In T. Dodge, & R. Higgot (Eds.), *Globalization and the Middle East* (pp. 36–56). The Royal Institute of International Affairs.

Harrigan, J., Wang, C., & Said, H. (2005). The economic and political determinants of IMF and world bank lending in the Middle East and North Africa. *World Development, 34*(2), 247–270.

Henry and , & Springborg. (2010). *Globalization and the politics of development in the Middle East*. CUP.

Hibou, B. (2003). *Le Partenariat en réanimation bureaucratique*. Special Issue 'Les faces cachées du Partenariat euroméditerranéen', Critique internationale, 18.

Hibou, B., Meddeb, H., & Hamdi, M. (2011, June). *Tunisia after 14 January and its social and political economy*. The issues at stake in a reconfiguration of European policy.Euro-Mediterranean Human Rights Network (EMHRN) (p. 92).

Higgot, R., & Dodge, T. (2002). Globalisation and its discontents: the theory and practice of change in the Middle East. In R. Higgot, & Dodge, T. (Eds.), *Globalization and the Middle East: Economy, society and politics*. Brookings Press. ISBN 9781862031340.

Hirst, P., & Thompson, G. (1999). *Globalisation in question*. Polity Press.

Hirst, P., Thompson, G., & Bromely, S. (2009). Globalisation in question. *The international economy and the possibilities of governance* (3rd Ed). Cambridge, UK: Polity Press.

IMF [International Monetary Fund]. (2003, September). *World economic and financial surveys*. World Economic Outlook Public Debt in Emerging

Markets. At: http://www.imf.org/external/pubs/ft/weo/2003/02/index.htm. Accessed on 20 November 2005.
Mittelman, J. H. (2011). *Contesting global order: Development, global governance, and globalisation*. London: Routledge.
Honohan, P., & Lane, P. R. (2000). *Will the Euro trigger more monetary unions in Africa?* World Bank Policy Research Working Paper, no. 2393.
Hudson, M. C. (Ed.) (1999). *Middle Eeast dilemma: The politics and economics of Arab integration*. New York: Columbia University Press.
Ibrahim, S. E. (1982). *The new Arab social order: A study of the social impact of the oil wealth*. Westview Press.
International Labour Organization (ILO). (2008). http://laborsta.ilo.org/cgi-bin/brokerv8.exe. Accessed 22 July 2008.
International Office for Migration (IOM). (2005). *World migration 2005*. IOM.
International Office for Migration (IOM). (2011). *Egypt after January 25: Survey of youth migration intentions, Cairo, May 2011*. IOM.
IOM [International Organization for Migration]. (2003). *Egyptian migration 2003*. IOM.
Kee, H. L., Nicita, A., & Olarreaga, M. (2009). Estimating trade restrictiveness indices. *Economic Journal, 119*, 172–199.
Kepel, G. (1999). Toward a social analysis of Islamist movements. In L. Binder (Ed.), *Ethnic conflict and international politics in the Middle East*. University Press of Florida.
Kitchen, N., Dodge, T., Lawson, G., El Issawi, F., Stein, E., Ulrichsen, K. C., Alaaldin, R., Phillips, C., Thiel, T., Rafati, N., & Voller, Y. (2012). *After the Arab Spring: Power shift in the Middle East?* IDEAS reports—special reports. Kitchen, N. (Ed.). (SR011). *LSE IDEAS*. London School of Economics and Political Science.
Kubursi, A. A. (1999). Prospects for Arab economic integration after Oslo. In Hudso.
Lixi, L. (2019). The ambivalent drivers of migration governance relations between the EU and Tunisia. In A. Geddes, M. V. Espinoza, L. H. Abdou (Eds.), *Dynamics of regional migration governance* (pp. 91–108). Leiza Brumat Publisher: Edward Elgar Publishing.
Masson, P. (2007). Currency unions in Africa: Is the trade effect substantial enough to justify their formation? *The World Economy*. https://doi.org/10.1111/j.1467-9701.01028.x,1-15
Meddeb, H. (2008). *Ambivalence de la politique migratoire en Tunisie*. FASOPO, mimeo.
Mensard, A. (2004). Temporary migration and self-employment: Evidence from Tunisia. *Brussels Economic Review, 47*(1).
Mittleman, J. H. (2000). *The globalisation syndrome: Transformation and resistance*. Princeton University Press.

Nassar, H. (2005). *Migration, transfers and development in Egypt*. European University Institute, RSCAS, Research Report 2005/01.

Natteur, K. (2015, May 28). *Revolution and political transition in Tunisia: A migration game changer?*. Migration Information Source, the Online Journal of the Migration Policy Centre. http://www.migrationpolicy.org/article/revolution-and-political-transition-tunisia-migration-game-changer.

Overbeek, H. (2000). Globalisation, sovereignty and transnational regulation: Reshaping the governance of international migration. In B. Gosh (Ed.), *Managing migration: Time for a new international regime*. Oxford University Press.

Piore, M. J. (1979). *Birds of passage: Migrant labor and industrial societies*. Cambridge University Press.

Pouessel, S. (2017). Tunisia and its diaspora: Between protection and control. In A. Weinar (Ed.), *Emigration and diaspora policies in the age of mobility* (pp. 205–220). Springer.

Ribas-Mateos, N. (2016). *Migration, mobilities and the Arab Spring spaces of refugee flight in the eastern Mediterranean*. Edward Elgar.

Richards, A., & Waterbury, J. (1996). *A political economy of the Middle East*. Westview Press. Chapter Fifteen.

Russell, S. S. (1992). International migration and political turmoil in the Middle East. *Population and Development Review, 18*(4), 719–727.

Schaebler, B., & Stenberg, L. (Eds.). (2004). *Globalization and the Muslim world: Culture, religion, and modernity*. Syracuse University Press. DS35.62 .G58 2004.

Seccombe, I. J. (1985). International labor migration in the Middle East: A review of literature and research, 1974–1984. *International Migration Review, 19*(2), 335–352.

Sell, R. (1987). *Gone for good? Egyptian migration processes in the Arab world*. Cairo Papers in Social Sciences, American University in Cairo Press.

Subramanian, A., & Tamirisa, N. T. (2003). Is Africa integrated in the global economy? *IMF Staff Papers, 50*, 352–372.

Talani, L. S. (2003). *Why do migrants leave their countries? Motivations to migrate at the point of departure, the case of Egypt*. http://www.files.emigration.gov.eg/Upload/Publications/English/11/why_do_migrants.pdf. Accessed 18 August 2008.

Talani, L. S. (2004). *European political economy*. Ashgate.

Talani, L. S. (2006). *Motivations for migration at the point of departure: The case of Morocco*. Project financed by the British Academy, mimeo.

Talani, L. S. (2010). *From Egypt to Europe*. London: I. B. Tauris.

Talani, L. S. (2012). *Globalisation, hegemony and the future of the city of London*. London: Palgrave.

Talani, L. S. (2014). *The Arab spring in the global political economy*. London: Palgrave.

Tsourapas, G. (2019). *The politics of migration in modern Egypt—Strategies for regime survival in autocracies*. Cambridge University Press.

UN Economic and Social Commission for Western Asia [ESCWA]. (2007, January 4). Economic trends and impacts: Foreign aid and development in the Arab region, 4, United Nations E/Escwa/Ead/2007/1.

UNCTAD [United Nations Conference on Trade and Development]. (2008). http://www.unctad.org. Accessed on [date 2008].

Valbjorn, M. (2016). North Africa and the Middle East. In T. A. Börzel, & T. Risse (Eds.), *The Oxford handbook of comparative regionalism*. Oxford University Press.

World Bank. (2010, June). *Arab development assistance: Four decades of co-operation*.

Zahlan, A. B. (1999). Labor migration and economic integration in the Middle East. In M. C. Hudson (Ed.), *Middle East dilemma* (pp. 259–279). Tauris & Co.

Zohry, A. (2006, June 21–24) *Egyptian irregular migration to Europe*. Paper presented at the European Population conference 2006, University of Liverpool.

Zohry, A. (2007). Egyptian irregular migration to Europe. *Migration Letters, 4*(1), 53–63.

Zohry, A., & Harrell-Bond, B. (2003). *Contemporary Egyptian migration: An overview of voluntary and forced migration*. University of Sussex working papers, C3.

CHAPTER 4

The Paradox of Regionalisation Within Globalisation: Migration to the EU and to the US Compared

This chapter addresses the paradox of regionalisation within globalisation and, in particular, how regionalisation (or the lack of it) produces different migratory patterns, with reference to the cases of migration to the EU and to the US.

The paradox of regionalisation within globalisation refers to the phenomenon of increased regional economic integration occurring in parallel with globalisation.[1] Such a phenomenon however does not take place in all regions of the globe, but is highly uneven, leaving some regions marginalised.[2] In turn, people from non-integrated regions have an increased incentive to migrate, as their standards of living worsen relative to those of the more integrated countries. The consequences of the above described dynamics are both the mass migration of unskilled workers, but also, importantly, brain drain of skilled personnel. Furthermore, such migratory flows acquire often regional patterns, due to social, geographical, economic or political reasons. Therefore, migratory policies are increasingly devised at the regional level.

In the following sections, the paradox of regionalisation within globalisation is addressed first in theoretical terms, looking at how the paradox is

[1] See, for example, Breslin et al. (2002) and Hettne et al. (1999).

[2] On this subject, see, for example, Murshed (2002).

L. S. Talani, *The International Political Economy of Migration in the Globalization Era*, International Political Economy Series,
https://doi.org/10.1007/978-3-030-79321-0_4

connected with the debate on the creation of a global governance system for international migration. Then, migratory patterns to the EU and the US are compared, as they present different dynamics related to the paradox of regionalisation. The analysis will conclude that migration to the EU from the Global South is an instance of extra-regional permanent migration due to marginalisation from the global political economy and the lack of economic regionalisation. On the contrary, migration to the US is mostly a case of intra-regional migration following from economic regionalisation as Saskia Sassan would predict.

Economic Regionalisation and Institutional Regionalism in the IPE Debate: Towards a Global Migration Governance Regime?

As already discussed above, the impact of globalisation on migration is related on the one hand, to the two paradoxes of marginalisation within globalisation and the lack of regionalisation, favouring extra-regional migration to more developed countries. On the other hand, however, globalisation also seems to produce more regional integration, both de jure (regionalism) and de facto (regionalisation).[3] This favours the creation of more or less developed and successful regional migration governance schemes to deal with both intra-regional and extra-regional migration, as underlined in the relevant literature (Geddes, 2019). Below we will consider the question of the governance system adopted at the regional level to deal with migratory issues in connection with the IPE debate on the relation between globalisation and regionalisation.

Neo-Realism: The State as the Engine of Regional Migration Schemes

From the realist perspective, scholars such as Hirst and Thompson (1999a, 1999b, 2009) consider globalisation a convenient myth, utilised by the most powerful regional states to perpetuate their hegemonic condition. This, however, does not rule out the possibility that those same

[3] For a discussion of the notions of region, regionalism and regionalisation, see Tanja, A. B., & Risse, T. (Eds.). 2016. *The Oxford handbook of comparative regionalism* (pp. 457–486). Oxford University Press, Introduction.

states would foster regionalisation processes with the aim of further strengthening their hegemonic position.

Thus, if for realists the establishment of a global political economy is far from certain, there is indeed a clear indication that regional economic integration is taking place. Such a regional process of integration could happen both de jure, meaning the creation of formal institutional arrangements at the regional level (regionalism), and de facto, which refers to the emergence from below of economically integrated regional blocks, not necessarily corresponding to their formal institutionalisations (regionalisation) (Hirst et al., 2009: 159). What is more important from the realist perspective is that similar processes of regionalisation, both de jure and de facto, are a consequence of the proactive role played by hegemonic states in the region to promote their interests. As such, regionalisation cannot be taken for granted, even in the case of the EU, where it is rather advanced at the institutional level. On the contrary, the process can be reversed any time if it stops to meet the interests of the most powerful states. Moreover, apart from being reversible, for realists regionalism is not a necessary phenomenon as it is for institutionalists, meaning that it only takes place if it is in the interests of the regional hegemon(s). Finally, from this perspective, de jure regionalism can only happen if there is de facto economic integration (regionalisation) (Hirst et al., 2009: 189; Moravcsik, 1991, 1993a, 1993b).

Saying that, in the 1990s scholars from the realist camp noticed the tendency towards the creation of a 'triad', i.e. three highly integrated regional blocks where most of trade and investment would take place. As Hirst and Thompson (1999a, 1999b) noticed in the late 90s, the triad was represented by groupings of countries clustered around the US, the EU and Japan, where most of the regional FDI flows and stock would be directed. Each one of these regional blocks was dominated by one or more hegemon(s).

In sum, regionalism, from the realist point of view, is a phenomenon that happens only if hegemonic states have an interest in increasing economic integration to foster their dominance over smaller states within the regional context. Therefore, contrary to what neo-institutionalists and transnationalists maintain, regionalism (or regionalisation) is not a step towards globalisation, but a demonstration of force of hegemonic states. Indeed, in triad regions, FDIs are first directed to the hegemonic countries, and then between the dominant country and its groupings, but not among the countries in the grouping (Hirst & Thompson, 1999b: 120).

More recently, this dynamic has been substantially altered by the extraordinary economic progress of East Asian countries, especially China (Hirst et al., 2009: 131). Indeed, the Asian region accounts for more than 35% of FDIs flows globally and some 25% of global exports. Since the start of the twenty-first century, Asia has been responsible for over 50% of the world's economic growth (Hirst et al., 2009: 140). As a consequence, China can no longer be ignored when talking about the integration of the Asian region, with India also having an increasingly relevant role. This has led to the creation of regionally integrated supply chains and to a degree of intra-regional trade similar to that of NAFTA. Clearly China is emerging as a new hegemonic power in a region which is fast expanding. For example, from 1979 to 2009 China grew at the fastest rate of any country, at 9.4% of total GDP growth, while per worker GDP growth was 7.7% and per capita GDP growth was 8.1% (Hirst et al., 2009: 142). Hirst et al. commented that China is the 'dominant part of a general shift in the historical geography of industrial capitalism to emerging Asia' (Hirst et al., 2009: 143).

However, if the Asian region is integrating into the global political economy, other regions of the world are increasingly diverging from it. These are Africa, parts of the Middle East and Central and Western Asia (Hirst et al., 2009: 148). This divergence, noticed by realist scholars, corresponds to what we call the paradox of marginalisation within globalisation.

Overall, despite a clear trend of Asian economies, and China in particular, towards a convergence with more advanced countries of the globe, for realists this does not signify a step towards globalisation (Hirst et al., 2009: 156–157). On the contrary, realists only recognise the existence of a process of 'supranational regionalisation', based on the definition of a region as:

> a geographically contiguous area composed of the territories of nation states that have either combined in an integrative economic or monetary union, or whose economies have evolved into a closely interdependent entity, or who can empirically be shown to be advancing along those routes. (Hirst et al., 2009: 159)

Clearly the EU is the most notable example of such a process, where both de jure and de facto integration are well advanced. However, there are other cases of 'supranational regionalisation', such as the North American

Free Trade Agreement (NAFTA), Mercorsur (including Argentina, Brazil, Paraguay, Uruguay and Venezuela), the association of South-East Asian Nations (ASEAN) and the West African Economic and Monetary Union, among others.

Summing up, according to realists the establishment of regional supranational blocks does not lead to the achievement of a global political economy. Quite to the contrary, it is precisely the fact that regional economic blocks are created that excludes the possibility of ever intensifying globalisation processes. As realists put it, 'under current conditions, the supranational regionalism/regionalisation process looks to be more convincing than full globalism/Globalisation' (Hirst et al., 2009: 189). This starkly contrasts with the opinion of scholars from different theoretical approaches (Amin, 2006; Cox, 1987; Gamble & Payne, 1996: 16; Held & McGrew, 2007) who consider regionalism/regionalisation as part and parcel of the globalisation process and not distinct from it, as we will see in the following sections of this chapter.

As far as the creation of regional migration governance schemes is concerned, realists tend to be sceptical about the idea that these, even when they exist, are the first step towards the identification of a global regime for migration. Furthermore, where regional provisions exist, they are often in the interest of the most powerful countries in the region.

Indeed, the first instances of institutions dealing specifically with migration attempted to tackle the need to control, either in a restrictive or liberal way, cross-border movements of workers or regional citizens within specific regions and were mostly regional mobility regimes (Lavenex, 2019: 23).

Moreover, these attempts at organising intra-regional migration were mostly informal intergovernmental networks dominated by the strongest countries of the region and aimed at catering for their specific interests. While these initiatives have taken place in different regions of the world, their degree of formalisation and even their effective implementation greatly differ (Lavenex, 2019: 18; similarly, Geddes, 2012).[4]

In the case of the EU, as further elaborated below and in the next chapter, nation states remain in charge of their own immigration policies, while cooperation focuses on extra-regional migration with the aim of

[4] Geddes, A. (2012). Regions and regionalism. In M. R. Rosenblum & D. J. Tichenor (Eds.), *The Oxford handbook of the politics of international migration* (pp. 573–593). Oxford University Press.

avoiding the entry of "unwanted" third country nationals. This happens however in the context of an almost totally liberalised intra-regional mobility regime, due to the abolition of internal border controls. As Lavenex puts it, '[t]he EU is therefore indicative of the dialectic dynamics of intra-regional opening and inter-regional closure in the process of state-led regional integration' (Lavenex, 2019: 19).

In the case of Latin America, regional mobility schemes are pretty institutionalised but still follow an inter-governmentalist logic and do not provide for the integration of human rights and refugee agendas in it (Lavenex, 2019: 20).

Also, in Africa there is a strong, intra-regional institutional framework, mostly related to economic integration although not particularly successful in its implementation (Lavenex, 2019: 21).

On the other hand, it is fair to claim that the North American migration scheme is dominated by the security agenda of the US (Borjas, 2016). Here co-operation did not take place within the NAFTA (North American Free Trade Agreement), but through an ad hoc regional cooperation process called the Regional Consultations on Migration (RCM or Puebla Process). This includes the US, Canada, Mexico and the countries of Central America and is primarily aimed at fighting irregular migration, trafficking and smuggling (Kunz, 2011). As far as the NAFTA is concerned, contrary to the case of the EU and Latin America, provisions for intra-regional mobility are limited to instances of temporary mobility linked to trade and investment (Lavenex, 2019: 21).

Finally, in Asia, formal institutional co-operation on migration is relatively underdeveloped. Indeed, the 1967 founding document of the Association of South East Asian Nations (ASEAN) was not concerned with migration at all. Some labour mobility provisions were included in its 1995 Framework Agreement on Services (AFAS), later recapitulated in the 2012 Agreement on Movement of Natural Persons. However, these are aimed at facilitating only the movement of highly skilled personnel as related to investment flows and necessitated by Mutual Recognition Arrangements. More than in the case of NAFTA, the little regulation provided by ASEAN regards a tiny share of the otherwise mainly informal migration movements in the region, and even then, its implementation is low (Jurje & Lavenex, 2018; Lavenex, 2019: 22).

Based on the above considerations, realists would conclude that the little that has been done in terms of regional migration governance is broadly intergovernmental and is hugely influenced by the agendas of the most prominent states in the regions or even in their proximity. This is far from the conceptualisation of the regionalisation process proposed by the institutionalists, as further elaborated in the next section.

Liberal-Institutionalism: Regional Migration Governance as a Step Towards Global Governance

From the liberal-institutionalist standpoint, regionalisation is not inconsistent with globalisation. On the contrary, regional integration processes are considered as a step towards the creation of a global supranational community (Held, 2000: 12; Held & McGrew, 2007: 78–81).

While institutionalists recognise that regional trade blocks are emerging in the form of the triad mentioned above, they also believe that the process of regional liberalisation has been accompanied by a parallel process of global integration (Held & McGrew, 2007: 80 quotes Baldwin, 2006: 24). On account of this, regional liberalisation alone is not sustainable in the long run, and will be substituted by a process of global liberalisation, as the trend towards price convergence globally demonstrates. Indeed, according to Held (Held & McGrew, 2007: 81), regional trade integration does not mean market segregation. Furthermore, especially for the case of trade, there are already global institutions regulating it, such as the WTO, whose role is precisely to set the global regulatory regime for markets. This means that global markets are already organised according to the legal framework set by the WTO and by the existence of a growing body of transnational private merchant law (Held & McGrew, 2007: 81).

Thus, although regionalisation is a phenomenon recognised by institutionalists, they also maintain that:

> trade globalisation is not simply about trends in world trade but also the critical importance of global and transnational trade authorities in the construction of global markets. (Held & McGrew, 2007: 81)

Simultaneously, although sometimes financial markets show signs of volatility and regional trends in financial flows, there is plenty of evidence that financial markets have achieved a high degree of global integration since the 1980s (Held & McGrew, 2007: 86). This is clearly demonstrated by the trend towards the convergence of interest rates and the decrease in the correlation between savings and investments from the 1990s (Held & McGrew, 2007: 86–87).

Even more relevant, however, is the establishment of global institutions, including legal regimes. In primis, financial flows have been substantially liberalised worldwide from the 1970s. As a consequence, global capital flows have increased, leading to what institutionalists call financial globalisation, as defined in a quantitative way (Cohen, 1996, 2001; Talani, 2012). Furthermore, trans-border financial transactions are increasingly being regularised and organised by private or public institutions, such as for example, the IMF, the Bank for International Settlements (BIS), the Financial Action Task Force (FATF) and the International Accounting Standards Board (IASB) (Held & McGrew, 2007: 89).

Overall, from the neo-institutionalist standpoint, regionalisation is not a stand-alone process, but merely a step towards the establishment of an integrated global political economy and governance system. Both global and regional governance regimes, such as MERCOSUR or the EU, concur in achieving a multilevel governance system at the global level (Held & McGrew, 2007: 115; Rosenau, 1990, 1997). Thus, conceptualising the new global economic regime as one of triadisation or regionalisation is not correct. In fact:

> It is the very clustering of FDI around the three major economic regions, combined with the intensity of interregional flows, that reinforces the dynamic of global economic integration. (Held & McGrew, 2007: 92)

In sum, there is no competition between regionalisation and globalisation, as they are complementarity forces (Held & McGrew, 2007: 108). As a consequence, institutionalists rarely focus only on regional institutional integration processes, with the exception, perhaps of neo-functionalists and their analysis of the evolution of the EU. Generally, they prefer to study the dynamics of the creation of a global governance system informed by cosmopolitan values.

Cosmopolitanism is indeed a very widespread notion among liberal institutionalists, intended as the end-result of the process of global restructuring (Archibugi, 2010; Beck, 2005, 2006, 2009; Held, 1995, 2002, 2010; Kaldor, 2010). Held, for example, conceptualises cosmopolitanism as a set of principles:

> These are principles that can be universally shared and can form the basis for the protection and nurturing of each person's equal significance in the moral realm of all humanity. (Held, 2010a: 230)

These cosmopolitan values will need to be adopted by a cosmopolitan democracy, based on cosmopolitan rules and institutions (Held, 2010a: 240–241; 2010b: 306).

By definition, a cosmopolitan legal regime is superimposed on national, regional and even international regimes. Therefore, also political authority should shift at the cosmopolitan level of governance through the creation of a cosmopolitan democracy. This would include the state system, but also supersede it when addressing transnational issues (Held, 2010a, 2010b: 306). Thus, globalisation is fostering the adoption of a human rights code at the cosmopolitan level of governance (Held, 2010b). A similar process started in the aftermath of the Second World War, which saw the progressive detachment of authority and sovereignty from the national level and the shift to a 'liberal international sovereignty' system. These transformations allowed the state to modify its role in the international system acquiring a new legitimacy through the establishment of supranational, including regional, governance systems (Held, 1995, 2010c; Held & McGrew, 2007).

Nevertheless, regional supranational schemes are not enough to address the new challenges of globalisation. This is the case because regional supranational schemes rotate around state structures and cannot easily accommodate the new multilevel and multilayered forms of governance brought about by globalisation. In Held's analysis, globalisation has led to new 'overlapping communities of fate', meaning political communities transcending national and even regional boundaries (Held, 2010a: 244). Examples of these new communities of fate are environmental ones, those affected by regulation of the global financial system or those affected by the global spread of diseases such as HIV/AIDS or COVID-19. Clearly, the challenges that these communities are facing cannot be resolved only at the national level nor by intergovernmental institutions at the regional or international level. Therefore, a new cosmopolitan democracy is needed to address the needs of these various and overlapping political communities. In a nutshell, cosmopolitanism will inevitably follow from already established international and regional institutional systems in order to bring about a global governance regime (Held, 2002, 2010b).

Also, according to Beck, the 'cosmopolitanisation' of society will follow almost automatically from the increase of trans-border interactions and processes involving actors who are increasingly threatened by global phenomena, such as global diseases or ecological catastrophes (Beck, 2005, 2006, 2010: 219; similarly Giddens, 1990). Therefore, the prosperity of humanity would require the creation of a sort of Kantian Cosmopolitan Res-publica (Wallace Brown, 2010: 45), where cosmopolitanism is based on its Greek meaning of dual citizenship: citizenship of the Cosmos (world/globe) and citizenship of the polis (city/state) (Wallace Brown, 2010: 4; similarly Beck, 2005, 2006).

We have already seen in Chapter 1 how the notion of universal citizenship has been used by neo-institutionalists to claim the existence of an international regime for migration in terms of human rights. Soysal recurs to the notion of post-national membership, deriving from international human rights standards based in universalistic notions of personhood, to imply that international migrants are already enjoying cosmopolitan citizenship and can therefore claim particularistic rights in the public sphere (Soysal, 2000).

More recently, Geddes et al. reiterate the idea that regional governance systems of migration are already emerging, including various governance 'levels' ranging from cities, neighbourhoods and communities, to the global (Geddes, 2019: 1). In their conceptualisation, as it is the case for neo-institutionalist approaches, regions represent the link between the national and the international. As such, regional migration governance schemes all over the globe have the power to constrain state behaviour legally and/or to produce new ideas and practices that can change the behaviour of states towards migration from within (Geddes & Scholten, 2016; Lavenex et al., 2016; Margheritis, 2018). Regionalisation of migration, in turn, occurs due to regional patterns of migration justifying the creation of regional migration governance systems of various types across the world (Geddes et al., 2019: 1). As it is undeniable that the borders and boundaries crossed by international migrants are predominantly national, the space for inter-state cooperation on international migration at regional or global level is not necessarily pre-defined. Whereas for neo-institutionalists, international migration exposes interdependencies in the international and regional systems, these do not necessarily lead to formal cooperation (Geddes et al., 2019: 4). However, cooperation could also happen informally, through a process of diffusion. Regions are likely to take hints from other regions on problem-solving, decision-making or good practices. From this perspective, this would explain the co-existence of interdependence and divergence in patterns of regional migration governance and would lead to the identification of "an international migration 'regime complex' comprising loosely coupled systems of institutions without an overall architecture to structure the whole set" (Geddes et al., 2019: 10). Ultimately, this would testify to the existence of an international migration governance system, comprised of various migration governance regimes which coexist without a clear hierarchy (Keohane & Victor, 2010: 4).

More explicitly, Levenx and Piper (2019), using the conceptualisation of regional migration regime as the interaction between top-down and bottom-up regionalism from a multilevel perspective, extend the notion to the context of global initiatives on migration governance. They point to the 2016 New York Declaration on Large Movements of Refugees and Migrants of the UN General Assembly and the Global Compacts on Migration and Refugees negotiated in 2018, as not only reinforcing regional initiatives but configuring international institutions as actors in international as well as regional migration governance schemes (Levenex & Piper, 2019: 17; Lavenex, 2018a).

This confidence in the existence of international regimes' ability to govern the interdependencies created by globalisation is not shared by transnationalists who instead underline the contradictions of globalisation and their potential to create tensions and crises.

Transnationalism: Regional Labour Migration Governance as a Tool for Neoliberal Restructuring

The conceptualisation of regionalism and regionalisation from the transnationalist point of view stresses the contradictions of the global restructuring of production and its impact on regions. While some regions of the globe are favoured by the globalisation process, others are doomed to fall progressively further behind (Overbeek & Léon, 2015). This perspective is substantially different from the institutionalist one, which emphasises the reorganisation of the governance system at the regional level in a multilayered fashion as a step towards an inclusive cosmopolitan res-publica.

In the transnationalist view, globalisation is a process ridden with contradictions and paradoxes. One of those contradictions and paradoxes is the renewed importance of the 'local' within the 'global', in that globalisation increases the relevance of sub-national actors, especially in "global cities", as noticed by Sassen (1981, 1988, 1991). As well as the 'localising' tendencies of globalisation there are also the regionalising ones, which do not affect all regions of the globe, but only core ones. Transnationalist scholars, such as Overbeek and Léon, talk of the creation of macro-regions where the restructuring of production is more advanced (Overbeek & Léon, 2015). The engine of this process of regionalisation is neoliberal restructuring which requires the creation of new and more open governance forms at the regional level thus leading to 'neoliberal regionalism' (Overbeek & Léon, 2015; Overbeek, 2000; Pellerin, 2015).

As already noted in Chapter 1, transnationalists recognise as one of the salient features of globalisation the subordination of politics to economics. Hence the crisis of nation states (Van der Pijl, 2011), which national governments in the core countries of the global political economy overcome by fostering 'open regionalism' (Pellerin & Overbeek, 2001), i.e.:

> forms of regional integration aimed at combining further liberalisation of the flow of goods and capital with a revival of certain structures of governance of the market. (Pellerin & Overbeek, 2001: 7 mimeo version)

Neoliberal regions are, for example, the EU, NAFTA, MERCOSUR and similar free trade regional projects worldwide (Gamble & Payne, 1996; Hettne et al., 1999).

In the opinion of Mittelman (2000: 112), regionalism in the current capitalist era is mostly neoliberal. This was not the case, for instance, in the 1930s when there were forms of 'autocentric regionalism' (Mittelman, 2000: 112), whose main characteristic was to be protectionist. There are also instances of not very successful 'development regionalism', a form of regionalism stressing redistribution of resources (Mittelman, 2000: 117). However, the form of regionalism prevailing in the globalisation era is the neoliberal one, which is not protectionist, and interacts with the global political economy at three different levels: the macro-regional level, the national level and the micro-regional level. Consequently, it is a multi-layered process, constantly moving to higher degrees of 'regioness' and therefore, certainly, part and parcel of globalisation (Hettne, 1994: 7–8).

Behind the creation of macro-regions are the needs of capital that, as it expands beyond the national, requires political and economic stability of the regions where it is investing (Gamble & Payne, 1996). Harvey refers to this phenomenon as 'the spatial fix' (Harvey, 2005). On the contrary, the regions where capital is not moving to have no reasons to integrate further and to create a new 'spatial fix'. One of those areas, as seen in Chapter 3, is the Middle East and North Africa.

Other authors, such as Overbeek and Pellerin, talk about the 'mobility-fixity nexus', which they define as 'the logic determining both instances and places of mobility and fixity of global capitalism' (Pellerin & Overbeek, 2001). While in the Fordist mode of production, immediately after the Second World War, production processes required capital to concentrate only within the nation state, the new globalised mode of production, with the creation of global value chains regionally concentrated, requires a new regional mobility-fixity nexus (Dicken, 2011; Mittelman, 2000; Pellerin & Overbeek, 2001).

Thus, neoliberal regionalism coincides with the new regional productive dimension, including regionally clustered chains of production, regional markets and regional divisions of labour (Gamble et al., 1996; Mittelman, 2000: 123).

Also, for transnationalists capital mobility occurs through regional dynamics. Therefore, the organisation of production needs to happen at the regional level as well (Overbeek & Léon, 2015). Transnationalist scholars (Dicken, 2011; Mittelman, 2000; Overbeek, 2000; Pellerin, 2015; Pellerin & Overbeek, 2001) highlight how the activities of transnational corporations, both productive and financial ones, are increasingly taking place within the context of the three global regions incorporated by the realists in the triad, i.e. North America (NAFTA), Western Europe (EU) and East Asia.

So, the triadisation of the global political economy is a phenomenon which is recognised by all IPE perspectives. However, for realists this phenomenon of creation of three macro-regions is the consequence of the interest of the hegemonic countries within them (Hirst et al., 2009). For institutionalists, it is a step towards the achievement of a global supranational community and a cosmopolitan democracy, while for trasnationalists it is the spatial fix that neoliberal capital, embedded in Transnational Corporations (TNCs), needs to reorganise the productive and financial processes in the neoliberal globalisation era (Dicken, 2011: Chapter 7; Mittelman, 2000: 123; Overbeek, 2000; Strange, 1991, 2002). Indeed, the increase of mergers and acquisition (M&As), taking place for the vast majority in developed countries, has enormously enhanced the power of TNCs. Moreover, transnational capital is able to exert power on the global economy recurring to less detectable activities, i.e. strategic alliances and other non-equity arrangements (Overbeek & Léon, 2015).

The concentration of capital in the hands of transnational companies, however, is not happening evenly throughout the globe, but is a highly non-homogenous process, focusing on some macro-regions, the triad, and leaving swathes of the globe marginalised. It is a fact that TNCs are locating their integrated productive chains mostly within the three regions of North America, Western Europe and East Asia/Pacific. Therefore, following Amin (2006), we could say that transnational companies are spreading neoliberal globalisation within the triad, which then emerges as a new collective imperialism. Thus, globalisation is de facto, in the hands of the triad, while the rest of the world is more and more left in the periphery of this process (Amin, 2006).

According to Overbeek (2000), this happens for the following reasons. To start with, productive capital is not completely mobile as there are still many limits to its full mobility such as technological, logistical or skill-related ones. Moreover, these limits are not the same for all industries, and therefore, some industrial sectors are more mobile than others (Dicken, 2011). On top of this, countries are in competition with transnational capital by adopting suitable structural and macroeconomic policies, as already noticed by Cerny (1999).

Another reason why transnational corporations have a strong incentive to re-organise spatially within regions is related to the uncertainties ensuing from the globalisation process ranging from exchange rate fluctuations to volatility of financial markets. These uncertainties do not favour the process of capital accumulation but can hardly be managed any longer by the national level of governance because technological developments have made it possible to overcome the national border for production purposes and exploit regional economies of scale. For this reason, in the globalisation era, capital can thrive more within a regional governance scheme in the core areas of the global political economy, such as for example, the European Single Market and European Monetary Union (Overbeek & Léon, 2015; Talani, 2013). Similar regional institutional schemes in the core areas of the global system, or regionalisms, are associated with the implementation of a clearly neoliberal programme, including privatisation, liberalisation, flexibility of labour markets and austerity (Cafruny & Ryner, 2007; Mittelman, 2000, Talani, 2013; Van

Apeldoorn et al., 2003). As Gill underlines, this neoliberal project has an evident disciplinary intent (Gill, 1997).

Thus, transnationalists like neo-institutionalists believe that the process of regionalisation is contiguous to that of globalisation. It is one of its manifestations, or better, globalisation needs regionalisation to succeed (Mittelman, 2000; Overbeek, 2000; Pellerin & Overbeek, 2001).

In any case, the establishment of a new regional mobility-fixity nexus entails a number of contradictions. As noticed by Callinicos (2010), regionalism(s) leaves unresolved a number of questions relating to the relationship between national and supranational institutions, and to the contrasting interests of different fractions of capital.

Apart from the conflicts between capitalist groups, tensions are also arising in the population, for example tensions between citizens, workers and migrants (Overbeek & Léon, 2015). On the one hand, the high technological content of globalisation increases the need for highly skilled labour in developed countries. As in countries such as China and India there is a surplus of high-skilled workers, (Papademetriou et al., 2008[5]), capital and private actors are increasingly important in the negotiation and enforcement of migration regimes (Betts, 2013). On the other hand, the 'skill biased technological change' also increased the demand for lower skilled jobs in the Global North, especially in so-called global cities. This has created new migratory dynamics, with a rise in mass migration from the Global South, especially irregular migration, and with instances of the feminisation of work (Samers, 2008).[6]

Often these migratory waves take a regional dimension, which frequently leads to some form of regional co-ordination of migration. However, for transnationalists the issue of regional migration governance should be contextualised within the general restructuring of governance structures in the current global political economy (Overbeek & Léon, 2015). In general, the existing schemes include rules aimed at the disciplining of labour and at the selective freeing of the mobility of labour. This increase in migration due to globalisation has exposed the limits of

[5] Papademetriou, D. G., Somerville, W., & Tanaka, H. (2008). *Talent in the 21st century economy*. Migration Policy Institute. http://www.migrationpolicy.org/research/talent-21st-century-economy.

[6] Samers, M. (2008). At the heart of 'migration management' immigration and labour markets in the European Union. In C. Gabriel & H. Pellerin, *Governing international labour migration: Current issues, challenges and dilemmas*. Routledge.

the state to deal with the consequences of migration alone. However, attempts at creating formalised co-operation schemes at the regional level have had mixed results, while at the global level a formal regime to regulate cross-border movements of people has not materialised (Betts, 2013).[7] To be sure, there are some international organisations concerned with the international movement of people, such as the International Labour Organisation (ILO), the International Organisation for Migration (IOM) or UN High Commissioner for Refugees (UNHCR). However, they do not constitute a strong international migration regime (Overbeek & Léon, 2015). For transnationalists, this outcome is not one of chance. The reason why this is the case is that migration is, as Overbeek and Léon (2015) put it,

> only one, and numerically not the most important, mechanism to integrate an increasing proportion of the world population directly into capitalist labour markets and lock national and regional labour markets into an integrated global one. (Overbeek & Léon, 2015: 47)

In their opinion, whereas labour mobility is socially and politically controversial, capital mobility through FDIs is more convenient to create regional and global production chains. Consequently, the regulation of migration in the globalisation era entails two opposed tendencies. On one side, the attempt to stop asylum and illegal migration. On the other side is the desire to increase migration of economically deserving labour. Therefore, governments of industrialised countries try to keep their hands as free as possible as the fate of the 1990 UN Convention on the Protection of the Rights of all Migrant Workers and Members of their Families demonstrates. Indeed, although it achieved the threshold to enter into force in 2003, no industrialised country ratified it (Kunz et al., 2011).[8] Thus, efforts to create migration regimes usually take a regional or an interest-based form and respond to the needs of the neoliberal restructuring process.

[7] Betts, A. (2013). The migration industry in global migration governance. In T. Gammeltoft-Hansen & N. Nyberg Sørensen (Eds.), *The migration industry and the commercialization of international migration* (pp. 45–63). Routledge.

[8] Kunz, R., Lavenex, S., & Panizzon, M. (2011). *Multilayered migration governance: The promise of partnership*. Routledge.

Importantly, transnationalists point to a growing discrepancy between the core areas of the global political economy and peripheral ones. Indeed, the process of reorganising the 'mobility-fixity' nexus at the regional level does not interest marginal regions such as Africa and the Middle East. As very appropriately noted by Dicken, the global political economy displays a remarkable continuity in geographical concentration which goes in parallel with a growing peripherisation of non-core areas (Dicken, 2011: 25).

In sum, what Ruggie terms the 'unbundling of territoriality' (Ruggie, 1993) makes it necessary to devise new forms of political and economic organisation of state functions. The regional level is the locus where this process is more needed and, therefore, more developed, but this refers only to the core regions of the globe. Such unbundling is very advanced in the European Union. Therefore, most theories of regional integration and, especially, regional migratory co-ordination policies, concentrate on this area as further discussed in the following chapter (Similarly Hettne, 1994: 12; contrary Mittelman, 2000: 115).

It is thus clear that, from a transnationalist perspective, peripherisation is a relevant phenomenon in the globalisation era, affecting migratory dynamics in a way as to increase extra-regional movements of people. It is however also evident that regionalisation processes do facilitate instances of intra-regional migration, leading to their substantial increase.

To elucidate these dynamics, it is worth comparing the evolution of the US and EU migratory patterns in the last decades with the aim of ascertaining whether extra- or intra-regional migration has prevailed.

Intra-Regional vs Extra-Regional Migration: Comparing the Dynamics of Migration to the US and the EU

Regional Migration to the US

At the onset it is important to clarify which region we are considering when dealing with regional migration to the US. As pointed out in the literature, talking of North America as a region is challenging (Abdou, 2019: 147). The notion of North America is not only disputable, but also controversial in terms of definition. According to the UN DESA, the SDG North America region comprises only the following countries:

Bermuda Canada, Greenland, Saint Pierre and Miquelon and the US.[9] If we look at North America as such, the number of migrants from the region as a share of total migrants in the US was in 2019 a mere 1.7%, while intra-regional migration in the North America region was as low as 2.5%.[10]

In the academic debate (Abdou, 2019: 147), however, NAFTA is the core of North American regionalism, and, therefore, the North America region is, from the signing of the NAFTA agreement in 1992, composed of Mexico, Canada and the US (Long & Suarez-Mier, 2017: 272). Moreover, looking at the dynamics of intra-regional migration in the North America/Nafta region, it becomes important to distinguish between regionalisation de-facto and regionalism de jure. Indeed, contrary to the case of the EU, where an intra-regional mobility regime is highly institutionalised and liberal, yet low in numbers, intra-regional mobility within the NAFTA has not been facilitated de jure, while is de facto very significant (Abdou, 2019: 147; Bacon, 2008: Ch. 5).[11] Perhaps, one might identify a relation between the degree of liberalisation of the intra-mobility regime and the numbers implied. Indeed, while migration between the US and Canada is very small, immigration from Mexico to the US is phenomenal and has increased substantially over the last four decades. According to the data from UNDESA, in 2019 the stock of legal Mexican migrants in the US was almost 12 million which was almost double the number of Mexican migrants in the US in 1995 (see Fig. 4.1).

Mexico was by far the highest country of origin for international migrants in the US. China and India follow with around 3 and 2.6 million, respectively, and then the Philippines and Puerto Rico with around 2 million[12] (Fig. 4.2).

When it started, in 1994, NAFTA covered the largest free trade area in the world. However, with respect to international migration, both regional and extra-regional co-operation was very limited. In particular, despite the introduction of the free circulation of goods, services and

[9] https://unstats.un.org/sdgs/indicators/regional-groups/.

[10] https://www.un.org/en/development/desa/population/migration/data/estimates2/countryprofiles.asp.

[11] See Bacon, D. (2008). *Illegal people: Creates migration and criminalizes immigrants*. Beacon Press.

[12] https://www.un.org/en/development/desa/population/migration/data/estimates2/countryprofiles.asp.

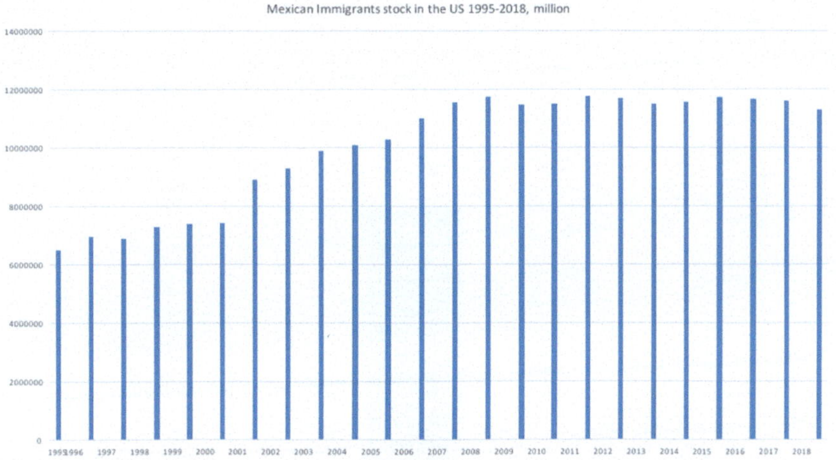

Fig. 4.1 Mexican immigrant stock in the US, 1995–2018, million (*Source* OECD, elaboration of the author)

capital within the region, the flow of labour was purportedly not liberalised. The NAFTA accord included only a few rules for temporary entry, which were taken from US immigration law already in force (Abdou, 2019). Even more, the agreement continued to differentiate between Mexican and Canadian migrants to the US. For example, the visa-exempt status which had already been guaranteed to Canadian citizens before the NAFTA by the Immigration and Nationality Act was not extended to Mexican citizens (Yost, 1996). However, this did not stop Mexicans from moving en-mass to the US. Indeed, as we have noticed above, from the entry into force of NAFTA, the stock of Mexican authorised migrants in the US has effectively doubled. This is in line with what the literature calls "regionalization from below" (Abdou, 2019: 149; see also Geddes et al., 2019; Lavenex, 2019). So, while regionalisation of migration is clearly happening within the NAFTA/North America region, formal regionalism is only taking place in the economic sphere. No doubt, however, the two phenomena are connected. One could even hypothesise that the informality of the management of intra-regional migration, and even its securitisation, with the ensuing increase of irregular flows, respond to the needs of the US or of its leading socio-economic sectors (Abdou, 2019: 150). From the Mexican side, considering that of the 11,796,178

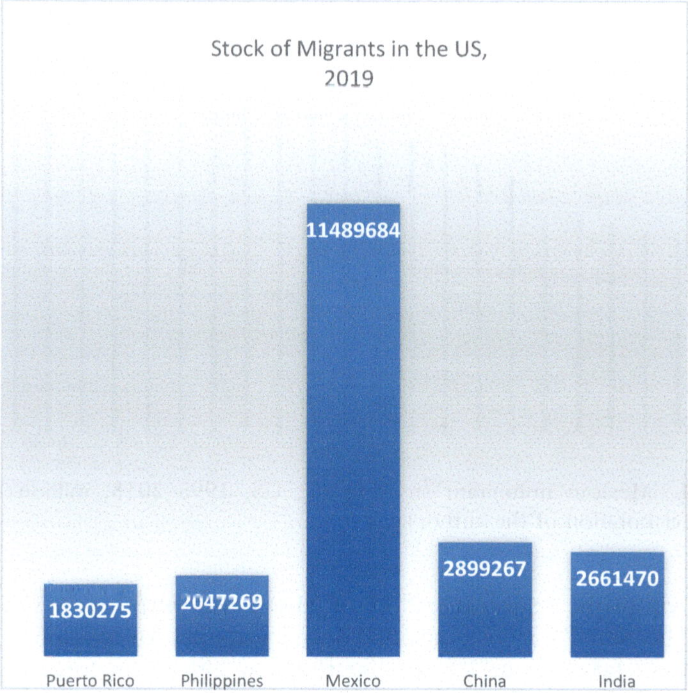

Fig. 4.2 Stock of Migrants in the US, 2019 (*Source* https://www.un.org/en/development/desa/population/migration/data/estimates2/countryprofiles.asp)

Mexican nationals living abroad in 2019, 97.4% were in the US, it is not a surprise that an essential part of the Mexican economy relies on remittances, which represent the third-largest source of foreign revenues in the country.

Mexican integration in the global political economy and in the region, however, is not to be doubted as data about the performance of FDIs inflows clearly demonstrates (Fig. 4.3).

Consequently, the Mexican approach to the issue of emigration to the US has traditionally been a 'policy of no policy' and more recently moved to a 'policy of depoliticised policy' (Espinoza, 2018: 166). This is based on technocratic bilateral migration cooperation dealing with

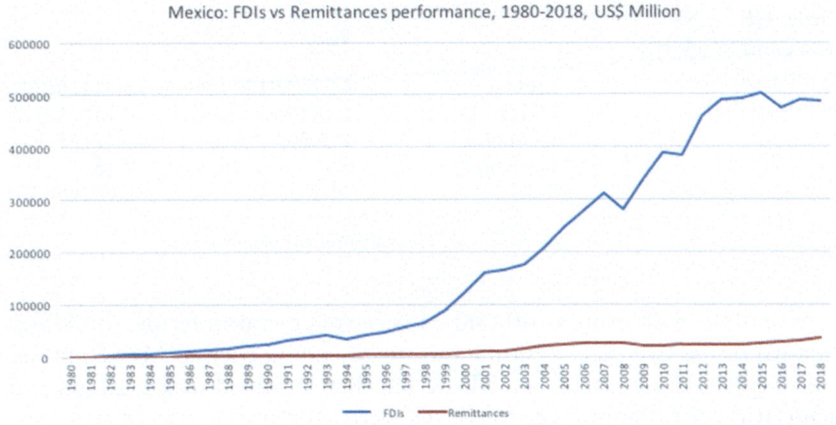

Fig. 4.3 Mexico: FDIs vs Remittances inflows (US$ Million) 1980–2018 (*Source* World Bank, and UNCTAD, elaboration of the author. https://www.worldbank.org/en/topic/migrationremittancesdiasporaissues/brief/migration-remittances-data)

both migration from, and return migration to, Mexico. Given its position as a 'weaker' state in the regional governance of migration, as well as in the NAFTA, Mexico's approach reinforces the status quo of weak regionalism that characterises North American intra-regional migration (Espinoza, 2018). This is extended also to other regional players, outside NAFTA. Indeed, some scholars (Brick et al., 2011) include in the notion of regional migration, apart from Mexico, also and the so-called Central America's "Northern Triangle", including El Salvador, Honduras and Guatemala. Thus defined, intra-regional migration has accelerated consistently with globalisation, to the extent that the US stock of authorised migrants from these four countries was more than 14 million in 2018, up from just 1 million in the 1970s. Overall, in 2018, these four countries represented 32% of all authorised migrants to the US (Table 4.1).

Looking at this broader definition of intra-regional migration, the evolution of the US migratory policy in the region followed three stages. Before the 1930s, the migratory regime was generally liberal, with very limited migration before the Second World War. Migration interested mainly Central Mexico and the southwest of the US and consisted mostly of seasonal workers in railroad construction and agriculture. Numbers

Table 4.1 US immigrant stock by country of birth, million, 2018

	Total	%
Mexico	11,269,913	25.31116693
El Salvador	1,401,832	3.148383112
Honduras	655,362	1.471881547
Guatemala	958,842	2.153469146
Total	14,285,949	32.08490073

Source OECD, elaboration of the author

were limited, with around 60,000 Mexicans per year entering the US at the turn of the twentieth century and usually going back to Mexico in the winter (Brick et al., 2011: 4). Those numbers doubled in the 1910s and again in the 1920s. Indeed, Mexicans were exempted by the restrictions imposed on Asian and European migrants until the late 1920s, when the Great Depression led to deportations to Mexico.

The 1930s are the only decade when migration within the region went from the north to the south. During and after the Second World War, the US introduced a large-scale Bracero Programme, targeting temporary migration from the region, which produced an increase in migratory flows. Initially Mexican workers, in high demand because of the needs of a country at war, were given extremely favourable terms. These included a guaranteed minimum wage, contrary to local farm workers, housing, transportation and health benefits. However, similar privileges were reverted in the 1950s, although the Bracero Programme lasted until 1964. With the elimination of the Bracero Programme and the new migratory regime introduced in 1965, with the Immigration and Nationality Act (INA), migratory flows from the region became mostly irregular. The law limited the number of visa available for migration to 20,000 per country, but more importantly limited the number of visa for employment to only 29,000 per year, including both workers and their families. However, this number grossly underestimated the needs of foreign, cheap labour of the US business sector. Moreover, while a law in 1952 made it illegal to help unauthorised migrants, it also explicitly exempted employers who hired them (the so-called Texas Proviso), thus creating a strong incentive to irregularly employ irregular migrants. Consequently, irregular flows persisted despite the introduction in the 1980s of a very strict immigration enforcement regime (Rosenblum & Tichenor, 2012). This culminated in the adoption of six laws between

2002 and 2006 mostly concerned with toughening immigration enforcement. But despite all this, the number of unauthorised immigrants has tripled in the last three decades, with Mexicans representing 60% of them and Central Americans around 12% (Brick et al., 2011: 6).[13]

Barak Obama promised to pursue a "comprehensive immigration reform (CIR)"[14] in the course of his 2008 presidential campaign. His approach included a strengthening of immigration support, a reform of the admission rules and a pledge to legalise undocumented migrants. This policy towards migration gained Obama 67% of the Latino vote and 64% of the Asian one, but once in office, he failed to implement his pledges with illegal migration remaining huge in the US. This is the case even after the toughening up of the US approach to unauthorised immigration implemented by Donald Trump.[15]

Obviously, estimating the size of the illegal alien population of the US is not an easy task. Most often, estimates of illegal aliens are based on US Census Bureau data[16] or survey data collected by private research organisations with their numbers sometimes differing by millions. Tichenor and Rosenblum (2012: 1) put the number of illegal aliens in the US at around 12 million in 2012. A detailed study by Yale/MIT[17] of 2018 proposed an estimate that ranges between 16.7 and 22.1 million. The Federation for American Immigration reform estimated 14.3 million in 2019 excluding an estimated 4.8 million American-born children of illegal aliens who, under the current interpretation of the Fourteenth Amendment, are US citizens.[18] Indeed, the definition of "illegal alien" is "anyone who

[13] Immigration to the US from Central America became significant only after the 1970s. Before, fewer than 40,000 Central Americans had obtained legal permanence in the US per year (Brick et al., 2011: 4).

[14] See Tichenor, D. J., & Rosenblum, M. R. (2012). Poles apart: The politics of illegal migration in America. In D. J. Tichenor & M. R. Rosenblum, *Oxford handbook of the politics of international migration* (p. 1). Oxford University Press.

[15] For more details, see https://www.nytimes.com/2017/02/21/us/trump-immigration-policies-deportation.html.

[16] https://www.dhs.gov/immigration-statistics/yearbook/2018/table10.

[17] Fazel-Zarandi, M., Feinstein J., & Kaplan, E. (2018). The number of undocumented immigrants in the United States: Estimates based on demographic modeling with data from 1990 to 2016. *PLoS ONE*, September 21, 2018. https://journals.plos.org/plosone/article?id=10.1371/journal.pone.0201193.

[18] https://www.fairus.org/issue/illegal-immigration/how-many-illegal-aliens-united-states#_edn3.

entered the United States without authorization or anyone who unlawfully remains in the United States once their authorized period of stay has expired".[19] In any case, the numbers are very high and do not seem to have decreased despite strict migratory policies implemented by various US administrations. It is not the place here to analyse why those policies have failed,[20] but instead it is underlined that illegal migration in the US continues unhindered and it is almost exclusively intra-regional (Brick et al., 2011: 6). The next section will deal with intra- and extra-regional migration to the EU.

Intra-Regional Mobility in the EU

There seems to be little doubt in the literature that, since Regulation 1612/68, the process of communitarisation of European migration policies has been accompanied by a growing consensus on the need to restrict entry of TCNs into the territories of the Member States and by a progressive securitisation of the issue (Huysmans, 2000; Guiraudon, 2000, Kostakopoulou, 2000). This is indeed reflected in the legislation regulating intra-regional migratory flows to the EU and extra-regional ones. Whereas migration from third countries has been progressively made more and more difficult and securitised, intra-regional migration in the EU, or, better, intra-EU mobility, is extremely liberal (see next chapter).

In fact, the free movement of workers within the EU is one of the fundamental rights guaranteed by Article 45 of the Treaty on the Functioning of the European Union, developed by EU secondary legislation and the case law of the Court of Justice. EU citizens, that is, the citizens of one of the Member States of the EU, have the right to work in any other EU country without the need for a work permit, reside in the country for the duration of their work activity and after, bring their

[19] https://www.fairus.org/issue/illegal-immigration/how-many-illegal-aliens-united-states#_edn3; FAIR offers a detailed explanation of who should be considered an illegal alien in a study titled "Why 'Illegal Alien' is the Correct Term. See O'Brien, M., Raley, S., & Ryan, C. (2018). *Why 'Illegal Alien' is the correct term*. Federation for American Immigration Reform, July 2018. https://www.fairus.org/sites/default/files/2018-07/IssueBrief_Illegal-Alien-is-the-Correct-Term.pdf.

[20] See Tichenor, D. J., & Rosenblum, M. R. (2012). Poles apart: The politics of illegal migration in America. In D. J. Tichenor & M. R. Rosenblum, *Oxford handbook of the politics of international migration* (p. 1). Oxford University Press.

extended family members, enjoy equal treatment with the citizens of the residing country in terms of access to employment, social and tax rights and conditions of work. EU citizens, thanks to the process of coordination of social security systems, may also enjoy the transfer of health and security coverage to the new EU country of residence (coordination of social security systems). Moreover, in some professions, there is the possibility of having professional qualifications recognised by other EU Member States (mutual recognition of professional qualifications).

Provisions relating to the free movement of workers apply also to the European Economic Area countries, i.e. Iceland, Liechtenstein and Norway.

Fair labour mobility in Europe is promoted by the EU removing barriers that hinder it. To improve it, the EU promotes the matching of job opportunities and jobseekers all over its territory, for example through the Posting of Workers directive, May 2014 (Official Journal of the EU). It also seeks to fight against irregular jobs (Decision on establishing a European Platform to enhance cooperation in tackling undeclared work) (Official Journal of the EU[21]).

Somehow paradoxically, however, intra-regional mobility, i.e. the percentage of EU citizens exercising their right to freely move within the EU, has increased only slightly, despite the 2004 and 2007 accessions, and remains relatively low in 2018.

Overall, considering EU citizens of working age (15–64), only around 3.75% in 2018 had established their residence in another Member State (Fig. 4.4). This has only slightly increased from 2008, when it was 2.55%. However, there are huge variations between different countries. In 2018, the share of EU mobile citizens out of the resident population was only 1.0% in Germany while it reached 21.3% in Romania.[22]

Romania and Croatia were, in 2018, the EU countries with the highest percentage of residents in another EU country over total residents, while Germany and the UK had the lowest percentage. However, looking at absolute numbers, in 2018 the most numerous national groups of mobile EU citizens aged 20–64 were from Romania (2,524,000 persons), Poland

[21] https://eur-lex.europa.eu/TodayOJ/.

[22] See EU citizens living in another Member State—statistical overview, https://ec.europa.eu/eurostat/statistics-explained/index.php?title=EU_citizens_living_in_another_Member_State_-_statistical_overview#Key_messages.

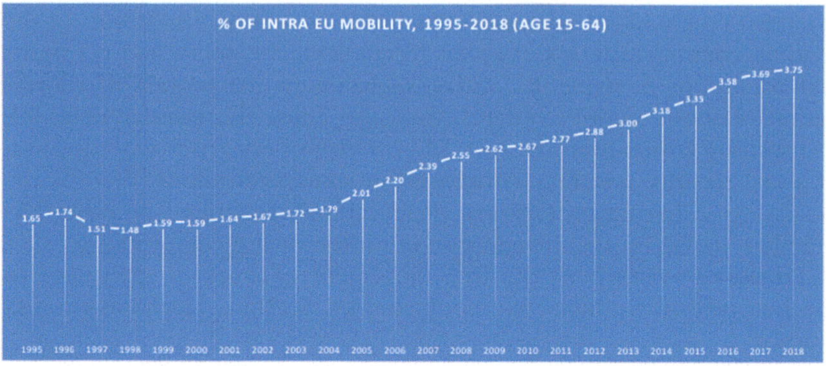

Fig. 4.4 Percentage of Intra-EU mobility 1995–2018 (age 15–64) (*Source* EUROSTAT Elaboration of the author)

(1,666,000 persons), Italy (1,133,000 persons), Portugal (824,000 persons) and Bulgaria (562,000 persons) (Fig. 4.5).[23]

Overall, the majority of EU citizens moving abroad have a higher tertiary level education than those remaining in their countries. This is even truer for countries that were part of the European Union before the 2004 accession (apart from Luxembourg and Portugal). In 2018, EU mobile citizens with tertiary education ranged from 67.5% for the Finnish to 16.4% for Portuguese citizens.

On the other hand, the proportion of low-skilled workers among mobile EU citizens decreased by 3.8% between 2008 and 2018, from 28.4% in 2008 to 24.6% in 2018 (Fig. 4.6).

In terms of employment rate, mobile EU citizens are on average more employed, with 77.1% in 2018, as compared to an EU average of 73.1%. These figures are often higher than the employment rates of their country of citizenship. To be sure, the employment rate of mobile EU citizens, as opposed to the EU average, has increased in the last decades (4.9%, compared to the total population with 2.9% see Figs. 4.7 and 4.8).

[23] See EU citizens living in another Member State—statistical overview, https://ec.europa.eu/eurostat/statistics-explained/index.php?title=EU_citizens_living_in_another_Member_State_-_statistical_overview#Key_messages.

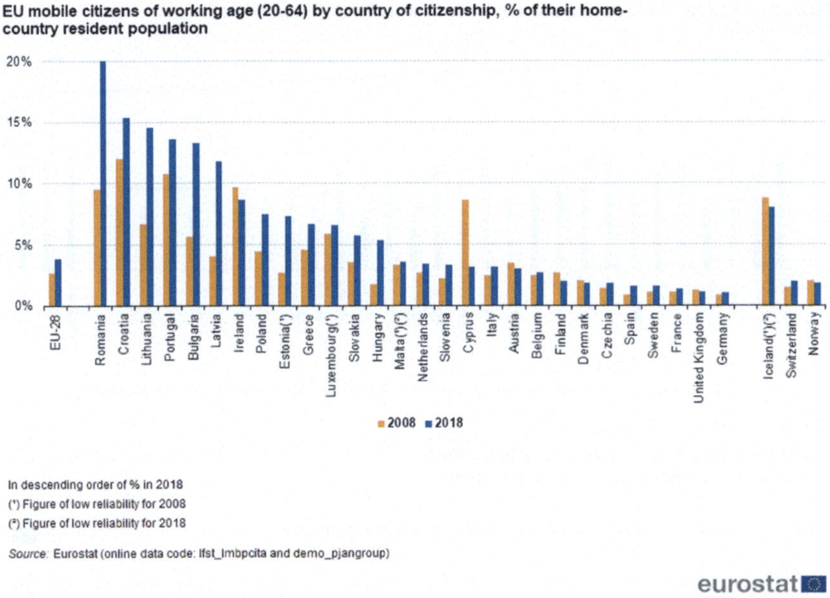

Fig. 4.5 EU mobile citizens of working age (20–64) by country of citizenship, % of their home-country resident population (*Source* Eurostat demo_pjangroup&language=en&mode=view (lfst_lmbpcita demo_pjangroup))

Migration of Third Country Nationals to the EU

If EU citizens moving or residing in another EU Member State are workers exercising their freedom to move within the EU territory ex Treaty, third country nationals who reside legally or illegally in the European Union are "migrants". Although still mainly regulated by the national legislation of each Member State, their treatment is likely to be very different from that of EU citizens. Also, their motivations for migration are likely to differ from those of EU nationals. In the next chapter, we will deal extensively with the characteristics and evolution of the EU migratory policy. Here we will look at some statistics relating to extra-regional migration to the EU.[24]

[24] https://ec.europa.eu/eurostat/statistics-explained/index.php?oldid=415659.

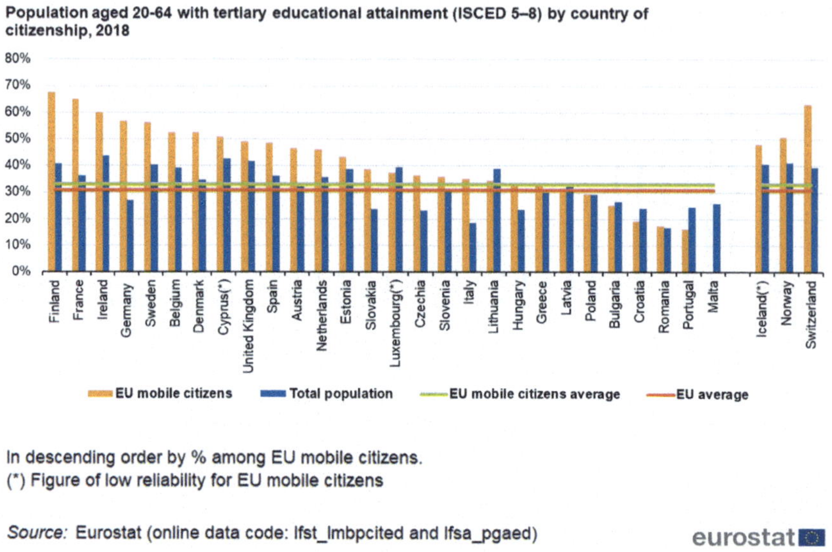

Fig. 4.6 Population aged 20–64 with tertiary educational attainment (ISCED 5–8) by country of citizenship, 2018 (*Source* Eurostat lfsa_pgaed&language=en&mode=view (lfst_lmbpcited lfsa_pgaed))

On the 1st of January 2018, around 4.4% (22.3 million) of the 512.4 million people living in the EU were third country nationals. In 2017, 825,000 persons were awarded nationality of one of the EU Member States. In 2017, 2.4 million immigrants entered the EU legally from non-Member countries, while the number of EU citizens moving to a different country from their own within the EU was only 1.3 million. Another 1.0 million people migrated to an EU Member State of which they had the citizenship (e.g. returning nationals or nationals born abroad), and there were around 11,000 stateless people.[25] Overall, in 2017, extra-regional migration was almost double (85% more) than intra-regional movement of people, and this is considering only legal migration (Fig. 4.9).

In terms of trends, in the last years, there has been a clear increase, by up to 50%, of the share of extra-regional vs intra-regional migration,

[25] Source: Eurostat (online data codes: migr_imm1ctz and migr_pop1ctz).

Population aged 20-64 with primary educational attainment (ISCED 0–2) by country of citizenship, %

Country of citizenship	EU mobile citizens 2008	EU mobile citizens 2018	Total population 2008	Total population 2018	% diff in EU mobile citizens 2018 vs 2008	% diff in Total population 2018 vs 2008
	%		%		%	
EU-28	28.4	24.6	27.9	21.3	-3.8	-6.5
Belgium	17.0	11.8	29.2	21.1	-5.1	-8.0
Bulgaria	23.6	34.0	21.8	17.2	10.4	-4.7
Czechia(1)	14.6	13.5	9.0	6.5	-1.1	-2.5
Denmark(1)	13.2	:	26.0	18.6	:	-7.4
Germany	13.3	7.2	15.7	14.2	-6.1	-1.5
Estonia(2)	:	26.0	12.4	11.3	:	-1.1
Ireland	19.9	11.7	26.7	15.2	-8.1	-11.5
Spain	31.0	19.0	48.0	38.9	-12.0	-9.1
Greece	44.7	31.4	36.7	24.8	-13.4	-11.8
France	15.3	8.7	29.1	19.7	-6.6	-9.3
Croatia	31.0	21.0	21.4	13.8	-10.1	-7.7
Italy	40.5	29.5	44.8	36.6	-11.0	-8.1
Cyprus(1)	26.9	:	25.7	16.8	:	-8.8
Latvia(1)	16.6	20.4	14.9	9.5	3.8	-5.4
Lithuania	19.6	16.0	9.8	5.5	-3.6	-4.4
Luxembourg	:	:	31.6	20.6	:	-11.0
Hungary(1)	10.5	12.9	20.0	15.1	2.4	-4.8
Malta	:	:	67.8	44.3	:	-23.5
Netherlands	18.2	15.2	28.0	20.3	-3.0	-7.7
Austria(1)(2)	12.4	8.9	18.8	14.4	-3.5	-4.3
Poland	17.7	17.5	12.4	7.7	-0.2	-4.7
Portugal	67.9	50.0	69.4	47.5	-17.8	-21.9
Romania	30.7	34.6	24.3	21.2	3.8	-3.1
Slovakia(1)	16.7	12.1	9.8	8.5	-4.6	-1.2
Slovenia(1)	29.9	:	17.1	11.6	:	-5.5
Finland(1)	22.1	:	18.4	11.0	:	-7.4
Sweden(2)	:	7.3	19.2	14.4	:	-4.8
United Kingdom	20.2	19.1	25.8	19.0	-1.2	-6.8
Iceland	:	:	36.9	23.8	:	-13.0
Norway(1)(2)	15.1	14.4	21.0	17.3	-0.7	-3.6
Switzerland(1)	11.1	:	13.6	11.6	:	-2.0

(1) Figure of low reliability for 2008
(2) Figure of low reliability for 2018
: data not available

Source: Eurostat (online data codes: lfst_lmbpcited and lfsa_pgaed)

eurostat

Fig. 4.7 Population aged 20–64 with primary educational attainment (ISCED 0–2) by country of citizenship (*Source* Eurostat lfsa_pgaed&language=en&mode=view (lfst_lmbpcited lfsa_pgaed))

which is in line with the paradox of regionalisation within globalisation (Fig. 4.10).

The foreign population was made of mostly non-EU citizens in most Member States. On 1 January 2018, Belgium, Ireland, Cyprus, Luxembourg, Malta, the Netherlands, Austria, Romania, Slovakia and the UK represented the only countries where the majority of non-nationals were EU citizens. Therefore, in the majority of EU Member States, third country nationals represented the majority of non-nationals (Fig. 4.9). The cases of Latvia and Estonia include a large percentage of citizens from outside the EU because of the high number of recognised non-citizens (mainly former Soviet Union citizens) (Fig. 4.11).

Looking at immigration by country of birth, instead of immigration by citizenship, can perhaps be considered more accurate because people can change citizenship once in a new country. With respect to those data, in

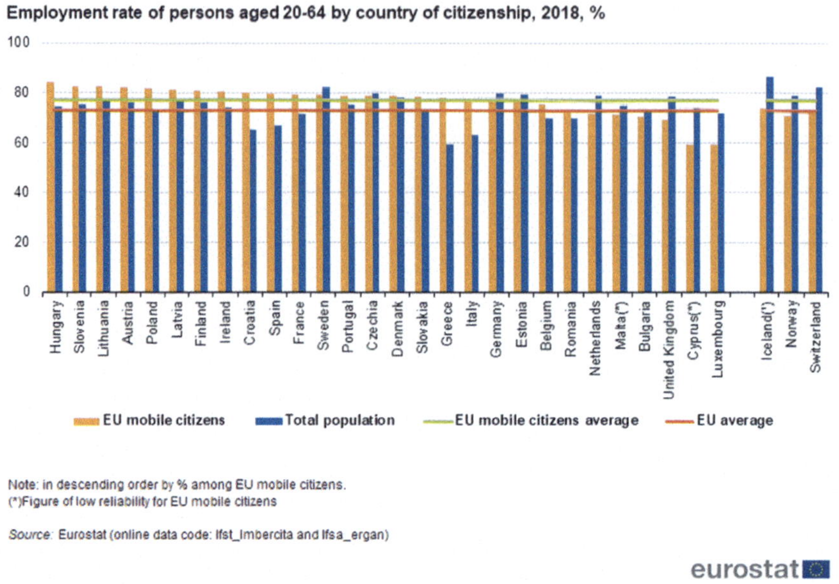

Fig. 4.8 Employment rate of persons aged 20–64 by country of citizenship, 2018 (*Source* Eurostat lfsa_pgaed&language=en&mode=view (lfst_lmbpcited lfsa_pgaed))

2017 the number of immigrants born in a non-member country of the EU was 2,350,248 while the number of EU28 born people exercising their right to move was almost half of it (55%), i.e. 1,310,787 (Fig. 4.12). Overall, there were 38.2 million people born outside of the EU-28 living in an EU Member State on 1 January 2018, while the number of persons born in a different EU Member State from the one where they were resident was only 21.8 million. In fact, the only countries where the number of people born in other EU Member States was higher than the number born outside of the EU-28 were Ireland, Cyprus, Luxembourg, Hungary, Malta and Slovakia (see Fig. 4.12).

As with the case of citizenship, the number of non-EU born immigrants with respect to those born in the EU has increased in the last years (Fig. 4.13).

Looking at the figures relating to the previous country of residence, Luxembourg was the EU country with the largest share of immigrants

Immigration by citizenship, 2017

	Total immigrants	Nationals		Non-nationals							
				Total		Citizens of other EU Member States		Citizens of non-member countries		Stateless	
	(thousand)	(thousand)	(%)	(thousand)	(%)	(thousand)	(%)	(thousand)	(%)	(thousand)	(%)
Belgium	126.7	17.5	13.8	108.5	85.6	60.2	47.5	48.3	38.1	0.0	0.0
Bulgaria	25.6	13.1	51.0	12.5	48.9	0.6	2.5	11.8	46.2	0.0	0.2
Czechia	51.8	4.5	8.7	47.3	91.3	16.6	32.0	30.7	59.3	0.0	0.0
Denmark	68.6	19.5	28.5	49.0	71.5	25.6	37.3	23.1	33.6	0.4	0.6
Germany (1)(2)	917.1	124.4	13.6	788.9	86.0	395.0	43.1	391.5	42.7	2.4	0.3
Estonia	17.6	8.5	48.5	9.1	51.4	4.6	25.9	4.5	25.5	0.0	0.0
Ireland	78.5	26.4	33.7	51.2	65.2	28.5	36.3	22.7	28.9	0.0	0.0
Greece	112.2	31.7	28.3	80.5	71.7	17.2	15.3	63.3	56.4	0.0	0.0
Spain	532.1	78.2	14.7	454.0	85.3	139.4	26.2	314.2	59.1	0.3	0.1
France	370.0	128.0	34.6	242.0	65.4	74.5	20.1	167.5	45.3	0.0	0.0
Croatia	15.6	7.9	50.9	7.6	49.1	2.2	14.1	5.4	35.0	0.0	0.0
Italy	343.4	42.4	12.3	301.1	87.7	61.1	17.8	240.0	69.9	0.0	0.0
Cyprus	21.3	4.0	18.6	17.4	81.4	9.3	43.7	8.0	37.7	0.0	0.0
Latvia	9.9	4.8	48.2	5.1	51.7	0.7	7.5	4.4	44.1	0.0	0.1
Lithuania	20.4	10.2	49.9	10.2	50.1	0.7	3.4	9.5	46.5	0.0	0.2
Luxembourg	24.4	1.2	4.9	23.2	95.0	16.7	68.3	6.5	26.6	0.0	0.0
Hungary	68.1	31.6	46.4	36.4	53.5	11.2	16.4	25.3	37.1	0.0	0.0
Malta	21.7	1.5	6.8	20.2	93.2	11.7	54.2	8.5	39.0	0.0	0.0
Netherlands	189.6	44.6	23.5	143.7	75.8	72.6	38.3	68.6	36.2	2.5	1.3
Austria	111.8	9.7	8.7	102.0	91.2	64.4	57.6	37.4	33.4	0.3	0.3
Poland (2)(3)	209.4	132.8	63.4	76.6	36.6	22.7	10.8	53.8	25.7	0.1	0.0
Portugal (2)	36.6	20.2	55.3	16.4	44.7	7.6	20.8	8.8	24.0	0.0	0.0
Romania (2)	177.4	146.3	82.5	26.8	15.1	9.2	5.2	17.5	9.9	0.1	0.1
Slovenia	18.8	3.3	17.5	15.5	82.5	3.3	17.6	12.2	64.9	0.0	0.0
Slovakia	7.2	4.3	59.5	2.9	40.5	2.3	32.4	0.6	8.1	0.0	0.0
Finland	31.8	8.1	25.4	23.1	72.6	6.5	20.3	16.5	51.8	0.2	0.5
Sweden	144.5	19.5	13.5	124.4	86.1	30.0	20.7	90.0	62.3	4.5	3.1
United Kingdom	644.2	80.9	12.6	563.4	87.4	242.7	37.7	320.7	49.8	0.0	0.0
Iceland	12.1	2.5	20.3	9.7	79.7	8.4	69.3	1.3	10.4	0.0	0.0
Liechtenstein	0.6	0.2	25.9	0.5	74.1	0.2	38.1	0.2	36.0	0.0	0.0
Norway	53.4	6.8	12.7	46.6	87.3	20.0	37.4	26.0	48.8	0.6	1.1
Switzerland	143.4	23.8	16.6	119.5	83.4	82.5	57.6	37.0	25.8	0.0	0.0

Note: The individual values do not add up to the total due to rounding and the exclusion of the 'unknown' citizenship group from the table.

(1) Break in series.
(2) Estimate.
(3) Provisional.

Source: Eurostat (online data code: migr_imm1ctz)

eurostat

Fig. 4.9 Immigration by citizenship, 2017 (*Source* Eurostat (migr_imm1ctz))

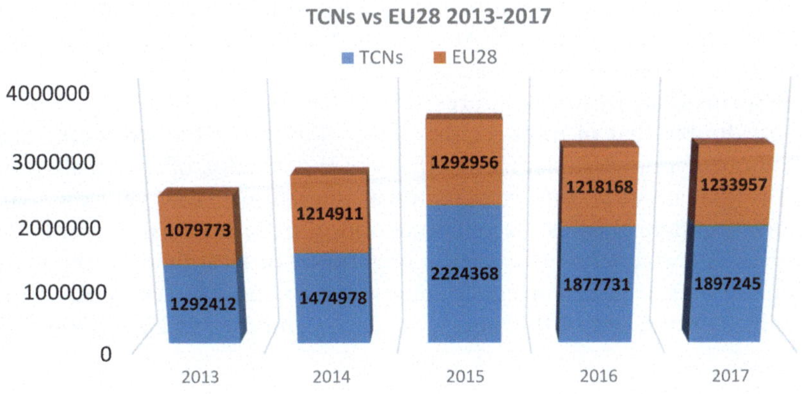

Fig. 4.10 EU 28 vs TCNs 2013–2017 (*Source* Eurostat (migr_imm1ctz))

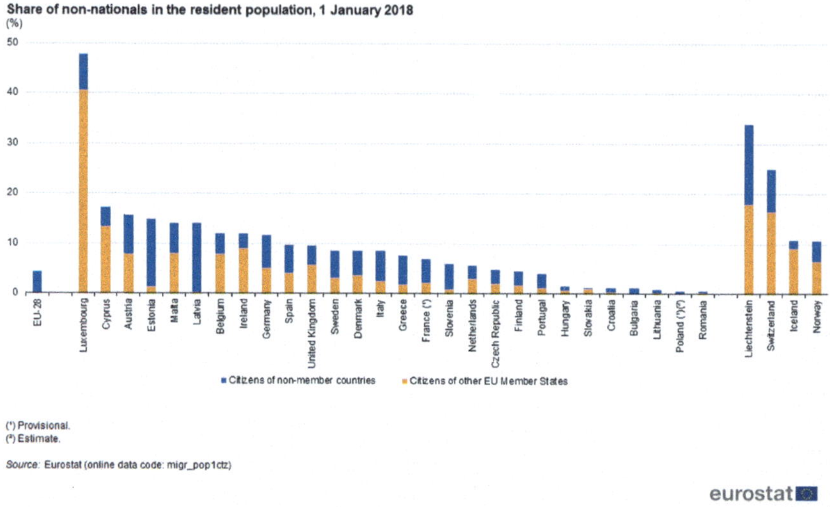

Fig. 4.11 Share of non-nationals in the resident population, 1 January 2018 (%) (*Source* Eurostat (migr_pop1ctz))

coming from another EU Member State (94% of its total number of immigrants in 2017), followed by Slovakia (at 79%) and Romania (at 69%). On the other hand, Italy had a high share of immigrants coming from a non-EU country, with only 22% of all immigrants coming from the EU. Slovenia, Sweden and Spain all had 28% of all immigrants coming from an EU Member State (see Fig. 4.14).

It seems clear from the figures above that extra-regional migration is almost double that of intra-regional migration in the EU, thus corroborating our hypothesis relating to effects of the paradox of marginalisation within globalisation on extra-regional migration. However, the data reported underestimates the phenomenon of extra-regional migration to the EU as it refers only to regular migration. In fact, there is abundant evidence of substantial irregular migration to the EU, especially from marginalised regions of the world, i.e. Africa and the MENA region.[26]

[26] I see African Migration to the EU: Irregular migration in context file:///Users/leilatalani/Dropbox/simona/migrationbook/IOMirregularmigration.pdf. https://gmdac.iom.int/sites/default/files/03_-_residence_permits-bbb.pdf.

Immigration by country of birth, 2017

	Total immigrants	Native born		Foreign-born Total		Foreign-born Born in another EU Member State		Foreign-born Born in a non-member country		Unknown	
	(thousand)	(thousand)	(%)	(thousand)	(%)	(thousand)	(%)	(thousand)	(%)	(thousand)	(%)
Belgium	126.7	14.5	11.4	111.7	88.1	53.0	41.8	58.7	46.3	0.6	0.4
Bulgaria	25.6	12.6	49.3	13.0	50.7	2.0	7.9	11.0	42.8	0.0	0.0
Czechia	51.8	6.8	13.0	44.6	86.1	12.8	24.6	32.3	62.3	0.0	0.0
Denmark	68.6	15.4	22.4	53.2	77.5	25.2	36.7	28.0	40.9	0.0	0.0
Germany (¹)(²)	917.1	90.7	9.9	821.7	89.6	371.8	40.5	449.9	49.1	4.8	0.5
Estonia	17.6	7.7	43.6	9.9	56.3	4.6	26.2	5.3	30.0	0.0	0.1
Ireland	78.5	27.0	34.4	50.6	64.5	28.0	35.6	22.7	28.9	0.9	1.1
Greece	112.2	27.7	24.7	84.6	75.3	20.0	17.8	64.6	57.5	0.0	0.0
Spain	532.1	38.4	7.2	493.7	92.8	122.5	23.0	371.2	69.8	0.0	0.0
France	370.0	84.8	22.9	285.2	77.1	75.6	20.4	209.6	56.6	0.0	0.0
Croatia	15.6	3.9	25.3	11.6	74.6	2.4	15.3	9.2	59.3	0.0	0.0
Italy	343.4	32.6	9.5	310.8	90.5	53.3	15.5	257.5	75.0	0.0	0.0
Cyprus	21.3	4.0	18.7	17.3	81.3	8.9	41.8	8.4	39.5	0.0	0.0
Latvia	9.9	4.1	41.7	5.8	58.2	1.4	13.6	4.4	44.6	0.0	0.1
Lithuania	20.4	8.8	43.4	11.5	56.6	1.5	7.3	10.0	49.3	0.0	0.0
Luxembourg	24.4	1.2	4.9	23.1	94.8	14.4	59.2	8.7	35.6	0.1	0.3
Hungary	68.1	21.3	31.3	46.8	68.7	14.9	21.9	31.8	46.8	0.0	0.0
Malta	21.7	1.7	7.8	20.0	92.2	10.7	49.2	9.3	43.0	0.0	0.0
Netherlands	189.6	29.2	15.4	160.4	84.6	68.6	36.2	91.9	48.4	0.0	0.0
Austria	111.8	7.9	7.0	102.8	91.9	59.2	52.9	43.6	39.0	1.2	1.1
Poland (²)(³)	209.4	81.8	39.1	127.6	60.9	64.4	30.8	63.1	30.2	0.0	0.0
Portugal (²)	36.6	13.8	37.7	22.8	62.2	8.6	23.3	14.2	38.9	0.0	0.0
Romania (²)	177.4	95.2	53.7	78.4	44.2	12.9	7.3	65.5	36.9	3.8	2.1
Slovenia	18.8	2.3	12.0	16.6	88.0	3.2	17.0	13.4	71.0	0.0	0.0
Slovakia	7.2	1.2	16.2	6.0	83.8	4.7	65.0	1.3	18.8	0.0	0.0
Finland	31.8	6.7	21.2	24.2	76.2	6.7	21.2	17.5	55.0	0.8	2.6
Sweden	144.5	14.4	10.0	129.8	89.9	29.0	20.1	100.8	69.8	0.2	0.1
United Kingdom	644.2	67.2	10.4	577.0	89.6	230.6	35.8	346.4	53.8	0.0	0.0
Iceland	12.1	1.9	15.9	9.6	79.4	7.8	64.5	1.8	14.9	0.6	4.7
Liechtenstein	0.6	0.0	6.5	0.6	93.5	0.3	39.5	0.3	54.0	0.0	0.0
Norway	53.4	4.4	8.2	49.0	91.8	18.9	35.4	30.1	56.4	0.0	0.0
Switzerland	143.4	17.1	11.9	125.1	87.2	73.1	51.0	52.0	36.2	1.2	0.9

Note: The values for the different categories of country of birth may not sum to the total due to rounding.

(¹) Break in series.
(²) Estimate.
(³) Provisional.

Source: Eurostat (online data code: migr_imm3ctb)

eurostat

Fig. 4.12 Immigration by country of birth, 2017 (*Source* Eurostat (migr_imm3ctb))

According to the IOM, the availability of quality and data on irregular migration dynamics from Africa and the MENA region, as well as other related topics, requires improvement.[27] The IOM defines irregular migration as "movement that takes place outside the regulatory norms of the sending, transit and receiving country".[28] This definition is comprised of: (a) irregular entry in a foreign country, (b) irregular residence in a foreign country or (c) irregular employment in a foreign country. Focusing

[27] http://gmdac.iom.int.

[28] See https://migrationdataportal.org/themes/irregular-migration.

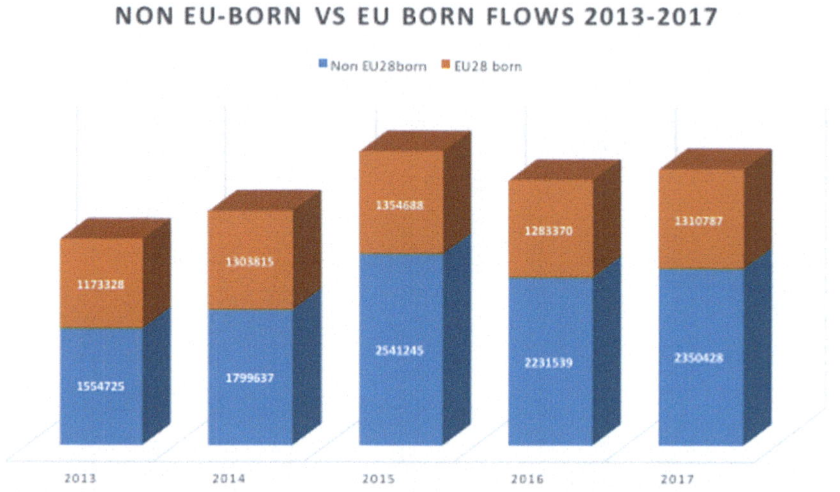

Fig. 4.13 Non-EU born vs EU born immigrants 2013–2017 (*Source* Eurostat (migr_imm3ctb))

on countries in Western and Northern Africa (UN regional classification),[29] only irregular arrivals (entries) by sea are considered here. The EU framework for regular migration to the EU of TCNs is the Global Approach to Migration and Mobility (GAMM). This comprises several legal tools and instruments, together with other cooperation frameworks, such as Mobility Partnerships for bilateral cooperation, and Common Agendas for Migration and Mobility (CAMM). Although the EU has pledged to create more legal migration pathways, including through the 2015 "Valletta Political Declaration and Action Plan", the 2016 framework "Towards a reform of the Common European Asylum System and enhancing legal avenues to Europe", currently legal migration to the EU is mostly allowed only for the highly skilled and for family reunification. On the other hand, irregular entries are generally being addressed within a securitisation agenda, as further elaborated in the next chapter. This includes search and rescue operations, law-enforcement-activities aimed

[29] Which are Algeria, Egypt, Libya, Morocco, Tunisia, Sudan (Northern Africa); Benin, Burkina Faso, Cabo Verde, Côte d'Ivoire, the Gambia, Ghana, Guinea, Guinea-Bissau, Liberia, Mali, Mauritania, the Niger, Nigeria, Senegal, Sierra Leone and Togo (Western Africa).

Immigration by previous country of residence, 2017

	Total immigrants	From an EU Member State of previous residence		From a non-member country of previous residence		From an unknown country of previous residence	
	(thousand)	(thousand)	(%)	(thousand)	(%)	(thousand)	(%)
Belgium	126.7	73.0	57.6	53.2	42.0	0.4	0.3
Bulgaria	25.6	9.2	35.8	16.4	64.1	0.0	0.1
Czechia	51.8	16.8	32.5	35.0	67.5	0.0	0.0
Denmark	68.6	34.0	49.6	34.1	49.7	0.5	0.8
Germany (1)(2)	917.1	435.4	47.5	398.9	43.5	82.8	9.0
Estonia	17.6	11.6	66.0	5.3	30.2	0.7	3.8
Ireland	78.5	42.6	54.2	35.9	45.8	0.0	0.0
Greece	112.2	44.6	39.7	67.7	60.3	0.0	0.0
Spain	532.1	150.0	28.2	382.1	71.8	0.0	0.0
France	370.0	128.3	34.7	241.7	65.3	0.0	0.0
Croatia	15.6	5.9	37.7	8.9	57.4	0.8	4.9
Italy	343.4	74.9	21.8	268.6	78.2	0.0	0.0
Cyprus	21.3	12.2	57.4	9.1	42.6	0.0	0.0
Latvia	9.9	4.8	48.8	5.1	51.2	0.0	0.0
Lithuania	20.4	8.7	42.8	11.7	57.2	0.0	0.0
Luxembourg	24.4	23.0	94.4	1.4	5.6	0.0	0.0
Hungary	68.1	36.1	53.0	31.9	46.9	0.1	0.1
Malta	21.7	12.5	57.7	9.2	42.3	0.0	0.0
Netherlands	189.6	93.5	49.3	94.7	50.0	1.4	0.8
Austria	111.8	66.3	59.3	40.6	36.3	4.9	4.4
Poland (2)(3)	209.4	139.3	66.5	70.1	33.5	0.0	0.0
Portugal (2)	36.6	17.1	46.6	19.6	53.4	0.0	0.0
Romania (2)	177.4	123.0	69.3	53.3	30.1	1.2	0.7
Slovenia	18.8	5.2	27.6	13.6	72.4	0.0	0.0
Slovakia	7.2	5.7	78.8	1.5	21.2	0.0	0.0
Finland	31.8	12.2	38.3	18.3	57.6	1.3	4.0
Sweden	144.5	40.6	28.1	99.8	69.1	4.1	2.8
United Kingdom	644.2	260.7	40.5	383.6	59.5	0.0	0.0
Iceland	12.1	9.5	78.6	2.5	20.5	0.1	0.9
Liechtenstein	0.6	0.2	28.7	0.5	71.0	0.0	0.3
Norway	53.4	24.6	46.1	28.8	53.9	0.0	0.0
Switzerland	143.4	90.4	63.1	48.6	33.9	4.4	3.1

Note: The values for the different categories of previous residence may not sum to the total due to rounding.

(1) Break in series.
(2) Estimate.
(3) Provisional.

Source: Eurostat (online data code: migr_imm5prv)

eurostat

Fig. 4.14 Immigration by previous country of residence, 2017 (*Source* Eurostat (migr_imm5prv))

at curbing irregular migration, and voluntary and forced returns schemes from both transit and destination countries. The EU also provides for various resettlement schemes, including, for example, provisions included in the EU-Turkey accord of 2016, as well as Member States' individual resettlement mechanisms. With respect to the African region, in 2016 the EU started an EU Emergency Trust Fund for Africa (EUTF) of €3.4 billion. This was devoted at producing job creation opportunities and vocational training, combatting human trafficking and other similar initiatives in various African countries (IOM, 2015).

To assess how many irregular entries by sea to the EU have taken place in recent years, it is worth looking at the Central Mediterranean route to Europe as this has been the most used. Of course, this is by no means a sufficient indicator of irregular migration to the EU as, for example, it does not include those who entered the EU regularly, but became irregular after the loss of employment or expiry of a visa. Unfortunately, however, no reliable data exists on these migrants, apart from the ones referring to regularisation processes taking place from time to time in various EU Member States. In Italy for example during the period between 2011 and 2017, there were 324,587 irregular arrivals from the top ten sending countries to Italy which were, in descending order: Nigeria, Tunisia, Mali, the Gambia, Cote d'Ivoire, Senegal, Ghana, Morocco and Egypt (IOM, 2015) (Fig. 4.15).

As already underlined above, those numbers, although already significant, grossly underestimate the size of irregular extra-regional migration to the EU. First of all, the figures reported above only include irregular entries by sea through the Central Mediterranean Route of the ten main origin countries of irregular migration to Italy from the MENA region. In general, it is extremely difficult to monitor irregular migration. Any data on flows, stocks or characteristics of irregular migrants is at best patchy, and at worst, unreliable. The reasons are obvious, as irregular migrants are not registered in national administrative offices, censuses or surveys. Even information on legal migrants are difficult to access, as this can change depending on national and regional legislation. Moreover, data relating

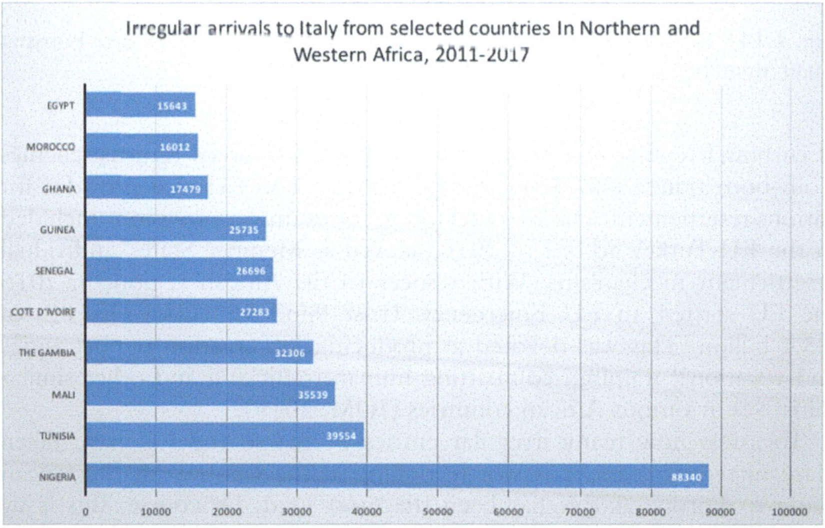

Fig. 4.15 Irregular arrivals to Italy from selected countries in Northern and Western Africa, 201–2017 (*Source* IOM, based on data from the Italian Ministry of the Interior)

to undocumented migrants' crossings of the Mediterranean is underestimating the real number, as they include only apprehended ones. Finally, reliable figures on any kind of migration from Africa are very difficult to obtain, as in Africa there are rarely reliable national statistics, including population censuses. Indeed, in 2017, 17% of countries in Africa had census data from before 2005. Not to mention that national statistical offices (NSOs) in Africa are not well funded, often migration-related data is not recorded or even shared within the country. As such, comparisons are difficult to make due to the differences across countries' data collection systems. Overall, therefore, it is very likely that the phenomenon of extra-regional migration to the EU is consistently more significant than shown by statistics on both regular and irregular migration of TCNs.

This is in line with the paradox of marginalisation within globalisation and its effects in terms of increasing extra-regional migration. In the next chapter, the characteristics and the evolution of the EU approach to migration from third countries will be discussed at length.

Conclusion

Globalisation-induced regionalism is a phenomenon that has been studied by different IPE approaches reaching different conclusions. Whereas all schools of thought recognise a tendency towards economic regionalisation and some forms of institutional regionalism, the interpretation of this trend changes substantially among the different intellectual perspectives.

For realists, regionalisation is an ad hoc phenomenon, unrelated to globalisation and following from the interests of the hegemonic states of the region to acquire more power. Quite to the contrary, both institutionalists and transnationalists connect the regionalisation process with globalisation. However, for the former, regionalisation and the creation of multilayered institutionally integrated regional governance schemes are just a step towards the creation of a global governance regime. For transnationalists, instead, this process is far from being a necessary one and entails several contradictions and paradoxes as it is highly uneven and unequal. Only regions that are progressively more integrated in the global political economy will see the creation of a new mobility-fixity nexus at the regional level. The others will be progressively more marginalised and less regionally integrated.

This applies also to the establishment of regional migratory governance schemes which present very regional dynamics. Therefore, intra-regional migration tends to be more regulated than extra-regional migration. This is clearly shown in the case of the EU migratory regime which is analysed in the next chapter.

Moreover, marginalisation and lack of regionalisation increase the incentives for extra-regional migration to more integrated regions, both under the form of brain drain and mass migration. On the other hand, intra-regional migration is usually favoured by the regionalisation process and happens de facto, if not de jure, as seen in the case of migration to the US.

References

Abdou, L. H. (2019). Immigrant integration: The governance of ethno-cultural difference. *Comparative migration studies, 7*(15). OnlineOnly[Migration Policy Centre]. https://orcid.org/0000-0002-5355-6286.

Amin, S. (2006). *Beyond US hegemony? Assessing the prospects for a multipolar world*. London: Zed books.

Archibugi, D. (2010). The architecture of Cosmopolitan Democracy. In G. Wallace Brown & D. Held (Ed.), *The Cosmopolitanism reader* (pp. 312–333). Polity Press.

Bacon, D. (2008). *Illegal people: How globalisation creates migration and criminalizes immigrants*. Beacon Press.

Beck, U. (2005). *Power in the global age: A new global political economy*. Polity Press.

Beck, U. (2006). *The cosmopolitan vision*. Polity Press.

Beck, U. (2009). *World at risk*. Polity Press.

Beck, U. (2010). The Cosmopolitan Manifesto. In G. Wallace Brown & D. Held (Eds.), *The Cosmopolitanism reader* (pp. 217–228). Polity Press.

Betts, A. (2013). The migration industry in global migration governance. In T. Gammeltoft-Hansen & N. Nyberg Sørensen (Eds.), *The migration industry and the commercialization of international migration* (pp. 45–63). Routledge.

Borjas, G. J. (2016). *We wanted workers: Unraveling the immigration narrative*. New York: W. W. Norton & Company.

Breslin et al. (2002). *New regionalisms in the global political economy: Theories and cases. Warwick studies in globalisation*. London, NY: Routledge. ISBN 9780415277679.

Brick, K., Challinor, A., & Rosenblum, M. (2011). *Mexican and central American immigrants in the United States*. Migration Policy Institute.

Cafruny, A., & Ryner, M. (2007). *Europe at bay*. Lynne Rienner Publishers.

Callinicos, A. (2010). *Bonfire of illusions*. London: Polity.

Cerny, P. (1999). Reconstructing the political in a globalising world. In F. Buelens (Ed.), *Globalisation and the nation state*. Edward Elgar.
Cohen, B. J. (1996). Phoenix risen: The resurrection of global finance. *World Politics, 48*(2): 268–296.
Cohen, B. J. (2001). Electronic money: New day or false dawn? *Review of International Political Economy, 8*, 197–225.
Cox, R. W. (1987). *Production, power, and world order: Social forces in the making of history*. New York: Columbia University Press.
Dicken, P. (2011). *Global shift: Mapping the changing contours of the world economy* (6th ed.). Guilford.
Espinoza, V. M. (2018). The limits and opportunities of Regional Solidarity: Exploring refugee resettlement in Brazil and Chile. *Global Policy, 9*(1), 85–94.
Gamble, A., & Payne, A. (1996). The political economy of regionalism and world order. In A. Gamble & A. Payne (Eds.), *Regionalism and world order* (pp. 1–20). Macmillan.
Geddes, A. (2012). Regions and regionalism. In M. R. Rosenblum & D. J. Tichenor (Eds.), *The Oxford handbook of the politics of international migration* (pp. 573–593). Oxford University Press.
Geddes, A. (2019). The Dynamics of Regional Migration Governance. European University Institute, Marcia Vera Espinoza, Queen Mary University of London, UK, Leila Hadj Abdou and Leiza Brumat, European University Institute, Italy Publication Date: 2019 ISBN: 978 1 78811 993 1 Extent: 256 pp.
Geddes, A., & Scholten, P. (2016). *The politics of migration and immigration in Europe* (2nd ed.). Sage.
Gill, S. (1997). An EMU or an Ostrich?: EMU and neo-liberal globalisation; limits and alternatives. In P. Minkkinen, & H. Potomaki (Eds.), *The politics of Economic and monetary Union*. The Hague: Kluwer Academic Publishers.
Guiraudon, V. (2000). European integration and migration policy: Vertical policy-making as venue shopping. *Journal Common Market Studies, 38*(2), 251–272.
Guiraudon, V. (2003). The constitution of a European immigration policy domain. *Journal European Public Policy, 10*(2), 263–282.
Harvey, D. (2005). *A brief history of neo-liberalism*. Oxford University Press.
Held, D. (1995). *Democracy and the global order: From the modern state to cosmopolitan governance*. Stanford, CA: Stanford University Press.
Held, D. (2002). Law of states, law of peoples. *Legal Theory, 8*(1), 1–44.
Held, D. (2010). *Cosmopolitanism: Ideas and realities*. Cambridge: Polity Press.
Held, D., & McGrew, A. (2007). *Globalisation/antiglobalisation*. Polity Press.
Hettne, B. (1994). The new regionalism: Implications for development and peace. In B. Hettne & A. Inotai, *The New Regionalism*. Implications for Global Development and International Security (UNU/WIDER).

Hettne, B., Inotai, A., & Sunkel, O. (1999). *Globalism and the new regionalism*. Macmillan Press Ltd.

Hirst, P., & Thompson, G. (1999a). *Globalisation in question* (2nd ed.). Polity.

Hirst, P., & Thompson, G. (1999b). The tyranny of globalisation: Myth or reality?. In F. Buelens (Ed.), *Globalisation and the nation state*. Edward Elgar.

Hirst, P., Thompson, G., & Bromely, S. (2009). *Globalisation in question: The international economy and the possibilities of governance* (3rd ed.). Polity Press.

Hoffmann, S. (1995). *The european sisyphus. Essays on Europe, 1964–1994*. Boulder Westview Press.

Huysmans, J. (2000). The European union and the securitization of migration. *Journal of Common Market Studies, 38*(5), 751–777.

IOM. (2015). *African migration to the EU: Irregular migration in context*. https://gmdac.iom.int/sites/gmdac/files/03_-_residence_permits-bbb.pdf.

Jurje, F., & Lavenex, S. (2018). Mobility norms in free trade agreements: Migration governance in Asia between regional integration and free trade. *European Journal of East Asian Studies, 17*(1), 83–117.

Kaldor, M. (2010). Humanitarian intervention: Towards a cosmopolitan approach. In G. Wallace Brown & D. Held (Eds.), *The Cosmopolitanism reader* (pp. 334–350). Polity Press.

Keohane, R. O., & Victor, D. G. (2010, January). *The regime complex for climate change* (Discussion Paper 10 33). The Harvard Project on International Climate Agreements.

Kostakopoulou, T. (2000). The 'protective union': Change and continuity in migration law and policy in post-amsterdam Europe. *Journal of Common Market Studies, 3*, 497–518.

Kunz, R. (2011). Depoliticisation through partnership in the field of migration: The Mexico–US Case. In R. Kunz, S. Lavenex, & M. Panizzon (Eds.), *Multilayered migration governance: The promise of partnership* (pp. 283–304). Routledge.

Kunz, R., Lavenex, S., & Panizzon, M. (Eds.). (2011). *Multilayered migration governance: The promise of partnership*. Routledge.

Lavenex, S. (2018a). Regional Migration Governance. Building Block of Global Initiatives? *Journal of Ethnic and Migration Studies*. https://doi.org/10.1080/1369183X.2018.1441606.34

Lavenex, S. (2018b). International migration. In A. Draude, T. A. Börzel, & T. Risse (Eds.), *Oxford Handbook of Governance in Areas of Limited Statehood*. Oxford University Press. https://doi.org/10.1093/oxfordhb/9780198797203.013.25

Lavenex, S. (2019). Special issue: Migration governance in an era of large movements regional migration governance – building block of global initiatives? *Journal of Ethnic and Migration Studies, 45*(8), 1275–1293. Published online: 06 Apr 2018.

Lavenex, S., & Nicola, P. (2019). Regional migration governance: Perspectives 'from above' and 'from below' 2019. In The Dynamics of Regional Migration Edited by Andrew Geddes, European University Institute, Marcia Vera Espinoza, Queen Mary University of London, UK, Leila Hadj Abdou and Leiza Brumat, European University Institute, Italy Publication Date: 2019 ISBN: 978 1 78811 993 1 Extent: 256 pp

Lavenex, S., Givens, T. E., Jurje, F., & Buchanan, R. (2016). Regional migration governance. In T. A. Börzel & T. Risse (Eds.), *The Oxford handbook of comparative regionalism* (pp. 457–486). Oxford University Press.

Long, T., & Suarez-Mier, M. (2017). Regional public goods in North America. In L. W. Goodman, & A. Estevadeordal (Eds.), *21st Century cooperation: Regional public goods, global governance and sustainable development* (pp. 265–286). New York: Routledge. ISBN 9781138722590.

Margheritis, A. (Ed.). (2018). *Shaping migration between Europe and Latin America: New approaches and challenges*. ILAS Publications.

Mathew, P., & Harley, T. (Eds.). (2016). *Refugees, regionalism and responsibility*. Edward Elgar Publishing.

Mittleman, J. H. (2000). *The globalisation syndrome: Transformation and resistance*. Princeton: Princeton University Press.

Moravcsik, A. (1991). Negotiating the single European act: National interests and conventional statecraft in the European community. *International Organisation, 45*(1), 19–56.

Moravcsik, A. (1993a). Preferences and power in the EC: A liberal intergovernmentalist approach. *Journal of Common Market Studies, 31*(4), 473–524.

Moravcsik, A. (1993b). Integrating international and domestic theories of international bargaining. In P. B. Evans, H. K. Jacobson, & R. D. Putnam (Eds.), *Double-Edged Diplomacy: International Bargaining and Domestic Policy*. Berkeley: University of California Press.

Murshed, S. M. (2002, July). Conflict, Civil War and Underdevelopment: An Introduction. *Journal of Peace Research, 39*(4), 387–393. Special Issue on Civil War in Developing Countries (Jul., 2002). Published By: Sage Publications, Inc.

Overbeek, H. (2000). Globalisation, sovereignty and transnational regulation: Reshaping the governance of international migration. In B. Gosh (Ed.), *Managing migration: Time for a new international regime*. Oxford: Oxford University Press.

Overbeek, H. W., & Léon, A. I. (2015). Neoliberal globalisation, transnational migration and global governance. In L. S. Talani, & S. McMahon (Eds.), *Handbook of the International Political Economy of Migration* (pp. 37–53). Edward Elgar. https://doi.org/10.4337/9781782549901.

Papademetriou, D. G., Somerville, W., & Tanaka, H. (2008). *Talent in the 21st Century Economy*. Migration Policy Institute. http://www.migrationpolicy.org/research/talent-21st-century-economy

Pellerin, H., & Overbeek, H. W. (2001). Neoliberal regionalism and the management of people's mobility. In A. Bieler, & A. D. Morton (Eds.), *Social forces in the making of the new Europe*. Palgrave.

Rosenau, J. N. (1990). *Turbulence in world politics*. Brighton: Harvester Wheatshef.

Rosenau, J. N. (1997). *Along the domestic-Foreign frontier*. Cambridge: CUP.

Rosenblum, M. R., & Tichenor, D. J. (Eds.). (2012). *Oxford Handbook of the Politics of International Migration*. Oxford University Press. Print Publication Date: Jun 2012. ISBN: 9780195337228. Published online: Nov 2012. https://www.oxfordhandbooks.com.

Ruggie, J. G. (1993). Territoriality and beyond: Problematizing modernity in international relations. *International Organisation, 47*(1), 139–174.

Samers, M. (2008). At the heart of 'migration management' Immigration and labour markets in the European Union. In C. Gabriel & H. Pellerin (Eds.), *Governing international labour migration: Current issues, challenges and dilemmas*. New York: Routledge.

Sassen, S. (1981). *Exporting Capital and importing labor: The role of Caribbean migration to New York City*. New York: NYU.

Sassen, S. (1988). *The mobility of labour and capital*. Cambridge University Press.

Sassen, S. (1991). *The global City: New York, London, Tokyo*. Princeton. NJ: Princeton University Press.

Soysal, Y. N. (2000). Citizenship and identity: Living in diasporas in post-war Europe?. *Ethnic and Racial Studies, 23*(1), 1–15.

Strange, S. (1991). An eclectic approach. In C. N. Murphy, & R. Tooze (Eds.), *The new International political economy* (p. 34). Boulder, CO: Lynne Rienner.

Strange, S. (2002). States, firms and diplomacy, 1992. In R. Tooze, & M. Christopher (Eds.), *Authority and markets, susan strange's writings on international political economy*. London: Palgrave.

Talani, L. S. (2012). *Globalisation, hegemony and the future of the city of london*. London: Palgrave.

Tichenor, D. J., & Rosenblum, M. R. (2012). Poles apart: The politics of illegal migration in America. In D. J. Tichenor & M. R. Rosenblum, *Oxford handbook of the politics of international migration* (p. 1). Oxford University Press.

Van der Pijl, K. (2011). *Arab revolts and nation-state crisis*. NLR 70, July—August 2011.

Wallace Brown, G., & Held, D. (Eds.). (2010). *The Cosmopolitanism reader*. Polity Press.

CHAPTER 5

The Paradox of Securitisation: Is There a Common Migratory Policy in the EU?

The debate on the implementation of a common migratory policy vis-à-vis third country nationals in the EU is a thriving one and one that does not seem to be easy to resolve. Much of the discussion focuses on the notion of 'Fortress Europe', defined in the literature as an area that enjoys internal mobility while erecting barriers to entry and stay with respect to non-EU citizens (Geddes, 2003). According to scholars, the notion of 'Fortress Europe' comes from the Council of European Communities or the Council (ECC) Regulation 1612/68, which for the first time distinguished between the movement rights of citizens of an EU country and the rights of movement of third country nationals (TCNs) (Huysmans, 2000; Ugur, 1995).

This divide was further intensified by the establishment of the freedom of movement for citizens of an EU country effected by the Single European Act in 1986, and even more by the 'EU citizenship' measures introduced by the Maastricht Treaty in 1992. Conversely, the rights of entry and stay for third country nationals were constantly restricted leading to a parallel increase of irregular migration. Thus, the EU common approach to migration has been progressively securitised and, according to some authors, after the refugee crisis of 2014–2015, it has come closer to the 'militarisation' of borders (Boswell & Geddes, 2011; Geddes, 2003; Guiraudon, 2018). The notion of 'securitisation" entails

L. S. Talani, *The International Political Economy of Migration in the Globalization Era*, International Political Economy Series,
https://doi.org/10.1007/978-3-030-79321-0_5

the conceptualisation and treatment of migration as a 'security issue', and its consequent management by security agencies, like the police or even, in the case of militarisation, the army (Guiraudon, 2000, 2018[1]; Huysmans, 2000).

This chapter reviews the developments leading to the creation of Fortress Europe and identifies the different migratory regimes existing within the EU for intra-regional and extra-regional migration. The chapter concludes by giving some insights on the explanations proposed to the securitisation of migratory policy and to its militarisation and externalisation, especially after the refugee crisis.

Does Fortress Europe Exist?

There is relative consensus in the literature on how the process of Europeanisation of EU migration policy has been constantly accompanied by the need to adopt a progressively more restrictive approach to the entry of TCNs and a related tendency to securitise the border, transforming the issue of migration into a security issue (Guiraudon, 2000, 2018; Huysmans, 2000; Kostakopoulou, 2000). At the same time, the communitirisation of migration policy has brought to the creation of 'Fortress Europe' through the passing of restrictive legislation at both national and EU levels. This chapter will be limited to addressing the developments in the EU approach to migration from third countries. As noticed by Geddes (2003), generally speaking European States have tendentially not considered the entry and stay of TCNs in their territories as a problem, and even less, a security threat, until the 1980s. Simultaneously, the question of the adoption of a common EU approach to migration had not been systematically addressed at the EU level as it was not considered salient (Baldwin-Edwards, 1999; Hollifield, 1992).

Indeed, the economic reconstruction of European countries in the 1950s and 1960s had made migratory flows from other countries, especially from southern European ones, a common phenomenon and one which was considered necessary to obtain much needed workers. Hence, national legislation tended to favour similar migratory dynamics instead of limiting them (Boswell, 2003: 10). Regularisation of migration, however, was not a problem, as migrants were often recruited in their home

[1] See Guiraudon, V. (2018). The 2015 refugee crisis was not a turning point. *European Political Science, 17*, 151–159.

countries and did not need to be regularised at the point of arrival (Boswell, 2003: 12). The adoption of a more restrictive approach to migration became more common only at the end of the 1960s and in the 1970s when economic crises started to hit European countries (Baldwin-Edwards, 1999: 9; Hollifield, 1992). Then, with mounting job losses, governments felt the urgency to protect national labour markets (Boswell, 2003: 15). In parallel, the question of migration acquired more saliency also at the European level, although this would remain at the state of infancy for a long time.

In fact, following the relevant literature, it is possible to identify three stages in the evolution of a common EU approach to migration (Boswell, 2003; Kostakopoulou, 2000; Stetter, 2000). The first phase is between 1957 and 1974; the second is from 1974 to 1985; and the third is represented by the period of 1985 to1992.

Given the circumstances mentioned above, the first period did not see much in terms of the communitarisation of migration policy. There were of course the treaty provisions connecting the free movement of people within the EEC to the creation of the internal market (art.3(1)© European Community Treaty (TEC)). According to Articles 39–55 TEC, such rights were reserved to economic actors, such as workers, self-employed persons and providers of services, who enjoyed the nationality of an EEC country. On the other hand, nation states were totally in charge of regulating migration from countries outside the EEC (Stetter, 2000). Still, Council Regulation 1612/68 sets the stage for the development of 'Fortress Europe' (Huysmans, 2000; Ugur, 1995: 967). In the words of Ugur:

> Fortress Europe has been with us since 1968 when intra-EU freedom of movement was established by regulation 1612/68. (Ugur, 1995: 974)

With this piece of legislation, the rights of free movement of the nationals of the EEC Member States were separated from the rights of free movement of third country nationals. This separation was then confirmed by the outcome of the Paris Summit of 1973, which marked a watershed in the common approach to migration. In Paris, for the first time, the question of migration of third country nationals received attention at the level of the European Community due to the need to create a more restrictive regime configuring 'Fortress Europe'. According to Stetter (2000), this was necessary as some measures facilitating the free circulation of Member

States' citizens would become effective already in 1974, while talks about the complete abolition of internal border controls were already taking place. Thus, there was a new urgency to adopt a common and more stringent approach to the question of entry and stay of TCNs.

This inaugurated the second phase of the development of 'Fortress Europe'. While nation states mostly retained competencies over migratory policy, the move towards the single market produced externalities with respect to border control. Hence, the need to cooperate on questions of migration from third countries emerged, although still exclusively through intergovernmental means.

Thus, in 1975, the so-called TREVI group was established, an intergovernmental gathering, addressing migration at the EU level for the first time. From the onset, its remit included competences relating to border control and security.

In particular, it was set up to facilitate coordination and cooperation of law-enforcement agencies devoted to internal security, especially anti-terrorist units. Later on, in 1985, its competences were extended to organised crime and illegal migration.[2]

However, the proper institutionalisation of 'Fortress Europe' occurred in the third period of the evolution of a common EU approach to migration. This period went from the passing of the Single European Act in 1985 to the Maastricht Treaty of 1992.

Clearly, the abolition of internal border controls and the establishment of the Single Market required addressing the issue of third country nationals. Thus, in March 1985, the European Commission issued a communication to the Council regarding 'Guidelines for a Community Policy on Migration' (Stetter, 2000: 86). This communication was aimed at: (i) updating the policy of the Commission on migrant workers and their families[3] and (ii) relaunching the question of how to address migratory issues at the level of EEC Institutions given the changing circumstances from 1974, when the first Commission Programme was adopted.[4]

[2] See http://ec.europa.eu/justice_home/fsj/customs/printer/fsj_customs_intro_en.htm as accessed on October 14, 2008.

[3] See Supplement 3/76 to the Bulletin of the EC.

[4] See web-site www.europa.eu as accessed on October 14, 2008.

Contextually, the Council of Ministers set up the ad hoc intergovernmental Working Group on Immigration in 1986. On this occasion, the Commission was invited to participate as an observer in discussions on migratory issues for the first time.

Still, little progress was achieved in terms of communitarising migratory policy at the European level (Stetter, 2000). On the contrary, the creation of 'Fortress Europe' was well advanced. This happened thanks to two important intergovernmental measures: the first was the Schengen Agreement of 1985 which covered migration policy; the second was the Dublin Convention of 1990 devoted to asylum and refugee policy. However, both of them did not enter into force until much later on, the Schengen Agreement after ten years from its original formulation in 1995 and the Dublin Convention after seven years, in 1997 (Guiraudon, 2000: 256).

Both agreements reinforced the idea that Europe was becoming progressively closer to a 'Fortress' as far as the rights of entry and stay of TCNs were concerned. This was achieved by adopting:

- External frontiers
- Common rules on carrier liability and punishment of those engaged in illegal migration
- A system of 'One State Checks' for asylum seekers with the Dublin Convention
- An executive Committee to monitor and harmonise policies on 'external frontiers'
- A database on illegal or undesirable migrants (Schengen Information System).[5]

The literature is consensual in considering both the Schengen agreement and the Dublin Convention as enforcing the restrictive nature of the European approach to extra-regional migration (Huysmans, 2000; Kostakopoulou, 2000; Miles & Thränhardt, 1995; Ugur, 1995). A clear example is the 1990 Convention Applying the Schengen Agreement of 14 June 1985 which creates a direct link between immigration and asylum

[5] See Thielemann 2005/06 lecture at the LSE.

and terrorism, transnational crime and border control.[6] More generally, the effect of the Schengen agreement is to insert the regulation of migration within the context of protection of internal security from external threats.[7]

Indeed, the Schengen agreement, which was signed by Belgium, Germany, France, Luxembourg and the Netherlands the same day the Commission passed its White Paper on completing the internal market (14 June 1985), postulated the progressive elimination of border checks between signatory states. In the preamble to the agreement, a commitment by the Member States involved the abolishment of all internal border checks at the Community level.[8]

More precisely, Article 30 of the agreement stipulated that the countries signing the Schengen Agreement were set to abolish, 'if possible, by 1 January 1990',[9] all checks at their common borders on both people and wares. Furthermore, the Schengen agreement included measures aimed at facilitating checks from its entry into force. This was compounded by a number of measures aimed at making the new frontier-free area as safe as possible.[10] Moreover, the latter measures needed to be implemented before internal borders could be opened up.[11]

It is revealing that the measures included in the Schengen Treaty to control its external border were very similar to those postulated in the Programme on Migration set out by the Commission's White Paper. In both cases, accompanying measures included among others, visa policies, right of asylum provisions, immigration controls, coordination and cooperation between police forces and provisions to combat drug trafficking.[12]

[6] For the text of the Convention see http://eur-lex.europa.eu/LexUriServ/LexUriServ.do?uri=CELEX:42000A0922(02):EN:HTML as accessed on November 11, 2008.

[7] See Huysmans (2000: 757).

[8] See Preamble of the Schengen Agreement available at http://www.garanteprivacy.it/garante/document?ID=33372 as accessed on November 11, 2008.

[9] See article 30 of the Schengen Agreement available at http://www.garanteprivacy.it/garante/document?ID=33372 as accessed on November 11, 2008.

[10] See articles 6/9 of the Schengen Agreement available at http://www.garanteprivacy.it/garante/document?ID=33372 as accessed on November 11, 2008.

[11] See article 30 of the Schengen Agreement available at http://www.garanteprivacy.it/garante/document?ID=33372 as accessed on November 11, 2008.

[12] See www.europa.eu as accessed on October 15, 2008.

Given the high political sensitivity of the Schengen agreement, particularly after the collapse of the German Democratic Republic at the end of 1989, its entry into force, as mentioned above, was delayed. Only on 19 June 1990 were the five signatory Member States able to issue the 'Schengen Convention' applying the Schengen agreement. The Convention's preamble states that:

> the Treaty establishing the European Communities, supplemented by the Single European Act, provides that the internal market shall comprise an area without internal frontiers' and that 'the aim pursued by the Contracting Parties (Schengen) coincides with that objective, without prejudice to the measures to be taken to implement the provisions of the Treaty.[13]

The Convention thus reiterates signatory states' commitment to eliminate all internal border controls on people and goods. It also includes a set of accompanying measures referred to above, aimed at protecting the newly acquired area of internal free movement from external threats to security.[14]

With respect to the Dublin Convention, its aim was to reduce the number of asylum applications by making it impossible for refugees to apply for asylum in more than one Member State. This would limit the possibility for an asylum seeker to be accepted as a refugee and was designed to discourage asylum applications. The latter objective was achieved by introducing the obligation to ask for asylum in the first EU country entered (Huysmans, 2000: 756).

The Dublin Convention was signed in Dublin on 15 June 1990. It was intended to cover a matter not addressed by the Geneva Convention on the status of refugees i.e. where to apply for asylum. According to the Dublin Convention, only the first Member State of entry of the asylum seeker had the jurisdiction to decide upon the application to refugee status, unless a 'safe' non EU country could be identified. The Dublin Convention specified the criteria to identify the state responsible

[13] See Preamble to the Schengen Convention available at http://eur-lex.europa.eu/LexUriServ/LexUriServ.do?uri=CELEX:42000A0922(02):EN:HTML as accessed on November 11, 2008.

[14] See various articles of the Schengen Convention available at http://eur-lex.europa.eu/LexUriServ/LexUriServ.do?uri=CELEX:42000A0922(02):EN:HTML as accessed on November 11, 2008.

for the filing of the asylum seeker's application. Similarly, Council Regulation (EC) No 343/2003 of 18 February 2003, replacing the Dublin Convention, contained detailed regulations on the mechanisms defining which Member State would take charge of lodging and deciding upon applications for refugee status by TCNs.[15] This was substituted by the so-called Dublin III regulation (Regulation (EU) No 604/2013 of the European Parliament and of the Council of 26 June 2013) containing similar provisions.[16]

The already security oriented and restrictive content of the Dublin Convention and the Dublin Regulation of 2003 and 2013 were further confirmed in 2003 by the introduction of the EURODAC Regulation. This entered into force on 15 January 2003 in the Member States of the European Union and in the third countries bound by the EURODAC Regulation (Norway and Iceland). The EURODAC system requires the gathering and comparison of the fingerprints of asylum seekers and illegal migrants. The system directly connects the entry of asylum seekers with illegal entry thus strengthening the restrictive nature of the EU approach to asylum seekers and refugees.[17]

Following the Single European Act and the Schengen Agreement, the European Union introduced with the Maastricht Treaty of 1992 the so-called third pillar on Justice and Home Affairs, where migration and security were again explicitly connected (Kostakopoulou, 2001: 130).

In the Maastricht Treaty, also known as the Treaty on the European Union (TEU), some elements of immigration and asylum policy were recognised as 'common interest' but not yet as 'common policies' (Geddes, 2003: 135). Art K1 of the TEU lists these issues as follows:

1. asylum policy;
2. rules governing the crossing by persons of the external borders of the Member States and the exercise of controls thereon;
3. immigration policy and policy regarding nationals of third countries:

[15] For the full text see http://eur-lex.europa.eu/LexUriServ/LexUriServ.do?uri=CELEX:32003R0343:EN:HTML as accessed on October 15, 2008.

[16] For the full text see https://eur-lex.europa.eu/legal-content/EN/TXT/PDF/?uri=CELEX:32013R0604&from=EN.

[17] See www.europa.eu as accessed on October 15, 2008.

(a) conditions of entry and movement by nationals of third countries on the territory of Member States;
(b) conditions of residence by nationals of third countries on the territory of Member States, including family reunion and access to employment;
(c) combating unauthorised immigration, residence and work by nationals of third countries on the territory of Member States;

4. combating drug addiction insofar as this is not covered by 7 to 9;
5. combating fraud on an international scale insofar as this is not covered by 7 to 9;
6. judicial cooperation in civil matters;
7. judicial cooperation in criminal matters;
8. customs cooperation;
9. police cooperation for the purposes of preventing and combatting terrorism, unlawful drug trafficking and other serious forms of international crime, including if necessary certain aspects of customs cooperation, in connection with the organisation of a Union-wide system for exchanging information within a European Police Office (Europol).[18]

The literature considers the Maastricht Treaty's approach to migration as consistently 'security oriented and restrictive' (Guiraudon, 2000: 257).

Articles K 1-9 of the TEU are devoted to the establishment of the third pillar on Justice and Home Affairs.[19] The third pillar is highly intergovernmental. It is set outside of the acquis communautaire and requires unanimous decision-making by the Council of Ministers, although there was an entire group (GD1) of the K4 committee dedicated to asylum, visa and migration.[20]

[18] See art K1 of the TEU (Official Journal C 191, 29 July 1992) available at http://eur-lex.europa.eu/en/treaties/dat/11992M/htm/11992M.html#0001000001 as accessed on November 11, 2008.

[19] See Title 6 TEU (Official Journal C 191, 29 July 1992) available at http://eur-lex.europa.eu/en/treaties/dat/11992M/htm/11992M.html#0001000001 as accessed on November 11, 2008.

[20] See article K4 TEU (Official Journal C 191, 29 July 1992) available at http://eur-lex.europa.eu/en/treaties/dat/11992M/htm/11992M.html#0001000001 as accessed on November 11, 2008.

As a consequence, very few common positions were deliberated within the framework of the third pillar on Justice and Home Affairs in its first five years of its establishment. These included one joint position on the common definition of a refugee and five legally binding joint actions regarding school travel for children of third country nationals, airport transit procedures, a common format for resident permits, burden-sharing for displaced persons and human trafficking (Guiraudon, 2000: 257). On the other hand, the securitisation of policy progressed considerably outside the remit of the third pillar. Securitisation measures included the exchange of information and data through the Schengen Information System (SIS), the identification of third countries needing visas, technological cooperation in areas such as the fingerprinting of asylum seekers, document fraud and the operationalisation of cooperation between border police, liaison officers, intelligence personnel and magistrates.[21]

It is widely held among scholars that the Maastricht approach to migration failed to bring clarity and was 'confused and confusing' (Geddes, 2000a: 108; Kostakopoulou, 2000: 499). On the other hand, it finally institutionalised the security-oriented nature of the EU approach to migration and asylum strengthening the idea of the creation of 'Fortress Europe'. This was the case due to the lack of a clear supranational structure devoted to guaranteeing scrutiny and accountability of the Member States' regulation on migration. In the absence of such a supranational level of scrutiny, Member States could implement control-oriented policies which failed to introduce measures protecting migrants' rights of entry, residence, employment and the like from discrimination.

On the one hand, the Treaty on the European Union introduced some elements of communitarisation of asylum and migration. On the other hand, however, this happened through the institutionalisation of two opposite models of immigration policy: one for EU citizens, a very liberal model; and one for nationals of third countries, which on the contrary, was extremely restrictive. The latter revealed what Geddes terms a 'strong security impulse' (Geddes, 2000a: 108).

In the aftermath of the Maastricht Treaty, the strategy of EU Member States towards immigration evolved and a new externalisation phase

[21] See Title 6 TEU (Official Journal C 191, 29 July 1992) available at http://eur-lex.europa.eu/en/treaties/dat/11992M/htm/11992M.html#0001000001 as accessed on November 11, 2008.

began. The aim was to keep both asylum seekers and undocumented migrants outside the EU's borders. The policy's focus remained on security and border control, with limited, if any, attention to the rights of migrants (Geddes, 2000a: 108). At the same time, any elements of communitarisation of the policy were devoted to highly secretive intergovernmental committees, with no supervision by EU institutions or other public bodies. Moreover, the TEU did not provide for a clear definition of the evolution of the EU migratory approach in future.

Summing up, the Maastricht Treaty achieved the result of shifting the debate about migration, and the related legislation at the European Community level, from socio-economic integration to an 'internal security project' by linking the elimination of internal borders to the control of the external ones (Huysmans, 2000: 766). This approach was confirmed with the communitarisation of the Schengen Agreement after the Treaty of Amsterdam (1997), and the Council and Commission's action plan on the implementation of the Treaty of Amsterdam in an 'area of freedom, security and justice', the so-called AFSJ (Geddes, 2000a: 108; Van Munster, 2009: 65–97).

The Treaty of Amsterdam achieved a partial communitarisation of the third pillar of the TEU on Justice and Home Affairs, but it did so by institutionalising the security paradigm (Kostakopoulou, 2000: 498; Van Munster, 2009: 65–97). EU Member States were allowed to extend their security-oriented approach to migration even beyond the EU borders with the support of EU legislation through EU externalisation policy (Kostakopoulou, 2000: 498). This happened at the expense of the protection of the rights of TCNs. Indeed, with the move of some aspects of migration and asylum policy from the third to the first pillar of the EU, the logic of exclusion informing the Justice and Home Affairs pillar of the Maastricht Treaty was not abandoned, but rather was extended to the first pillar. Furthermore, the treaty of Amsterdam embedded within the acquis communautaire the security-oriented approach to migration of the Member States and ensured that their restrictive policies continued unhindered. The reasons why this happened have been identified in the literature as follows.

First, only a limited number of measures were communitarised in the Treaty of Amsterdam under Title IV (Articles 61–69) of the EC Treaty. These included a few provisions in the areas of free movement of persons, checks at external borders and customs cooperation, asylum, immigration and protection for the rights of nationals of non-member countries and judicial cooperation in civil matters.[22]

The Treaty of Amsterdam also included a clause according to which, despite the transfer of the above mentioned subjects to the acquis communautaire, the Council of the European Union would still be the main decision-maker on migration policy for the five years after the entry into force of the Amsterdam Treaty. At the expiry of the five year period the Council would remain central for policy, although it would no longer be the only institutional player.[23]

Indeed, during the first five years, the Council could take decisions by unanimity, based on the proposal of the EU Commission or a Member State. The Parliament only had a consultation role.

At the end of the transition period, only the EU Commission could make proposals to the Council, while the EU Commission was bound to consider any request by a Member State for a proposal to be put before the Council. After consultations with the European Parliament, the Council could decide by unanimous vote to apply the co-decision procedure and qualified majority voting when adopting measures under Title IV and to modify the clauses relating to the Court of Justice of the European Communities, according to the Treaty of Amsterdam.[24]

The Council of Ministers thus retained the role as the main decision-maker and also consolidated its position as the main coordinator between Member States and the EU Commission and within the relevant departments of the different states (Article 66).[25]

It is true that for the first time, the Treaty of Amsterdam attributed some jurisdiction on the subjects included in Title IV to the Court of Justice. However, the Court of Justice could act only if called upon by a national court, the Council, the Commission or a Member State and could only rule on matters of interpretation of the new title or of acts

[22] See articles 61/62/63 of the EC Treaty as amended by the Amsterdam Treaty website http://eur-lex.europa.eu/en/treaties/dat/11997E/htm/11997E.html#0173010078 as accessed on October 27, 2008.

[23] See http://europa.eu/scadplus/leg/en/s22000.htm as accessed on October 27, 2008.

[24] See http://europa.eu/scadplus/leg/en/s22000.htm as accessed on October 27, 2008.

[25] See http://eur-lex.europa.eu/en/treaties/dat/11997E/htm/11997E.html#0173010078 as accessed on October 27, 2008.

adopted on its basis. Importantly, any measures or decision taken by a Member State pursuant to Article 62(1), pertaining to the maintenance of law and order and the safeguarding of internal security, did not fall under the jurisdiction of the Court of Justice.

In fact, Member States retained full responsibility to guarantee internal security, law and order and had the prerogative to take into account foreign policy considerations. This was reflected in the possibility to restrict freedom of movement or entry of TCNs in cases of emergency for a maximum of six months. To this aim, the Council would vote by qualified majority on a Commission proposal in the interests of the Member State concerned (Article 64).[26]

The UK and Ireland obtained an opt-out from the application of Title IV. Thus, they could still implement border controls on anyone seeking to enter their territory. They could also take decisions on whether or not to let people enter their territory, including on citizens of states which are contracting parties to the Agreement on the European Economic Area or to any agreement by which the United Kingdom and/or Ireland is bound. Equally, other Member States could control anyone coming from the United Kingdom or Ireland.

Denmark was also exempted from the application of almost all provisions under Title IV apart from those relating to visa requirements for non-member countries and measures on the same format for visas.[27]

It must also be underlined that there were still many areas which were not included in Title IV of the Amsterdam Treaty and which remained under the intergovernmental Title VI of the TEU ('Provisions on police and judicial cooperation in criminal matters', Articles 29–42). These areas related to preventing and combating migration flows and were:

- racism and xenophobia;
- terrorism;
- trafficking in persons and offences against children;
- drug trafficking;
- arms trafficking;

[26] See http://eur-lex.europa.eu/en/treaties/dat/11997E/htm/11997E.html#0173010078 as accessed on October 27, 2008.

[27] See http://europa.eu/scadplus/leg/en/s22000.htm as accessed on October 27, 2008.

- corruption and fraud.[28]

For the accompanying measures to obtain results in these areas, they were geared towards more securitisation of migration. This included increasing the level of co-operation between police forces, customs authorities and other competent authorities in Member States, both directly and through Europol; increasing cooperation between judicial and other competent authorities of Member States, both directly and through Europol; and approximating, where necessary, rules on criminal matters in Member States.[29]

The responsibility for increased cooperation in the areas listed by Title VI of the TEU rested almost exclusively with Member States. The involvement of EU institutions, mechanisms and procedures could be summoned on a voluntary basis, but Member States could also decide to coordinate their action with one-another without including the EU.

Moreover, the Schengen Agreement was incorporated in the acquis communautaire by the Treaty of Amsterdam. This included a protocol annexed to the Treaty establishing the European Community, which regulated asylum for Member States nationals of the European Union.[30]

Overall, the Treaty of Amsterdam reveals the limitations to the delegation of competences the Member States were prepared to relinquish on Asylum and Migration policy. This is clear when looking at the integration in the acquis communautaire of the Schengen Treaty without changing its essentially 'excluding' content. Furthermore, the fact that many of the mechanisms introduced by Title VI of the TEU were retained in Title IV of the TEC demonstrates a reluctance to regulate migration at the supranational level. These mechanisms ranged from the retention of unanimity voting, to the limits imposed to the EU Commission's right of initiative; from the merely consultative role of the European Parliament, to the restrictions to the ECJ's jurisdiction. Hence, the widespread belief amongst scholars that the new title made it possible for Member States to reinforce their regulatory capacity in a more controlling way and

[28] See art 29 of the TEU web-site http://eur-lex.europa.eu/en/treaties/dat/11997M/htm/11997M.html#0145010077 as accessed on October 27, 2008.

[29] See art 29 of the TEU web-site http://eurlex.europa.eu/en/treaties/dat/11997M/htm/11997M.html#0145010077 as accessed on October 27, 2008.

[30] For further information see http://europa.eu/scadplus/leg/en/s22000.htm as accessed on October 27, 2008.

to deepen the restrictive elements of 'Fortress Europe' (Kostakopoulou, 2000: 505; Ucarer, 2002: 29).

Thus, while EU citizens were increasingly able to enjoy a liberal model of migration, third country nationals were to endure an increasingly more restrictive and securitised regime. In fact, it was institutionalised in the Treaty of Amsterdam that the liberties of movement and establishment of EU citizens could only be guaranteed by adopting two separate migratory regimes, an intra-EU and extra-EU one. Thus, internal freedom was associated with appropriate anti-crime and anti-terrorist measures, external border controls and similar actions. In the language of the EU, internal mobility could be achieved by establishing what became to be known as an Area of Freedom, Security and Justice (AFSJ), a leitmotif in all EU communications on migration and asylum policy which made it possible to realise a 'normalisation of security' (van Munster, 2009: 91).

More generally, the Amsterdam Treaty continued alongside the path established by the Treaty of Maastricht by adopting a rather confused and confusing approach to immigration and asylum policy at the EU level (Geddes, 2000b: 129; Huysmans, 2000: 756; Ucarer, 2002: 29).

However, the lack of democratic and judicial control of TCN's rights, perpetuated by the provisions of the Treaty of Amsterdam limiting the role of the ECJ and of the EP on the matter, made it possible for EU Member States to be in charge of the regulation of migration at the EU level without being subject to the judicial scrutiny that they would have to abide to at a national level. In sum, the shift of the level of governance of some areas in migration and asylum policy from the national to the community level reduced the scope for the protection of the 'social and political spaces' of migrants from outside the EU (Geddes, 2000b: 129). This was not compounded by a real supra-nationalisation of the policy, as the Treaty of Amsterdam mostly institutionalised intergovernmental practices rendering states' interests in migration policy easier to achieve (Huysmans, 2000: 756). As we have noticed already, states' interests were, and still are, to protect the external border through security measures creating an Area of Freedom, Security and Justice for EU citizens without much attention to the rights of third country nationals (Geddes, 2000b: 130).

This conception of migration and asylum policy did not change with later iterations of the policy following the Amsterdam Treaty. Quite to the contrary, the latest developments confirmed and even accentuated the security-oriented outlook of 'Fortress Europe'.

Indeed, the Council and the Commission's Action Plan on how best to implement the provisions of the Amsterdam Treaty (1998) explicitly made the link between intra-EU freedom of movement and the securitisation of extra-EU migration by stating:

> Freedom loses much of its meaning if it cannot be enjoyed in a secure environment and with the full backing of a system of justice in which all Union citizens and residents can have confidence[31]

The European Council in Tampere (15–16 October 1999), further added to the securitisation of the border by specifying the aim of reconciling internal freedom of movement within the Union with high standards of legal protection and guarantees for all.[32]

Also the Seville European Council in June 2002, while reiterating the need to adopt a common policy in the areas of asylum and immigration, underlined its relation with the adoption of effective measures to combat illegal migration and control external borders.

Finally, at the Hague European Council of 4–5 November 2004, the Hague programme was approved. This built upon the results of the Tampere programme to achieve an Area of Freedom, Security and Justice (AFSJ) by tightening external border controls. The proposals included in the Hague Programme referred to:

- initiating a debate on the possible creation of a European corps of border guards;
- setting up the Schengen Information System II (or SIS II) and the Visa Information System (VIS);
- establishing the Internal Security Committee;
- introducing the European evidence warrant by 2005;
- setting up a European police record information system.[33]

[31] European Council and European Commission. (1998). Action Plan on how best to implement the provisions of the Treaty of Amsterdam establishing an area of freedom, security and justice. 12 July, 1998 web-site http://ue.eu.int/jai/article.asp?lang=en&id=39813844, pp.1–2 as quoted in Kostakoupolou (2000: 507).

[32] See http://www.europarl.europa.eu/facts/4_11_1_en.htm as accessed on October 27, 2008.

[33] For more information see http://ec.europa.eu/justice_home/news/information_dossiers/the_hague_priorities/index_en.htm as accessed on October 27, 2008.

The Hague programme was translated into action by the Hague Action Plan, approved by the European Council meeting on 16–17 June 2005 and updated at the end of 2006.[34]

Later on, the Stockholm Programme, aimed at covering the period of 2010–2014, emphasised again the aim to establish an area of Freedom, Security and Justice guaranteeing EU citizens with a protected environment, secure from external threats.[35]

This was the background to the new provisions on migration and asylum policy introduced by the Lisbon Treaty. When at the end of 2009, the Lisbon Treaty was fully ratified and entered into force, the establishment of an Area of Freedom Security and Justice (AFSJ) represented the core element of the EU's approach to migration and asylum policy. The Lisbon Treaty confirmed this tendency, emphasising the focus of EU action as a way to limit migration 'unwanted by Member States policies' (Boswell & Geddes, 2011: 9). Therefore, although migration and asylum policy were finally totally integrated in the acquis communautaire, including with co-decision with the EU Parliament, qualified majority voting by the Council and full involvement of the ECJ, Articles 77–80 made clear that the stress of EU action on migration is on border control dynamics. More explicitly, Article 79(5) of the Lisbon Treaty states that EU rules on immigration 'do not affect the right of Member States to determine volumes of admission of TCNs coming from third countries to their territory, in order to seek work, whether employed or self-employed' (Boswell & Geddes, 2011: 11).

The 2014/2015 Refugee Crisis in the EU and the Militarisation of the Border

The EU's approach to migration faced new challenges with an extraordinary inflow of refugees from Syria in the period 2014–2015, in what came to be known in the literature and mass media (e.g. FT, 2015a) as Europe's 'migrant or refugee crisis'. As well as a social crisis, the

[34] See http://www.europarl.europa.eu/facts/4_11_1_en.htm as accessed on October 27, 2008.

[35] See the European Council Stockholm Programme, web-site http://eur-lex.europa.eu/LexUriServ/LexUriServ.do?uri=OJ:C:2010:115:0001:0038:en:PDF as accessed on October 4, 2012.

refugee crisis became an institutional one, with the widespread perception that the EU was unable to manage it either in a consistent or coherent way. Indeed, EU Member States failed to find a suitable agreement to reform the Dublin approach to refugee policy. They also failed to properly establish an EU wide resettlement scheme and reverted to the re-adoption of border controls within Europe, notably by suspending Schengen prerogatives.[36]

Much of the confusion about whether the crisis was a 'migrant' or 'refugee' one comes from the fact that asylum seekers, unable to enter the first safe country within the EU by regular means, had to revert to irregular entry into the EU, often using the same dangerous routes of irregular migrants. Indeed, whereas the United Nations Convention relating to the Status of Refugees (1951) and the Dublin Convention (later the Dublin Regulations) gave the right for refugees to apply for asylum, both failed to specify "how" precisely a safe country could be reached. As regular entry is usually allowed to those showing regular documents, it becomes very difficult for someone fleeing a situation of civil war or political prosecution in their own country, to obtain a regular visa, leaving no other choice than to revert to irregular entry as irregular migrants.[37] Hence the confusion in the public opinion, press and sometimes even the literature, between migrants and refugees (similarly Guiraudon, 2018).

Irregular routes to the EU, as detailed by FRONTEX, are those through Western African, the Western Mediterranean, the Central Mediterranean, the Eastern Mediterranean, the circular route from Albania to Greece, the Western Balkan, the Black Sea and the Eastern Land Borders (see Chapter 6). During the 2014 and 2015 'refugee crisis', the routes mostly used to enter the EU were the Mediterranean ones, especially the Central and the Eastern Mediterranean and the Western Balkan. The Western Mediterranean route, through Spain and its enclaves of Ceuta and Melilla, has acquired a more relevant role recently, especially for the smuggling of drugs through organised crime, but was less traversed during the refugee

[36] Asylum policy: the EU's 'crises' and the looming policy regime failure, Florian Trauner, Journal of European Integration, Volume 38, 2016–Issue 3: EU Policies in Times of Crisis, Pages 311–325 | Published online: 24 March 2016.

[37] The UN convention on refugee actually explicitly recognises the possibility that asylum seekers enter a safe country illegally, see https://www.unhcr.org/3b66c2aa10.

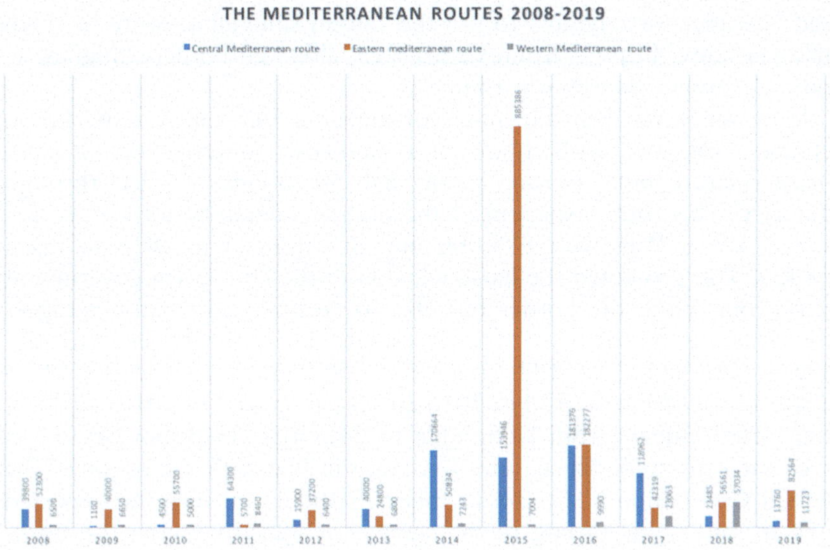

Fig. 5.1 The Mediterranean routes 2008–2019 (*Source* FRONTEX https://frontex.europa.eu/along-eu-borders/migratory-routes/centralmediterranean-route/, elaboration of the author)

crisis (Frontex, https://frontex.europa.eu/along-eu-borders/migratory-routes/centralmediterranean-route/) (Fig. 5.1).

The Central Mediterranean route leading to Malta and Italy had been well established for a long time. The refugee crisis' novel feature was the increase in the number of migrants choosing to take it. In 2014, the number of arrivals who had taken the Central Mediterranean route amounted to 170,000, a figure that represented a record high, but was soon surpassed in 2016 when around 180,000 were detected crossing it illegally. The numbers in 2015 were still high, although slightly less than in 2014 and 2016, with 153,946 migrants using it to reach the EU (Frontex, https://frontex.europa.eu/along-eu-borders/migratory-routes/centralmediterranean-route/).[38] The chaos created by the Libyan civil war provides an explanation for such an upsurge, as it

[38] See Frontex https://frontex.europa.eu/along-eu-borders/migratory-routes/central-mediterranean-route/.

made it easier for migrants to use the Libyan smuggling route to reach Italy (Frontex, https://frontex.europa.eu/along-eu-borders/migratory-routes/centralmediterranean-route/).

However, given the difficulties of pursuing this route, most Syrian refugees, together with Iraqis and Afghans, preferred the Eastern Mediterranean route, moving to the Balkans to enter the EU through Hungary. This route experienced the highest migratory wave since the Second World War, with 885,000 migrants using it in 2015 to reach the EU. This prompted the harsh reaction of the Hungarian government to not only close the border but also to erect a wall to stop refugees from entering its territory.[39] The unsustainability of the Eastern Mediterranean route led to an EU agreement with Turkey, costing the EU 6 billion Euros (3 billion paid in 2016 and 3 billion paid in 2018) to effectively stop refugees from being able to cross the Greek border and to force them to apply for asylum in Greece in line with the terms of the Dublin Convention.[40] Economic migrants should, in turn, be deported back to Turkey. According to the agreement signed between the European Council and Turkey on 18 of March 2016, for every Syrian returned to Turkey, another Syrian could be resettled in the EU. Apart from the debate over the legality of the agreement which is not recognised in the literature as an international treaty and whose negotiation did not follow the EU procedure (or maybe because of this), the agreement was not respected as only a handful of Syrian refugees were returned to Turkey from Greece (den Heijer & Spijkerborer, 2016; Guiraudon, 2018: 158). However, since the signing of this agreement, the number of irregular arrivals on this route substantially diminished. On the other hand, the impact of a similar arrangement on the Italian route was to bring new arrivals to the Italian coasts.[41]

Overall, Italy is mainly a transit country for migratory flows to the rest of the EU. Its transitory role is testified by the images of migrants waiting

[39] http://www.express.co.uk/news/world/648269/Hungary-plan-fence-border-Romania-migrants-refugees-crisis-Viktor-Orban-Schengen as accessed on May 13, 2016.

[40] See EU web-site http://europa.eu/rapid/press-release_MEMO-16-963_en.htm as accessed on May 13, 2016.

[41] See La Repubblica web-site http://video.repubblica.it/rubriche/rispostarep/rispostarep-quanti-migranti-sbarcheranno-questa-estate/239361/239274?ref=HRER2-2 as accessed on May 13, 2016.

at Milan station to embark trains to Germany,[42] or those who are blocked in Mentone, by the Franco-Italian border.[43]

It is a well-known fact that, before the refugee crisis, Italian and Greek authorities experienced difficulties in implementing the EURODAC regulation, according to which asylum seekers should have their fingerprints taken in the first country of entry into the EU (Trauner, 2016: 316).[44] The implementation of fingerprinting was necessary in order to identify where i.e. in which Member State, had an asylum seeker entered the EU, thus requiring them to ask for asylum there, as required by the Dublin regulation. As Greece and Italy's practice of not taking fingerprints produced an increased flow of asylum seekers to other EU countries, such as France and Hungary, in 2015 in the midst of the refugee crisis, some camps in Greece and Italy were designated as 'hotspots'. Here all the operations relating to the registration, fingerprint gathering and sorting of migrants, had to take place in front of personnel from other EU countries, staff from FRONTEX, Europol, Eurojust and European Asylum Support Office (EASO), apart from all other national and local authorities and NGOs. This measure was taken ex Article 78(3) TFEU, the same legal basis as the relocation of refugees, according to which the European Council could adopt additional measures in case "one or more Member States (are) being confronted with an emergency situation characterised by a sudden inflow of country nationals of third countries" (Guiraudon, 2018: 158).

Indeed, data on the number of people asking for the refugee status shows that neither Italy nor Greece had been the destination countries most wanted by refugees before the creation of the hotspots, with refugees clearly preferring other destinations, such as Germany, Sweden or France (EUROSTAT, 2018). For example in 2010, only 10,000 migrants had claimed asylum in Italy and only 10,275 in Greece, whereas the number of applications in Germany were 48,475, in France 52,725, in Sweden 31,850 and in the UK 24,335 (Fig. 5.2).

[42] See La Stampa web-site http://www.lastampa.it/2015/06/11/italia/cronache/migranti-le-stazioni-di-milano-e-roma-sembrano-campi-profughi-mmdKvo2kXzB5HdYTbZZcSM/pagina.html as accessed on May 13, 2016.

[43] See Il Secolo XIX web-site http://www.ilsecoloxix.it/p/imperia/2015/09/15/ARKqHuxF-migranti_arrestati_fermati.shtml as accessed on May 13, 2016.

[44] Trauner (2016: 316).

	2010	2011	2012	2013	2014	2015	2016	2017	2018	2019
European Union - 28 countries (20	259,630	309,040	335,290	431,095	626,960	1,322,845	1,260,910	712,235	647,165	417,975
Belgium	26,080	31,910	28,075	21,030	22,710	44,660	18,280	18,340	22,530	27,460
Bulgaria	1,025	890	1,385	7,145	11,080	20,390	19,420	3,695	2,535	:
Czechia	775	750	740	695	1,145	1,515	1,475	1,445	1,690	:
Denmark	5,065	3,945	6,045	7,170	14,680	20,935	6,180	3,220	3,570	:
Germany (until 1990 former territor	48,475	53,235	77,485	126,705	202,645	476,510	745,155	222,560	184,180	165,615
Estonia	35	65	75	95	155	230	175	190	95	:
Ireland	1,935	1,290	955	945	1,450	3,275	2,245	2,930	3,670	:
Greece	10,275	9,310	9,575	8,225	9,430	13,205	51,110	58,650	66,965	77,280
Spain	2,740	3,420	2,565	4,485	5,615	14,780	15,755	36,605	54,050	117,795
France	52,725	57,330	61,440	66,265	64,310	76,165	84,270	99,330	120,425	:
Croatia	:	:	:	1,080	450	210	2,225	975	800	:
Italy	10,000	40,315	17,335	26,620	64,625	83,540	122,960	128,850	59,950	:
Cyprus	2,875	1,770	1,635	1,255	1,745	2,265	2,940	4,600	7,765	:
Latvia	65	340	205	195	375	330	350	355	185	:
Lithuania	495	525	645	400	440	315	430	545	405	:
Luxembourg	780	2,150	2,050	1,070	1,150	2,505	2,160	2,430	2,335	:
Hungary	2,095	1,690	2,155	18,895	42,775	177,135	29,430	3,390	670	:
Malta	405	1,890	2,080	2,250	1,350	1,845	1,930	1,840	2,130	:
Netherlands	15,100	14,590	13,095	13,060	24,495	44,970	20,945	18,210	24,025	:
Austria	11,045	14,420	17,415	17,500	28,035	88,160	42,255	24,715	13,710	:
Poland	6,540	6,885	10,750	15,240	8,020	12,190	12,305	5,045	4,110	:
Portugal	155	275	295	500	440	895	1,460	1,750	1,285	:
Romania	885	1,720	2,510	1,495	1,545	1,260	1,880	4,815	2,135	:
Slovenia	240	355	295	270	385	275	1,310	1,475	2,875	3,820
Slovakia	540	490	730	440	330	330	145	160	175	:
Finland	3,085	2,915	3,095	3,210	3,620	32,345	5,605	4,990	4,500	:
Sweden	31,850	29,650	43,855	54,270	81,180	162,450	28,790	26,325	21,560	26,005
United Kingdom	24,335	26,915	28,800	30,585	32,785	40,160	39,735	34,780	38,840	:
Iceland	40	75	115	125	170	370	1,125	1,085	775	:
Liechtenstein	105	75	70	55	65	150	80	150	165	50
Norway	10,015	8,990	9,675	11,930	11,415	31,110	3,485	3,520	2,660	2,260
Switzerland	15,425	23,615	28,400	21,305	23,555	39,445	27,140	18,015	15,160	:
Total	285,215	341,795	373,550	464,510	662,165	1,393,920	1,292,740	735,005	669,025	420,290

Fig. 5.2 Asylum requests 2010–2019 (*Source* EUROSTAT, https://appsso.eurostat.ec.europa.eu/nui/show.do?dataset=migr_asyappctza&lang=en)

The declaration by Angela Merkel, on the 31st of August 2015, that Germany could accept up to 800,000 refugees, the famous "Wir Shaffen das!", demonstrated that refugees were correct in believing that Germany was a more convenient country of destination. Indeed, in 2016, around 750,000 migrants applied for asylum in Germany, whereas almost 500,000 had applied in 2015.

The increase in the arrivals of 2014–2015 produced a marked change in the composition of migratory flows to Italy and Greece, with a substantial increase of asylum seekers. For example, the composition of migratory flows through the Eastern Mediterranean route changed dramatically during the 2014–2015 crisis, with an extraordinary increase in the number of Syrian refugees (Fig. 5.3).

The change in the nationality of arrivals by sea was by far less dramatic for the Central Mediterranean route, but the increase in the number of Syrians in 2014 was still noticeable (Fig. 5.4).

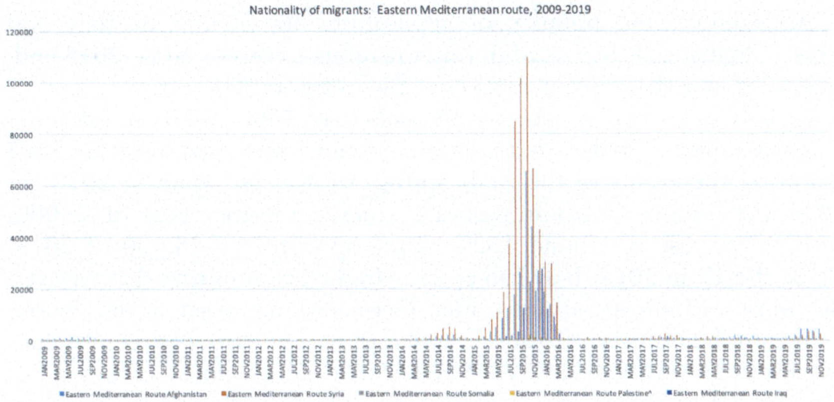

Fig. 5.3 Top Nationality of Migrants: Eastern Mediterranean Route, 2009–2019 (*Source* FRONTEX, https://frontex.europa.eu/along-eu-borders/migratory-map/, elaboration of the author)

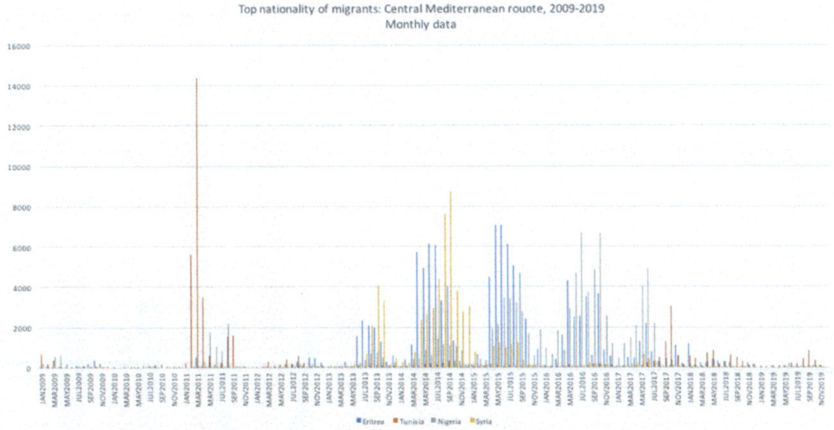

Fig. 5.4 Top nationality of migrants: Central Mediterranean route, 2009–2019 (*Source* FRONTEX, https://frontex.europa.eu/along-eu-borders/migratory-map/, elaboration of the author)

Accordingly, the number of applications for asylum in Italy and Greece dramatically increased during the refugee crisis of 2014–2015 and, especially after the implementation of the hotspots (Fig. 5.5).

In Italy in particular, whereas the period of 2008–2010 recorded the lowest number of asylum requests, this number increased in 2011–2012 following the events of the Arab Spring, to decrease again in 2013. In 2014, the amount of asylum requests achieved a historic peak of 65,000, with an increase of around 132% with respect to the 28,000 of 2013. Thus, already in 2014, Italy had entered in the five countries that received the most asylum applications after Germany and Sweden and before France and Hungary. However, looking at the ratio between asylum requests and inhabitants, it was Sweden that had the most asylum applications (8.4 applicants per thousand inhabitants) followed by Hungary (4.3) and Austria (3.3) (Eurostat, 2018).

In 2016 and 2017, after the creation of the hotspots, the number of asylum requests to Italy had doubled with respect to 2014, to achieve

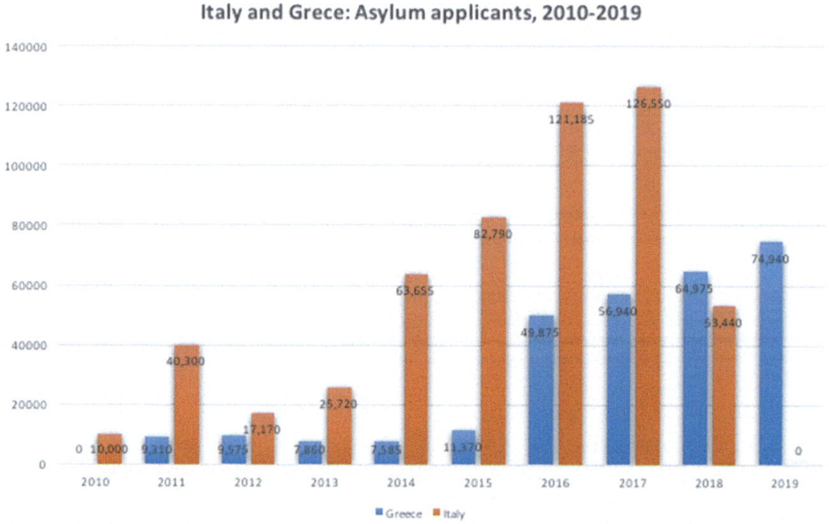

Fig. 5.5 Italy and Greece: Asylum applicants, 2010–2019 (*Source* EUROSTAT, https://appsso.eurostat.ec.europa.eu/nui/show.do?dataset=migr_asyappctza&lang=en)

more than 120,000 applications, second only to Germany (Eurostat, 2018).

At the peak of the crisis, the European Commission tried to implement a more balanced distribution of asylum seekers, relocating on a voluntary basis 160,000 refugees stuck in camps in Greece and Italy. In September 2015, the EU Council passed a plan providing for the relocation of 160,000 asylum seekers from Greece and Italy to other Member States in the following two years. This had to happen on the basis of the GDP, population and previous numbers of asylum applications. The asylum seekers exceeding this quota, would have to follow the regular Dublin procedure, including registration with EURODAC and asylum application in the first country of entry. Eventually, the plan failed, with only 5% of the refugees in the quota having actually been distributed to other countries from Italy and Greece by the end of 2016 (Guiraudon, 2018: 157).

On the other hand, the dangerousness of the Mediterranean route became evident when tragedy came to the island of Lampedusa in October 2013, when more than 360 migrants drowned (Trauner, 2016: 318).[45] In reaction to those dramatic events, the Italian Government established a search and rescue operation in the Mediterranean known as Mare Nostrum, saving many lives in 2014.[46] However, while 2014 and 2015 were record years for the number of asylum seekers reaching Italy and the EU, records were also broken for the number of those who died in the Mediterranean, estimated to be around 3,300 in 2014 and 2,872 until September 2015 (IOM, 2015).

On 1st November 2014, Mare Nostrum was substituted by the much less ambitious, and less costly, EU joint operation Triton, managed by FRONTEX.[47] Despite the fact that Triton did not have a mandate for the search and rescue of migrants, nor could it operate in international waters (with a reduced service based within 30 miles of the Italian coastline) its

[45] See La Repubblica web-site http://www.repubblica.it/cronaca/2013/10/03/news/lampedusa_brucia_un_barcone_strage_di_migranti-67817611/ as accessed on May 13, 2016.

[46] See Italian navy web-site http://www.marina.difesa.it/cosa-facciamo/operazioni-concluse/Pagine/mare-nostrum.aspx as accessed on May 13, 2016.

[47] FRONTEX is the EU border agency, see web-site http://frontex.europa.eu as accessed on May 13, 2016.

budget was tripled in 2015 (Trauner, 2016: 318).[48] In June 2015, as a part of the EU Common Security Policy, a new operation, called Sophia (or EUNAVFOR MED) was launched with the aim of intercepting and stopping human traffickers and migrant smugglers at sea, including in international waters.[49] This was a proper military operation, under the orders of an Italian Admiral in Rome. Finally, drones started to be used to monitor migratory routes within the so-called EUROSUR system of surveillance.[50]

Conclusion

The measures described above account for a de-facto militarisation of the border. However, they had little or no impact on reducing the number of victims at sea. In 2016, the IOM reported that 2,977 persons were found dead in the Mediterranean only in the first half of the year (Guiraudon, 2018: 157). Therefore, the question remains why did the EU decide to react to the refugee crisis of 2014/2015 by further increasing the securitisation approach, thus adding to the paradox of securitisation of globalisation?

As outlined above, scholarly interventions on refugee policy have traditionally adopted institutionalist approaches by, for example, focusing on the Europeanisation of national asylum systems (e.g. Guild, 2006; Lavenex, 2001; Toshkov & de Haan, 2013), on the decision-making procedures and the content of EU asylum law (e.g. Kaunert & Léonard, 2012; Ripoll Servent & Trauner, 2014), on the attempt by the EU to externalise refugee protection (e.g. Andrijasevic, 2010; Gammeltoft-Hansen, 2007; Klepp, 2010) and on how burden sharing has played a role in shaping EU asylum policy (e.g. Bendel, 2015; Thielemann & Armstrong, 2013; Uçarer, 2006). In particular, institutionalist accounts have sought to explain the EU approach to the 'refugee crisis' by identifying the actors that compose its 'hierarchical and horizontal' levels

[48] For the differences between Mare Nostrum and Triton, see Rai website http://www.rainews.it/dl/rainews/articoli/Scheda-Mare-Nostrum-e-Triton-le-differenze-4fedb886-58d9-48c7-88d8-e5bb2a58b8d3.html as accessed on May 13, 2016.

[49] See Italian Navy web-site http://www.marina.difesa.it/cosa-facciamo/operazioni-in-corso/Pagine/EUNAVFORMED.aspx as accessed on May 13, 2016.

[50] See Italian navy web-site http://www.marina.difesa.it/cosa-facciamo/operazioni-concluse/Pagine/mare-nostrum.aspx as accessed on May 13, 2016.

(Geddes & Lixi, 2018). Security-oriented actions were mainly implemented within the hierarchical level by actors such as the Interior Ministries in EU, Member States and the DG HOME in the European Commission. With the 2014/2015 'crisis', the EU had to come to terms with the increased politicisation of migratory issues at the Member State level, especially by populist parties (see Chapter 9). Given the difficulties reported above in identifying common solutions to the crisis, the EU sought to externalise border controls and refugee protection to third countries. This approach is notable in the various deals made with Turkey and even with Libya despite its on-going political turmoil (ECRE, 2017). When externalisation could not be proposed, such as, for example in the case of post-Arab Spring Tunisia, the EU reverted to the 'fight against migrant smuggling' in parallel with a renewed effort to achieve higher rates of return and readmission of irregular migrants (see COM, 2015a, 2015b, 2016, 2017a).

Both approaches reflect the aim of securitising the prevention of irregular migration, in line with the EU's traditional external migration governance regime. This hierarchical approach to the EU's external governance of migration/refugee policy aligns with a more global trend towards the migration of third country nationals (Lixi, 2017: 95). Guiraudon explicitly talks about the militarisation of the EU's approach to migration and refugee policy after the 2014 and 2015 'crisis' (Guiraudon, 2018: 156–157). Overall, the literature suggests that the EU's reaction to the 'crisis' is a continuance of the security-oriented framework of previous approaches to refugee and migration policy at the EU level. The EU basically passed only "emergency measures", such as a new military operation, as well as two Council decisions never being fully implemented to derogate from the Dublin regulation and relocate refugees on the basis of other criteria, and a statement between the EU and Turkey which was outside any international and legal framework. All this suggests that the EU failed to effectively address the problem. This applies in particular to the humanitarian aspect of the 'crisis' (Guiraudon, 2018: 158; Trauner, 2016: 313).[51]

However, Lixi notes that since the 'European migration crisis' of 2015, the horizontal level of governance of migration/asylum policy has grown. This includes networks of actors working mainly on foreign policy issues

[51] Trauner, F. (2016). Asylum policy: The EU's 'crises' and the looming policy regime failure. *Journal of European Integration, 38*(3), 311–325.

and international development, which have been the recipients of huge funds as compared to before the 'crisis' (Den Hartogh, 2016; Lixi, 2017: 93). According to this author,

> The 'crisis' triggered the development of an architecture of funding instruments through which the EU reached out to its external partners, with the emergence of a constellation of projects that resulted in a 'fragmented and incoherent' funding landscape. (Lixi, 2017: 93)

This kind of horizontal intervention, although certainly a step in the right direction, is unlikely to reverse the condition of marginalisation from globalisation leading to the increasing trend in migration from the global south to the Global North and to the EU in particular. In this respect, it is debatable whether the securitised, if not militarised, responses to the EU 'refugee crisis' are the most appropriate to increase security or if, by adding to irregular migration, they indeed achieve the opposite result of increasing insecurity. This will be the subject of the next chapter.

References

Andrijasevic, R. (2010). The international organization for migration and the international Government of borders, William Walters First Published December 1, 2010. Research Article, 28(6), 977–999. https://doi.org/10.1068/d1509.

Baldwin-Edwards, M. (1999). Where free market reigns: Aliens in the Twilight Zone. In M. Baldwin-Edwards & J. Arango (Eds.), *Immigrants and the informal economy in Southern Europe* (pp. 1–16, p. 9). Frank-Cass.

Bendel, P. (2015). But it does move, doesn't it? The debate on the allocation of refugees in Europe from a German point of view. *Border Crossing*, *5*(1–2), 25–32. https://doi.org/10.33182/bc.v5i1-2.512.

Boswell, C. (2003). *European migration policies in flux: Changing patterns of inclusion and exclusion*. Blackwell and RIIA.

COM. (2015a). *EU Action Plan against Migrant Smuggling (2015–2020)*. COM (2015) 285 final.

COM. (2015b). *A European Agenda for Migration*. COM (2015) 240 final.

COM. (2016). *Communication on Establishing a New Partnership Framework with Third Countries under the European Agenda on Migration*. COM (2016) 385 final.

COM. (2017a). Communication of the Commission on a More Effective Return Policy in the European Union—A Renewed Action Plan.

den Heijer, M., Rijpma, Jorrit J., & Spijkerboer, T. (2016). Coercion, prohibition, and great expectations: The continuing failure of the common European asylum system. *Common Market Law Review, 53*, 607–642. Available at SSRN: https://ssrn.com/abstract=2756709. Accessed 30 March 2016.

Den Hartogh, L. (2016). *EU budgetary responses to the "refugee crisis": Reconfiguring the funding landscape.* CEPS Paper in Liberty and Security in Europe.

ECRE. (2017). *EU externalisation efforts set the scene for escalation of rights violations by Libyan and Turkish Coast Guard.* ECRE News. https://www.ecre.org/eu-externalisation-efforts-set-the-scene-for-escalation-of-rights-violations-by-libyan-and-turkish-coast-guard/. Accessed 1 October 2018.

Eurostat. (2018). *Migr_eipre databases.* http://appsso.eurostat.ec.europa.eu/nui/show.do?dataset=migr_eipr. Accessed 17 May 2017.

Gammeltoft-Hansen, T. (2007). *The extraterritorialisation of asylum.* DIIS Working Paper, No. 2.

Geddes, A. (2000a). *Immigration and European integration: Towards the fortress Europe?* Manchester University Press.

Geddes, A. (2000b). Lobbying for migrant inclusion in the European Union. *Journal of European Public Policy, 7*(4), 632–649.

Geddes, A. (2003). *The politics of migration and immigration in Europe.* Sage.

Geddes, A. (Ed.). (2011). *International migration.* Sage.

Geddes, A., & Lixi, L. (2018). New actors and new understandings in European Union external migration governance? The case of EU–Tunisian migration relations. In C. Carrera, D. Kostakopoulou, & M. Panizzon (Eds.), *The external faces of EU migration, borders and asylum policies: Intersecting policy universes* (pp. 60–80). Brill Nijhoff.

Guild, E. (2006). The Europeanisation of Europe's asylum policy. *International Journal of Refugee Law, 18*(3–4), 630–651.

Guiraudon, V. (2000). European integration and migration policy: Vertical policy-making as venue shopping. *Journal Common Market Studies, 38*(2), 251–272.

Hollifield, J. F. (1992). *Immigrants, markets and states: The political economy of postwar Europe.* Harvard University Press.

Huysmans, J. (2000). The European Union and the securitization of migration. *Journal of Common Market Studies, 38*(5), 751–777.

IOM. (2015). *African Migration to the EU: Irregular migration in context.* Web-site: https://gmdac.iom.int/sites/gmdac/files/03_-_residence_permits-bbb.pdf.

Kaunert, C., & Léonard, S. (2012). The development of the EU asylum policy: venue-shopping in perspective. *Journal of European Public Policy, 19*(9).

Klepp, S. (2010). A contested asylum system: The European uniofn between refugee protection and border control in the mediterranean sea. *European Journal of Migration and Law, 12*, 1–21. https://doi.org/10.1163/157181610790931128.

Kostakopoulou, T. (2000). The 'Protective Union': Change and continuity in migration law and policy in post-Amsterdam Europe. *Journal of Common Market Studies, 3*, 497–518.

Kostakopoulou, T. (2001). *Citizenship, identity and immigration in the European Union*. Manchester University Press.

Lavenex, S. (2001). Migration and the EU's New Eastern Border. *Journal European Public Policy, 8*(1), 24–42.

Lixi, L. (2017, November). *Beyond transactional deals: Building lasting migration partnerships in the Mediterranean*. MPI Europe. https://www.migrationpolicy.org/research/beyond-transactional-deals-building-lasting-migration-partnerships-mediterranean. Accessed 1 December 2017.

Miles, R., & Thränhardt, D. (Eds.). (1995). *Migration and European integration*. Pinter.

Ripoll Servent, A., & Trauner, F. (2014). Do supranational EU institutions make a difference? EU asylum law before and after 'communitarization'. *Journal of European Public Policy, 21*(8), 1142–1162.

Stetter, S. (2000). Regulating migration: Authority delegation in JHA. *Journal European Public Policy, 7*(1), 80–103.

Thielemann, E., & Armstrong, C. (2013). EU internal security as a collective action problem understanding European asylum cooperation under the Schengen/Dublin system: A public goods framework. *European security, 22*(2), 148–164. Retreived 13 Feburary 2012, accessed 31 May 2012, Published online 11 October 2012.

Toshkov, D., & de Haan, L. (2013). The europeanization of asylum policy: An assessment of the EU impact on asylum applications and recognitions rates. *Journal of European Public Policy, 20*(5), 661–683.

Ucarer, E. M. (2002). Guarding the borders of the European Union: Paths, portals and prerogatives. In S. Lavenex & E. M. Ucarer (Eds.), *Migration and the externalities of European integration* (pp. 15–33). Leximgton Books.

Uçarer, E. M. (2006). Burden-shirking, burden-shifting, and burden-sharing in the emergent European asylum regime. *International Politics, 43*, 219–240.

Ugur, M. (1995). Freedom of movement vs. exclusion: A reinterpretation of the "insider"–"outsider" divide in the European Union. *International Migration Review, 29*(4), 964–999.

Van Munster, R. (2009). *The politics of risk in the EU*. London: Springer.

CHAPTER 6

Towards a Political Economy of Irregular Migration: Theory and Practice

This chapter deals with IPE approaches to irregular migration building on the theoretical paradigms presented in previous chapters. This discussion is tightly connected to the explanation of the "policy gap" or "policy failure" that different IPE approaches are able to provide (Boswell & Geddes, 2011; Guiraudon & Lahav, 2006; Rosina, 2020). Indeed, if for the realists, the state is firmly in control of migratory policy-making in the age of globalisation, the existence of a "policy gap" and therefore, the existence of irregular migrants on its territory, must somehow serve the interests of the host country. As Borjas explicitly claims, 'the existence and persistence of a black market for immigrants implies that all parties participating in these exchanges benefit from these voluntary transactions' (Borjas, 1990: 58). The "policy gap" is therefore, only a rhetorical one, a "discoursive gap" (Czaika & de Haas, 2013: 487).[1] The state only claims to be willing to control migration, while what it wants in reality is to extract the benefits of both regular and irregular migration. Such benefits span from the political to the economic, as further elaborated in this chapter.

[1] Czaika, M., & de Haas, H. (2013). The effectiveness of immigration policies. *Population and Development Review, 39*(3), 487–508.

L. S. Talani, *The International Political Economy of Migration in the Globalization Era*, International Political Economy Series,
https://doi.org/10.1007/978-3-030-79321-0_6

On the other hand, from the institutionalist point of view, the policy gap is a reality that cannot be avoided, as all national migratory policies restricting migratory flows are destined to fail. Indeed, migration is a transnational phenomenon and, as such, it cannot be controlled by national institutions. Moreover, migration often acquires regional patterns because of historical, geographic, social or cultural factors. This is why, from the institutionalist point of view, there is a tendency to focus on regional migratory regimes, as seen in Chapter 4. While it is true that regional migratory regimes, like US or EU immigration policies, are becoming stricter and less liberal (Geddes, 2000a, 2000b; Thielemann, 2003), it is, however, also true that migrants, both regular and irregular, do acquire rights in western democracies. These rights are protected by national and supranational bodies of law which are implemented by the judiciary. This is by itself a strong pull factor, making it impossible to completely stop migration and explains why there is a "real" policy gap, or "implementation gap" (Czaika & de Haas, 2013), at both the national and regional level. There is, however, scope for the supranational regulation of migration (Bacon, 2008: 246).

Finally, in the transnationalist analysis, policy failure is inevitable as international migration is a structural phenomenon and, as such, it cannot be regulated by national or even supranational institutions. By trying to stop migration, the only result is to transform regular immigration into irregular immigration. However, in the dichotomy between regular and irregular immigration, the latter is clearly favours the neoliberal order. This happens because it allows for the flexibilisation of the domestic labour market, while legal immigrants should be integrated in the existing welfare state system. This leads to the various political economy advantages of irregular migration, including further decreasing the costs of both migrant and local labour, decreasing the bargaining power of organised labour and providing a weak and "blackmailable" working force (Bacon, 2008). Moreover, as irregular migrants can only work in the irregular or underground economy, this entails a vicious cycle of marginalisation and stigmatisation of immigration which eventually leads to political backlashes against immigration (Chapter 9).

These three IPE approaches to migration will be analysed in the following sections to conclude with a more detailed identification of the modalities of irregular migration.

IPE Approaches and Irregular Migration: Does the Policy Gap Exist?[2]

If we define, following Czaika and de Haas (2011, 2013), immigration policy as, "(…) rules (i.e., laws, regulations, and measures) that national states define and implement with the (often only implicitly stated) objective of affecting the volume, origin, direction, and internal composition of immigration flows" (Czaika & de Haas, 2013: 489) then the sheer existence and persistence, as well as the generalised increase in the age of globalisation, of irregular migration is, by itself, a demonstration that immigration policy is somehow failing.

Indeed, the very existence of irregular migration is itself clear, unconfutable evidence that the effort made by any nation state, or any supranational institutions for that matter, to "regulate" migration has failed somehow. Unless the idea was to regulate migration with the aim of obtaining irregular migrants, which is a paradoxical use of regulation. Even more paradoxically, if states did not try to regulate migration, irregular migration would not exist.

Similar considerations somehow undermine the discussion in the literature on whether or not the "policy gap" exists and whether or not we can talk about "policy failure". We can legitimately talk about "policy failure" because there are irregular migrants.

We could, and will, however, take a more nuanced approach to the discussion about the failure of the states to regulate migration, by reverting to Czaika and de Haas' (2011, 2013) claim that by immigration policy we do not only mean the "rules" de jure adopted by the State (jus in civitate positum). They also include within the notion of "immigration policy" the "rhetoric" or "discourse" about those "rules" stating that:

[2] The Age of Migration: International Population Movements in the Modern World (Inglese) Copertina rigida—9 dic 2019, Hein De Haas (Autore), Stephen Castles (Autore), Mark J. Miller (Autore); Czaika, M., & de Haas, H. (2011). *The effectiveness of immigration policies: A conceptual review of empirical evidence* (International Migration Institute Working Paper 33). IMI; de Haas, H. (2006).*Turning the tide? Why 'development instead of migration' policies are bound to fail* (IMI Working Paper); de Haas, H. (2011). *The determinants of international migration: Conceptualising policy, origin and destination effects* (International Migration Institute Working Paper 32). IMI.; de Haas, H. (2015, January 7). Borders Beyond Control? Hein de Haas Blog. http://heindehaas.blogspot.co.uk/2015/01/borders-beyond-control.html (last accessed 25 August 2015).

> There is often a considerable discrepancy between publicly stated and "real" objectives of migration policy, resulting in a wide gap between policy rhetoric and actual policy objectives and policies on paper. (Czaika & de Haas, 2013: 487)

Moreover, we also need to distinguish between the rules on paper and their implementation, and finally between the implementation of the law and its results. This means moving the debate from whether the state fails to regulate migration, which is undeniable, to the degree of this failure, or, seen from the opposite perspective, the degree of "effectiveness" of migration policy.

This is an even more elusive discussion as the definition of "effectiveness" is debatable. In the literature policy, "effectiveness" is defined as the ability of a policy to obtain the "desired" effects (Czaika & de Haas, 2011, 2013: 491). This is different from policy effects. Hence, we can assume this means that policies might also obtain undesired effects, or results which are different from the desired ones.

So, overall, we can have the following instances of "policy failure" or "policy gap":

(1) Policy failure only at the rhetorical level (discoursive gap), when the real objectives pursued by the State with a specific piece of migration policy are not the ones declared in the public discourse.
(2) Policy failure at the implementation level (implementation gap), when the state fails to implement the policies adopted.
(3) Policy failure at the result level (efficacy gap), when the effects of the policies implemented are different from the desired ones (Czaika & De Haas, 2013: 491).

Whether or not migration policies have become more restrictive,[3] it is difficult to claim that they have, at least in the public discourse, become more favourable to irregular migration. In fact, there is widespread evidence of the criminalisation of irregular migrants, which means making irregular migration a criminal offence (Bacon, 2008: 78; Rosina, 2020). How would the different IPE approaches explain the existence, persistence and increase of irregular migration?

[3] E.g., Ortega and Peri 2009, 2013 Challenge this assumption.

As already discussed at length in previous chapters, from the realist perspective, the state is not only willing but also capable of controlling immigration. From this standpoint, nation states are able to achieve the volume and type of migration they desire by enacting proper legislation but also by failing to enact it, if the real objectives they want to achieve cannot be disclosed publicly.

The aims of the state can be various, according to the different realist scholars. For Weiner (1995) and his national security thesis, the state has the primary objective of achieving security within its territory and this implies that any nation state can host only a limited number of immigrants, but this number does not need to be represented only by regular migrants. If this number is exceeded, threatening national security, the state can use any means to stop migrants from reaching its territory, including intervening in conflicts and criminalising migration. Thus, irregular migration exists because irregular migrants are included in calculation of the sustainable number of migrants the state can host.

Borjas (1990) adopts a more economic realist approach. Based on the open labour market thesis, he advocates competition between states to attract the best migrants. However, also for him:

> the existence and persistence of a black market for immigrants implies that all parties participating in these exchanges benefit from these voluntary transactions. (Borjas, 1990: 58)

Finally, for Freeman and his socio-economic interests approach, there are many different interests the state needs to consider when adopting migration policies and this is why there can be a discrepancy between what the state claims it wants to achieve and what the policies' objectives effectively are (Freeman & Kessler, 2008: 671). This "discursive gap", however, does not conceal a real policy failure. As Czaika and de Haas (2013) explicitly claim:

> governments of countries with restrictive migration policies accept officially "unwanted" (legal and irregular) migrants, particularly if they are perceived to fulfil a useful economic role in agriculture, construction, catering, domestic work, or other low-skilled service. (Czaika & de Haas, 2013: 494)

Within this context, migration policy becomes the remit of interest coalitions, each of them with different disclosed and undisclosed aims. These interests include some objectives that can be better pursued by allowing irregular migration such as, for example, lowering the costs of both local and immigrant labour, reducing the power of organised labour and a more expendable and flexible working class (Bacon, 2008: 76).

Overall, although there are different interest coalitions that immigration may produce, still migration, both regular and irregular, is beneficial to the host country:

> migration is economically advantageous for leading economic sectors. Those for which it is not beneficial can be compensated for their losses, or if not, typically wield limited influence in national politics. (Freeman, 2008: 11)

Others have focused more on "discourse coalitions" to explain why migration policies are apparently incoherent and irrational (Cziaka & de Haas, 2013; Hajer, 1993; Jobert, 2003; Pian, 2010). A "discourse coalition" is created precisely with the aim of concealing the real interests of its stake holders. The example proposed by Czaika and de Haas (2013: 491) is very relevant for this discussion:

> The idea is that all stakeholders in such a coalition agree upon a common, publicly stated definition of a situation or a policy objective—for instance, to "combat" illegal migration from Africa to Europe. However, the formation of a discursive coalition around such belligerent rhetoric does not necessarily imply the genuine sharing of a system of values or beliefs, since each stakeholder is focused on the pursuit of its own interests. (Czaika & de Haas, 2013: 491)

Thus, from the realist perspective, even if the government in its public discourse proposes tough policies against irregular migration, in reality, the laws that are passed are more permissive because different politicians' constituencies have different interests on migration, including irregular migration.

This conceals a degree of hypocrisy in the rhetoric of the state against unwanted migrants. Therefore, policies against irregular migration are the outcome of a compromise between the need to gain and keep political support from different political constituencies of the government and the real interest the different socio-economic groups want to obtain.

In a sort of two level game Putnam's style,[4] the government and its supporting allies are trying to maximise the interests of two opposing coalitions. On the one hand, the political coalition's support from an electorate which is more and more anti-migrant, partly due to the magnitude of the irregular component of migratory flows (see next section). On the other hand, its socio-economic constituencies whose interests are disparate and include those who would gain from irregular migration. Therefore, because interests and objectives are multiple and often not explicitly stated, it is frequently impossible to identify a singular "real objective of a given policy" (Czaika & de Haas, 2013: 494).

Consequently, as the real and stated policy objectives do not coincide, and might even be opposite, it is not possible to speak about "policy failure". The government, in the realist perspective, is able to obtain its real aims. These, as already discussed, include allowing irregular migration. However, politicians do not necessarily want to disclose their real intentions in their public discourse as this could alienate important sectors of their electorate. Therefore, they form "discourse coalitions" against migration.

In a few words, for realists, irregular migration is an intended, if unspoken, outcome of migration policy, and therefore, the policy gap is only a "discourse one". As Sakia Sassen reminds us:

> those who deal with the real political economy of the city already know this: the rhetoric of the law-and-order Republican mayor here in New York City is surprisingly friendly to illegal immigrants. (Sassen, 1998: xiii)

Contrary to realists, neo-institutionalists do indeed believe that immigration policies can only have a limited impact on the long-term number and characteristics of immigrants. Therefore, for them the policy gap is real. This is for two reasons. First of all, according to the 'liberal state thesis' (Brettell & Hollifield, 2008), the ability of liberal states to control immigration is constrained by laws and institutions (ibid.: 195; Hollifield et al., 2014: 4–5). Moreover, migration is a transnational phenomenon, which is often organised at the regional level because of geographical, historical or cultural reasons. This has often led institutionalists to concentrate on the regional dimension of migratory governance schemes (Geddes, 2011). While on the one hand, regional migratory regimes like the US or the

[4] Putnam, R. (1991).

EU are becoming stricter and less liberal (Geddes, 2000a, 2000b; Thielemann, 2003), on the other hand, migrants, both regular and irregular, still enjoy rights in western democracies.

It follows that irregular migration is the consequence of the impossibility by the state to completely control the number of migrants entering and residing in the country. This is due to the fact that liberal democracies guarantee, even to them, some social, political and civil rights. This configures the so-called "liberal paradox" (Hollifield et al., 2014: 8), implying that liberal democracies cannot deny liberal rights to migrants, otherwise they would not be liberal democracies any longer. This is considered by institutionalists a sort of pull factor for migration, both regular and irregular. As Hollifield states:

> Civil rights– based policies help immigrants not only to get in but also to remain and settle. At the same time, human rights and refugee conventions have underscored the rights of asylum seekers, migrant workers, and their families. (Hollifield et al., 2014: 8)

If Hollifield's perspective relies on the role of the nation state to extend rights to migrants, other scholars underline that migrants acquire rights also at the supranational and even international level. For example, Geddes (2003, 2015: 435) argues that EU institutions have guaranteed a significant degree of institutionalisation of migrants' rights at the EU level. According to Soysal, migrants have acquired post-national citizenship, deriving from the recognition of international human rights standards (Soysal, 1998, 2015).[5]

As migrants' rights at all levels are mostly protected by the judiciary and the courts, the failure of immigration policy to limit their numbers happens mainly at the level of its implementation configuring what Czaika and de Haas (2013) call an "implementation gap".

Indeed, the implementation gap is considered in the literature to be more significant when there is scope for discretion and interpretation by the courts, such as in the case of refugee and migrants' rights (Czaika & de Haas, 2013).

More specifically, according to Massey (1999: 315), the state's ability to implement restrictive migratory policy is constrained by the degree to

[5] Soysal, Y. N. (2015). *Transnational trajectories in East Asia: Nation, citizenship, and region*. Routledge Taylor & Francis. 9781138819351.

which political rights of citizens and non-citizens are protected by the judiciary and by its relative independence.

Finally, from the transnationalist perspective, the failure of states' policies to influence the volume, timing, direction and composition of migration flows is not due to the role of the judiciary in limiting their implementation, but by the structural nature of migration in the age of globalisation. The uneven nature of globalisation, due to the regional patterns of the restructuring of the production process and to the related marginalisation of excluded regions and countries, leads to a new Global Division of Labour and Power (GDLP) (Mittelman, 2000, 2011: 188). Alongside the creation of global commodity chains,[6] a global market for labour is created, which produces brain-drain and mass migration from regions not included in the globalisation and regionalisation processes (Mittelman, 2011; Overbeek & Léon, 2015).

Thus, restrictive migration policies are, by definition, destined to fail. They can only modify the modalities of entry and stay of migrants, for example, changing the routes, increasing family reunification, and most importantly, transforming regular migration into irregular migration. Eventually, the desired effects of limiting the numbers of migrants and of controlling the border cannot be achieved, thus leading to an "efficacy gap".

In fact, policies of border control or limitations to the volume of migrants only produce unintended and/or unwanted consequences, what the literature calls "substitution effects" (de Haas, 2011). These make immigration restrictions ineffective, revealing the structural nature of migration. Indeed, one of the four "substitution effects" identified in the literature is "categorical substitution" when regular migration becomes irregular migration (de Haas, 2011).

Thus, from the transnationalist point of view, by attempting to control the volume and characteristics of migratory flows in the globalisation era, states necessarily incur an efficacy gap, a structural failure of the policy which obtains, amongst various other unwanted effects, the one of transforming regular migration into irregular migration. This transformation fosters a "real irregular migration industry", including organised crime recruiters, smugglers and other intermediaries: the so-called "Dark Side of Globalisation" (Talani & Roccu, 2019).

[6] As originally defined by Hopkins and Wallerstein (1986: 159), a commodity chain is "a network of labor and production processes whose end result is a finished commodity."

In the next section, we will elaborate more on the consequences of irregular migration for both migrants and host societies and the role that a thriving underground economy plays in favouring it.

The Underground Economy and Irregular Migration

This section aims to assess the consequences on receiving societies of integrating migrants generally, and irregular migrants in particular, in the "underground" economy of host countries.[7]

There is often confusion as to the definition of the terms "underground economy" and "shadow economy" and how the two of them differ. According to Schneider and Williams, the term "underground economy" usually refers to "currently unregistered economic activities that would contribute to the officially calculated gross national product if the activities were recorded" (2013: 23). For Smith, the "shadow economy" comprises "market-based production of goods and services, whether legal or illegal, that escapes detection in the official estimates of GDP" (1994: 18). Thus, in general use, the terms "underground" and "shadow" economy are interchangeable. In Table 6.1, the activities usually included in the broadest definition of the underground (or shadow) economy are listed.

There is, however, a more specific use of the two notions, as suggested by Schneider and Williams (2013), which distinguishes between the underground economy, defined in general terms as above, and the shadow economy, in which illegal or criminal activities are not included. In the opinion of Schneider and Williams (2013: 25) (similarly Bühn

[7] See Reyneri, E. (1999). *Immigration and the underground economy in new receiving South European countries: Manifold negative effects, manifold deep-rooted causes*. Final Report of the Project "Migrants' insertion in the informal economy, deviant behaviour and the impact on receiving societies"—CE/DGXII-Science, Research And Development, Tsr Program, Contract No. Soe2-Ct95-3005, Mimeo, p. 16. Reyneri, E. (2001). *Migrants' involvement in irregular employment in the Mediterranean countries of the European Union* (International Migration Papers (IMP) Working Paper Series).

Reyneri, E. (2009, September 17–19). *Immigration and the economic crisis in Western Europe*. Presented at the VI Conference on migrations in Spain, A Coruna.

Reyneri, E. (2003). Immigration and the underground economy in new receiving South European countries: Manifold negative effects, manifold deep-rooted causes. *International Review of Sociology*, 13(1), 117–143. https://doi.org/10.1080/0390670032000087023.

Table 6.1 Taxonomy of types of underground economic activities

Type of activity	*Monetary transactions*		*Non-monetary transactions*	
Illegal actives	Trade in stolen goods; drug dealing and manufacturing; prostitution; gambling; smuggling; fraud; human trafficking, drug trafficking and weapon trafficking		Barter of drugs, stolen goods, smuggling, etc., producing or growing drugs for own use; theft	
	Tax evasion	*Tax avoidance*	*Tax evasion*	*Tax avoidance*
Legal actives	Un reported income from self-employment; wages, salaries and assets from unreported work related to legal services and goods	Employes discounts; fringe benefits	Barter of legal services and goods	All do-it-yourself work and neighbour help

Source Schneider and Williams (2013: 24)

et al., 2010: 444) the notion of a "shadow economy" has a narrower meaning than a "underground economy":

> The shadow economy (...) includes all market-based production of legal goods and services that are deliberately concealed from public authorities. (Schneider & Williams, 2013: 25)

Moving to the subject of this section, it is worth noting that there is now a thriving literature on the relationship between the underground economy and migration.[8] One of the first and most comprehensive primary research projects dealing with the integration of irregular migrants in the

[8] Ambrosini, M. (2013a). *Irregular migration and invisible welfare*. Palgrave Macmillan. Crossref.; Ambrosini, M. (2013b). 'We are against a multi-ethnic society': Policies of exclusion at the urban level in Italy. *Ethnic and Racial Studies, 36*(1), 136–155. Ambrosini. M. (2015). Irregular but tolerated: Unauthorized immigration, elderly care recipients, and invisible welfare. *Migration Studies, 3*(2), 199–216; Ambrosini, M. (2016). From 'illegality' to tolerance and beyond: Irregular immigration as a selective and dynamic process. *International Migration, 54*(2): 144–159; Ambrosini, M. (2017). Why irregular migrants arrive and remain: The role of intermediaries. *Journal of Ethnic and Migration Studies, 43*(11), 1813–1830; Anderson, B. (2008). *'Illegal Immigrant': Victim or Villain?* (COMPAS, Working Paper No. 64). University of Oxford (WP-08–64).

underground economies of Southern European countries was promoted by the European Commission and evaluated by the author of this book. The title of the project was "Migrants' Insertion In The Informal Economy, Deviant Behaviour And The Impact On Receiving Societies"—(CE/DGXII-Science 1999) and it provides relevant theoretical as well as empirical insights on this topic (EC, 1999).

Triandafyllidou, A. (Ed.). 2010. *Irregular migration in Europe: Myths and realities*. Ashgate.

Triandafyllidou, A. (Ed.). (2017). Beyond irregular migration governance: Zooming in on migrants' agency. *European Journal of Migration and Law, 19*, 1–11. Triandafyllidou, A., & Maroukis, T. (2012). *Migrant smuggling: Irregular migration from Asia and Africa to Europe*. Palgrave Macmillan.

Triandafyllidou, A., & Ambrosini, M. (2011). Irregular migration control in Italy and Greece: Strong fencing and weak gate-keeping strategies serving the labour market. *European Journal of Migration and Law, 13*, 251–273. Triandafyllidou, A., & Dimitriadi, A. (2014). Deterrence and Protection in the EU's Migration Policy. *The International Spectator, 49*, 142–162.

Triandafyllidou, A., & Maroukis, T. (2012). *Migrant smuggling: Irregular migration from Asia and Africa to Europe*. Palgrave.

Triandafyllidou, A., & Maroukis, T. (2008). *The case of the Greek Islands: The challenge of migration at the EU's Southeastern Sea Borders* (Documentos CIDOB, Serie: Migraciones 17, pp. 63–82). CIDOB. Walters, W. (2010). Imagined migration world: The European Union's anti-illegal immigration discourse. In M. Geiger & A. Pécoud (Eds.), *The politics of international migration management: Migration, minorities and citizenship* (pp. 73–95). Palgrave Macmillan. See Ambrosini, M. (2001). Immigrati e lavoro indipendente. In G. Zincone (Ed.), *Secondo rapporto sull'integrazione degli immigrati in Italia*. Bologna, Il Mulino. Cavounidis, J. (1998). *The immigrant labour force in the informal economy in Greece*. Paper presented at the Third International Metropolis Conference, Israel, November 1998. Magatti, M., & Quassoli, F. (2000, June). The *Italian case: Socio-economic characteristics of immigrant businesses in Italy*. Second Conference of the Network 'Working on the Fringes: Immigrant Businesses, Economic Integration and Informal Practices', Jerusalem. Portes, A. (1994). The informal economy and its paradoxes. In N. J. Smelser & R. Swedberg (Eds.), *The handbook of economic sociology*. Princeton University Press. 08 reyneri (jk/d).fm Page 142 Monday, May 12, 2003 10:11 AM Immigration and the Underground Economy 143. Portes, A. (1995). Economic sociology and the sociology of immigration: A conceptual overview. In A. Portes (Ed.), *The economic sociology of immigration*. Russell Sage Foundation. Reyneri, E. (1999). Unemployment patterns in the European countries: A comparative view. DML-online, No. 1 (www.lex.unict.it/DML-online). Rowlands, D. (1998, June 29–July 3). Poverty and environmental degradation as root causes of international migration: A critical assessment. In UN-IOM, *Technical Symposium on International Migration and Development*. United Nations, World Population Monitoring. (1997). *Issue of International Migration and Development: Selected Aspects*. Williams, C. C. & Windebank, J. (1995). Black market work in the European Community: Peripheral work for peripheral localities? *International Journal of Urban and Regional Research, 19*(1), 23–39.

From the theoretical point of view, the project manages to overcome the disciplinary boundaries in the study of migration by adopting an interdisciplinary analytical perspective within an interdisciplinary theoretical context. In particular, the study emphasises the "mirror effect" of immigration on receiving societies. This means that immigration allows the inner problems of host societies to emerge (EC, 1999).

To verify how host societies reveal their shortcomings in dealing with immigrants, especially irregular ones, the project rejects the adoption of a rationalistic, economic approach to migration, based on a pure calculation of costs and benefits. The decision to migrate influences the life of the migrant in a holistic way as it is a "crucial" (non-reversible) event. As a consequence, it affects the entire personality and identity of the migrant, and the whole social and value context in which their personality and identity were defined. This does not mean that the approach used is a relativistic one i.e. one that does not permit a clear identification of the aims and the motivations of the emigration process. It is true that on many occasions, the motivations for migration are of a structural nature and remain unclear to the migrants at the point of departure. However, migrants often rationalise them after the migratory process has taken place and therefore are deeply influenced by the way in which the host society has treated them. In particular, migrants can form an opinion on their motivations to leave based on any factors influencing their experiences, such as for example, the chances of staying, the possibilities of earning or of improving their standard of life. This is why the behaviour of the receiving society, and its ability to integrate migrants or not, is a particularly important one, because it also influences, in a "mirror effect", the behaviour of the migrants regardless of what their original motivations were or even whether they were aware of them (EC, 1999).

Thus, the study by the European Commission importantly emphasises how the decision to migrate is "relived" by the migrants on the basis of their experiences in the receiving countries and this deeply influences their behaviour in all societal and economic realms. That is to say that those migrants who have a very negative experience in the host society are more likely in turn to behave badly and engage in illegal or illicit activities. In sum, migrants' behaviour is not independent from the conditions in which they find themselves in receiving societies and this makes the latter co-responsible for their deviant behaviour.

This approach informed the fieldwork of the EC project and is substantiated in a series of interviews and surveys where migrants are asked to describe their motivations for migration at the point of arrival. The research is conducted on people who have been living for some time in the receiving country, which allows them to reconstruct their migratory projects and their motivations for migration, after they have experienced life in their host societies. This allows the researchers to better grasp how the conditions at the point of arrival can influence the future direction of migrants' behaviour (EC, 1999).

Thanks to this interdisciplinary approach and to the methodology described above, the project overcame the cleavages between temporary and permanent emigration, short or long distance emigration and economic or non-economic motivations. In fact, with this approach, the main determinants of emigrants' orientation to action were on one side the modalities in which they re-elaborated their experience after arrival and, relatedly, the modalities in which the host country treated them after entry. Therefore, it is possible to hypothesise that migrants can have more than a single life project or a set of intentions or aims to achieve through the migratory project. It is also likely that these multiple sets of intentions and aims might exert contradictory influences on migrants' behaviour (EC, 1999).

Empirically, the research project focuses on the economic and social integration of migrants in Southern Europe, in particular Italy, Spain, Greece and Portugal. Importantly, the authors of the research underline the role played by the underground or informal economy in attracting and integrating migrants, as well as the reactions of the host societies to their inclusion in it.

Mediterranean countries are relevant case studies in order to understand the relationship between the underground economy and migration for a number of reasons. One reason is the size of the informal economy which in some cases, like in Italy, also includes substantial illegal activities (Talani, 2018; Talani & Roccu, 2019). Moreover, Mediterranean countries are relatively new to immigration problems as they have traditionally been countries of emigration. As such, unlike countries such as Germany and France, they did not experience the same level of immigration inflows in the '60s and '70s (Geddes, 2003). For this reason, the analysis of Mediterranean societies' reaction to immigration is particularly interesting as they act as a sort of "pure" case study. Finally, a so-called "Mediterranean" welfare state model prevails in Southern Mediterranean countries (Esping-Anderson, 1999).[9] This is defined as a familial type of welfare state, characterised by the presence of a family male as the main breadwinner. The consequence of this are extremely high levels of female and youth unemployment coupled with very low levels of social protection and relatively high levels of labour protection, at least until recently (Talani, 2017). In fact, the family is the main provider of social protection, with the wife being dependent on the income of the husband, as are children up to a late age. Thus, protecting the job of the male breadwinner becomes necessary to protect the entire society. However, with the globalisation era, and its ensuing flexibilisation and precaritisation of labour, the rigid labour market structure of Mediterranean countries facilitates the development of an underground economy (Talani, 2017). Conversely, in the Anglo-Saxon liberal model of the welfare state, the flexibility of the labour market is pursued by the state, thus limiting exclusion but also increasing inequality (Martin, 2015; Rhodes, 1997).[10]

The presence of a relevant underground sector with a substantial informal labour market is therefore embedded in the Mediterranean welfare state model (Martin, 2015).[11] Moreover, the waves of labour markets liberalisation that are associated with globalisation rather than reduce the size of the informal economy, in fact make it more salient (Martin, 2015; Talani, 2017).

The EC project's methodology relies on field work as well as on secondary literature and statistical sources, especially EUROSTAT.

Field research was of a qualitative nature including interviews with migrants. In particular, the stress was on the following areas of enquiry. First, the study addressed the question: to what extent are irregular migrants integrated into the labour market? It aimed, in particular, to identify instances of competition with the local workforce. Then, the research was interested in assessing whether migrants would engage in deviant behaviour, from micro-criminality to contributing to the illegal

[9] See Esping-Anderson, G. (1999). *Social foundations of post-industrial economies*. Oxford University Press.

[10] See Rhodes, M. (1997). *Globalisation, labour markets and welfare states: A future of 'competitive corporatism?'* (EUI Working Papers, No. 97/36).

[11] Following OECD manual, irregular or underground employment is meant to be "Employment concealed by the enterprises choosing not to respect employment regulations or immigration laws by hiring labour off the books". See OECD (2002, p. 38).

economy. Finally, the analysis moved to the reactions of receiving societies to irregular migration focusing on instances of migrant segregation, exclusion and social hostility, including the creation of urban movements aimed at defending local communities from migrants' criminal behaviour.

From field micro-analysis, it is then possible to gather insights on other more general issues, such as: the identification of the underground economy as a pull factor for irregular migration; the revival of the shadow economy in Southern European countries; the integration of migrants in the informal sector; how the insertion of irregular migrants in the underground economy contributes to both their social exclusion and their deviant behaviour, as well as how host societies react to this (EC, 1999).

These issues, discussed in the project reports, represent valuable tools for both policy-makers and scholars. The final report presents a summary of the findings by the various contributors to the project and provide a wealth of data and information which is useful to address the question of the impact of irregular migration on both migrants' lives and host societies. For this reason we will refer to them below.

Reyneri and Maria Baganha (1999) contribute to the final report with a study about "New migrants in South European Countries and their insertion in the underground economy". This is concerned with assessing the kind of migration in the countries considered, the characteristics of their labour markets and the integration of migrants in them. Their study includes issues of competition with the local labour force and the role played by the underground economy in attracting both regular and irregular migrants (Reyneri & Baganha, 1999).[12]

Salvatore Palidda, Maria Frangoulis and Antonios Papantoniou (1999) in their section of the report titled "Les conduits deviantes et la criminalisation des immigres" analyse the extent to which the social exclusion of migrants can result in the adoption of deviant and criminal behaviour and how this contributes to the criminalisation of migrants by host societies. The topic is addressed primarily at the theoretical level by referring to the legal implications of this phenomenon whose political consequences can hardly be underestimated.

The issue of the societal reaction of the different Southern European societies to immigration is studied by Carlota Sole' in the report "The Impact of immigration on South European new receiving societies". The

[12] See Reyneri and Baganha (1999).

report pays particular attention to the cases of Spain, Greece, Italy and Portugal and focuses on the material factors conditioning the life of immigrants in their receiving countries, such as their position in occupational structures or immigration legislation (Sole', 1999).[13]

Finally, the contribution by Czarina Wilpert and Smain Laacher (1999), "New forms of immigration and the informal labour market in old receiving countries: France and Germany" concentrates on the informal economy's role in countries of old migration, such as France and Germany. After sketching out the history of migration in both nations, the research addresses their responses to the new waves of migration. The comparison between the two countries is effected alongside the following dimensions: what kind of migratory flows, types and policies are there; what is the relationship between the formal and informal labour market and, how new migratory dynamics impact on the informal economy.[14]

Summing up, the research commissioned by the EC concluded that the underground economy has a major impact on the characteristics of migration to the South of Europe (similarly Ambrosini, 2017; Reyneri, 2007).[15] Most of the migrants involved in the field work had been employed in the underground at some point in their migration experience. Importantly, however, migrants are not responsible for the existence and size of the underground economy in these countries, which predates their arrival (Baldwin-Edwards & Arango, 1999; Reyneri, 2007).[16] In fact, the analysis proves how it is precisely the presence of a thriving underground economy, with its work opportunities for irregular migrants, which acts as a strong pull factor for migrants' decisions to enter Southern European countries without proper documents. This allows migrants to stay even if their chances of legalisation after entry are limited (Reyneri, 1999: 21; Reyneri, 2007).[17]

[13] See Solé, C. (1999). *The impact of immigration on South European new receiving societies*. Mimeo.

[14] See Wilpert, C., & Laacher, S. (1999). New forms of immigration and the informal labour market in old receiving countries: France and Germany, mimeo.

[15] See also Baldwin-Edwards, M., & Arango, J. (1999). *Immigrants and the informal economy in southern Europe*. Frank Cass.

[16] See also Reyneri, E. (1999b). The mass legalisation of migrants in Italy: Permanent or temporary emergence from the underground economy? In M. Baldwin-Edwards & J. Arango, *Immigrants and the informal economy in southern Europe* (p. 84). Frank Cass.

[17] Reyneri (1999, p. 21).

Indeed, after entering those countries illegally, very few migrants are able to gain a legal status. Looking at the regularisation schemes implemented by Southern European countries (Table 6.2), it is possible to estimate the percentage of migrants who stayed in the countries studied without proper documents. These were 75% in Italy, more than 50% in Spain, 25–30% in Portugal and 10% in Greece (Reyneri, 1999a: 2; 2003: 117).

Overall, the report proves the existence of a vicious cycle of unauthorised entry and irregular work in all countries considered (Reyneri, 1999: 20–22). The explanation for this phenomenon is that without proper documents to enter, migrants cannot obtain regular work. Even when they do have a work permit, this can be restricted in terms of geographical validity, jobs allowed or expiry time. In fact, work permits are usually not renewed automatically but through lengthy and time consuming processes. In some occasions, when a work permit is renewed, the migrant might have been undocumented for quite some time. In other instances, due to the long process of regularisation, the person might lose their job, thus becoming irregular again. Moreover, work permits last for a limited time period, after which the migrant's position is again made irregular as losing their work permit means losing their right to legally reside in the country. A migrant who either entered the country without a work permit or who lost it after entry for the reasons underlined above is extremely likely to enter the informal labour market (Reyneri, 1999b, 2003: 118).

On the other hand, a migrant can legally enter the country through a tourist or a student visa, and then start working in the informal economy as such kinds of visa do not usually permit lawful employment. Finally, family reunification visas often limit the ways in which the arriving family members may work in the host country. Indeed, the family member requesting reunification must show that they are in a financial position to maintain their family. Moreover, it should be considered that the definition of family can be different in different countries.

Overall, the research concludes that the underground economy often represents the only opportunity for migrants to find gainful employment. As such, it acts as a powerful pull factor. Migrants are more likely to move irregularly to countries, such as Southern European states, where

Table 6.2 Unauthorised immigrants filing for regularisation schemes (thousands)

Italy population	*57,000*	*Spain population*	*39,000*	*Portugal population*	*9,000*	*Greece population*	*10,000*	*France population*	*58,000*
1986–1987	118.7	1985–1986	43.8*	1992–1993	39.2	1998–1999	373.0	1981–1982	121.1
1990	217.7	1991	108.3	1996	21.8			1998–1999	80.6
1995–1996	238.2	1996	21.3	2001	147.5				
1998–1999	193.2	2000	200.0*						
2002	695.0*								

Source EMILIO REYNERI Illegal immigration and the underground economy, National Europe Centre Paper No. 68 Paper presented to conference entitled The Challenges of Immigration and Integration in the European Union and Australia, 18–20 February 2003, University of Sydney

the existence of a thriving underground economy allows them to find a job, despite lacking proper work and residence permits (EC, 1999).[18]

The next section will analyse how the connection between the underground economy, irregular migration and the criminalisation of migrants unfolds in the case of Italy.

Irregular Migration, the Underground Economy and the Migrants' Deviant Behaviour: The Case of Italy

In this section, the theoretical background set out above, regarding the relationship between the underground economy and irregular migration, will be discussed with a focus on the case of Italy.[19] In particular, the following dimensions of this relationship will be studied:

a. Whether there is competition between the local and foreign labour force and how migrants are included in both the formal and informal labour market.
b. Whether the underground economy acts as a pull factor for migration and what role migrants have in it.
c. Whether the inclusion of mostly irregular migrants in the underground economy results in both their social exclusion and their deviant behaviour.
d. How migrants are perceived by receiving societies and how the latter react to them.

The evidence for the analysis comes from statistical data and reports gathered by the author of this book.[20]

[18] Ibid.

[19] Similarly, Reyneri, "Immigration and the Underground Economy", "The mass legislation of migrants in Italy", "Illegal Immigration and the Underground Economy", and "Immigration in Italy".

[20] ISMU, "Ventunesimo Rapporto sulle migrazioni".

Competition Between the Local and Foreign Labour Force and How Migrants Are Included in Both the Formal and Informal Labour Market

Looking at the first question, and focusing in particular on the developments of the Italian labour markets after the global financial markets and the Eurozone crisis, it is relevant to underline how the number of migrants active in Italian labour markets increased dramatically during this period (Zanfrini, 2015a). To start with, in the manufacturing sector, the generational turn-over of the workforce has been guaranteed by regular and irregular migrant inflows. Thanks to them, and their lower costs due to their lower skills as explained below, companies which had lost competitiveness because of the Eurozone crisis, were able to survive. In fact, the Italian manufacturing sector experienced a higher increase in its value than the service sector which requires a higher level of skills (De Arcangelis et al., 2015). This occurred precisely because of the employment of more migrants who, on average, have a lower level of education and skills than the Italian workforce. From this perspective, it is possible to rule out any competition between the local and migrant labour contingents. This is in line with the conclusions in the literature as detailed in a next chapter. However, the effect on the Italian productive system was a negative one. Indeed, these dynamics exasperated a very detrimental characteristic of the Italian model, which is its over-reliance on low-skilled production (Talani, 2017).

Overall, immigrants who arrive in Italy present a lower educational level and skills than the Italian population. This is despite the fact that Italians are not particularly well placed on those two indicators in comparison with other EU and OECD countries. Looking at the distribution of immigrants in Italy per level of education, it transpires that in 2015, the share of immigrants with low education was more than 11% higher than that of Italians. In 2010, it had only been 7% higher (OECD, 2015, 2018).

Contextually, the share of highly educated migrants moving to the country has decreased. In fact, in 2010, the share of immigrants with high education was only 5% less than that of Italian, while in 2015, it was 10% less (OECD, 2015, 2018). Not enjoying brain gain limits Italy's opportunities to improve in terms of innovation, which inhibits the chances of the country to compete effectively in the global political economy over the medium and long term (Chapter 7; Zanfrini, 2015a: 103).

The predominance of unskilled labour in the Italian migrant labour force is also favoured by the presence of a thriving irregular migrant labour market, which in turn, is facilitated by the existence of a substantial underground economy (Talani, 2017). Unlike other Southern European countries, irregular migrants in Italy can find gainful work in the manufacturing sector (Reyneri, 2003). However, irregular migrant employment is often concentrated in the personal and domestic service sector (Zanfrini, 2015a: 104). Interestingly, the Euro-zone crisis hugely impacted this sector, by limiting the ability of Italian families to afford other solutions than to employ low-cost informal immigrant labour for their domestic and personal services (Zanfrini, 2015a: 104).

Many irregular migrants in Italy find employment in the agricultural sector. Here, the percentage of the labour force represented by migrants increased by three times during the crisis, reaching the level of 14% of the total work force. This trend did not seem likely to be reversed in the aftermath of the economic crisis. Unioncamere estimates that seasonal work in agriculture represents between 68 and 77% of total work required.[21] As most seasonal activities are carried out by migrants, this makes them essential for the survival of the sector. Moreover, in the South of Italy, seasonal agricultural immigrant workers, whose main task is picking fruit and other agricultural products, are overwhelmingly irregularly employed. They often endure unacceptable working conditions and very low working standards (Zanfrini, 2015a: 106). This makes their competition with the local working force limited (Zanfrini, 2015b).

Unfortunately, such irregular working activities often see the involvement of local organised crime, not only in the employment of irregular migrants, but also in their trafficking and smuggling into the country. Needless to say, this results in serious tensions with the local population and creates difficulties of coexistence within interethnic communities.[22]

Also ethnic entrepreneurs were essential to support the growth of Italy's entrepreneurial communities during the crisis years. The number of Italian enterprises with at least one owner born outside the EU reached 335,452 in 2014. Again, this is not a good development for the Italian productive system as it reinforces one of its most negative traits. It

[21] https://www.unioncamere.gov.it/.

[22] See the case of Rosarno, https://www.avvenire.it/attualita/pagine/rosarno-dopo-la-rivolta-nulla.

limits the size of companies to a small, even micro-level which allows for reduced costs but also less productivity, profits and lower skills.

The size of the companies owned by third country nationals is, indeed, only one person in 80% of cases. Their establishment is favoured by low access barriers in sectors with problems for the generational turn-over (Zanfrini, 2015b: 106). This accounts for the lack of competition with the local labour force. This is, however, a serious problem for Italian competitiveness due to the low cost/skill model of migrants' mostly individual companies and their limited entrepreneurial competence.

Overall, it is fair to conclude that third country nationals' integration in the Italian labour market did not pose a competitive threat to the national labour cohort. It did, however, reinforce some of the most detrimental features of the Italian productive model i.e. the low-skill content of production and the permeability of the border between the informal and the formal economy, especially in the south of Italy.

In fact, it is possible to identify a north/south divide in the Italian labour market. In the north, in Emilia Romagna, Tuscany and Lombardy, migrants face similar conditions of labour as the native population with two-thirds of migrants being regularly employed. Instead, in the south of Italy, incidences of irregular labour are much higher, particularly in agriculture, for both Italian and foreign workers.

Clearly, irregular work also exists in the north of the country. In fact, irregular migrants can only be employed in the irregular labour market. On the one hand, this degrades the life and work experience of migrants, but on the other, it is less likely that migrants will compete with locals for the same jobs. The result of these dynamics exacerbates the exclusion and marginalisation of migrant communities and exasperates the existing weaknesses of the Italian productive structure (De Arcangelis et al., 2015).

The Underground Economy as a Pull Factor for Migration and the Role of Migrants in It

As previously highlighted in this chapter, there is now substantial literature on the relevance of the underground economy in shaping the characteristics of migratory flows to the south of Europe.[23] As far as

[23] Ambrosini, M. (2013a). *Irregular migration and invisible welfare*. Palgrave Macmillan. Crossref.; Ambrosini, M., (2013b). 'We are against a Multi-ethnic society':

Italy is concerned, there is plenty of evidence of migrants' involvement in the informal sector (Ambrosini, 2013a, 2013b; Ambrosini, 2015). This does not mean, however that, they are responsible for the existence of the underground economy (Reyneri, 2003). Quite the contrary is true. It is because of the existence of a thriving underground economy that irregular migrants can find gainful employment in the country. This represents an important element of attraction to access Italy in particular and the South of Europe in general, given the difficulties migrants find in obtaining regular documents (Kosic & Triandafyllidou, 2004); Triandafyllidou, 2010, 2017).

There is indeed a vicious cycle between irregular entry and stay in a country and insertion into the informal or even illegal labour market. Such a cycle can only be broken by allowing migrants to enter and reside regularly in a country (Reyneri, 2003: 20–22).

With respect to Italy, it is common knowledge that the size of the underground economy is very substantial. Bühn et al. (2010) estimate

Policies of exclusion at the urban level in Italy. *Ethnic and Racial Studies, 36*(1), 136 155. Ambrosini. M. (2015). Irregular but tolerated: Unauthorized immigration, elderly care recipients, and invisible welfare. *Migration Studies, 3*(2), 199–216; Ambrosini, M. (2016). From 'illegality' to tolerance and beyond: Irregular immigration as a selective and dynamic process. *International Migration, 54*(2), 144–159; Ambrosini, M. (2017). Why irregular migrants arrive and remain: The role of intermediaries. *Journal of Ethnic and Migration Studies, 43*(11), 1813–1830; Anderson, B. (2008). *'Illegal immigrant': Victim or villain?* (COMPAS, Working Paper No. 64). University of Oxford (WP-08–64).

Triandafyllidou, A. (Ed.). (2010). *Irregular migration in Europe: Myths and realities*. Ashgate.

Triandafyllidou, A. (Ed.). (2017). Beyond irregular migration governance: Zooming in on migrants' agency. *European Journal of Migration and Law, 19*, 1–11. Triandafyllidou, A., & Maroukis. T. (2012). *Migrant smuggling. Irregular migration from Asia and Africa to Europe*. Palgrave Macmillan.

Triandafyllidou, A., & Ambrosini, M. (2011). Irregular migration control in Italy and Greece: Strong fencing and weak gate-keeping strategies serving the labour market. *European Journal of Migration and Law, 13*, 251–273. Triandafyllidou, A., & Dimitriadi, A. (2014). Deterrence and protection in the EU's migration policy. *The International Spectator, 49*, 142–162.

Triandafyllidou, A., & Maroukis, T. (2012). *Migrant smuggling: Irregular migration from Asia and Africa to Europe*. Palgrave.

Triandafyllidou, A., & Maroukis, T. (2008). *The case of the Greek Islands: The challenge of migration at the EU's Southeastern Sea Borders* (Documentos CIDOB, Serie: Migraciones 17, pp. 63–82). CIDOB. Walters, W. (2010). Imagined migration world: The European Union's anti-illegal immigration discourse. In M. Geiger & A. Pécoud (Eds.), *The politics of international migration management: Migration, minorities and citizenship* (pp. 73–95). Palgrave Macmillan.

that the shadow economy, not including illegal and criminal activities constituted on average around 27% of the national GDP between 1999 and 2007.[24]

According to Eurispes, in 2016, the shadow economy, estimated at 540bn Euros, accounted for one-third of the official GDP of around 1,500bn Euros. Moreover, another 200bn Euros were to be added to account for illegal activities, up from the 175bn Euros estimated in 2008. Thus, in total, the underground economy would represent 740bn Euros, namely half of the national income. To this figure, we should add 370bn Euros which represented the tax evaded on the underground economy based on a tax rate of around 50% (EURISPES, 2016).[25]

The Italian Official Statistical Office (ISTAT), given its governmental outlook, proposes more conservative figures relating to the size of the underground economy. Still, the value added of the underground economy had always been considered to be around 20% (ISTAT, 2015).

However, after adopting a new estimation system in 2010, called the SEC2010, allowing for a reduction of the weight of the underground economy, ISTAT estimated that the figure relating to the value added of the shadow economy in Italy dropped in 2013 to 11.9% of GDP, still higher than 11.7% in 2012, and 11.4% in 2011. Adding an estimate of the value added of illegal activities, around 1% of the GDP, it arrives at 12.9% of the Italian GDP in 2013 (ISTAT, 2015).

It must be noted that the drop in the official estimate of the non observed economy does not mean that there is less underground economy in Italy. As Schneider and Williams underline, the opposite is true (2013: 144). Given the growth of the value added by the shadow economy in recent years, some countries, as is the case of Italy, with the support of the OCED (OECD, 2011: 14), have included some segments of the shadow, or even the illegal economy into the official accounts. This automatically reduces the estimate of the underground economy that remains outside the official records.

Schneider and Williams report that the Italian official accounts have thus been boosted by between 14.8% and 16.7% of national income

[24] Source Bühn, A., Montenegro, C., & Schneider, F. (2010). New estimates for the shadow economies all over the world. *International Economic Journal*, *24*(4), 443–461.

[25] See also Il fatto quotidiano, https://www.ilfattoquotidiano.it/2016/01/28/evasione-eurispes-fenomeno-di-massa-sommerso-per-540-miliardi-un-terzo-del-pil/2412877/ as accessed on October 23, 2020.

(2013: 144). This was the case before they incorporated the value added of illegal activities, such as drug dealing, prostitution and smuggling, into the official numbers (similarly OECD, 2011).[26]

Against this backdrop, irregular labour is thriving. However, the quantification of the phenomenon is not an easy task as it can take various forms. Irregular work can include extra jobs outside or during the official working time; it can be a job done by people outside the official labour markets, or by people who cannot enter the official job market, such as irregular migrants. Work in the underground economy can be full time, part-time, self-employment with tax evasion or any job performed without a proper contract, from child minding to elderly care or bar tending. (Schneider & Williams, 2013: Chapter 2).

EURISPES notes that the highest percentage of irregular work in Italy is to be found among the following categories: childminders and babysitters (80%); tutors and educators (78.7%) and private household workers (72.5%). Other categories presenting a high degree of irregularity are carers, gardeners, builders, plumbers, electricians, wood workers and even medical doctors with around 50% irregularity (EURISPES, 2016).[27] The liberalisation of the job market effected through the so-called "Jobs Act" of 2014 seems to have worsened the phenomenon (Talani, 2017). After its adoption, EURISPES revealed that 28.1% of people interviewed in 2015 admitted to being willing to take up an irregular job, whereas in 2014, they had been only 18.6% (EURISPES, 2016).[28]

Looking at the composition of the Italian underground economy, according to ISTAT, in 2013, 47.9% of its value came from undeclared economic activity, while 34.7% was derived from the value added of irregular work. The other components were undeclared rents, tips and sales (9.4%) and straightforward illegal activities (8%) (ISTAT, 2015).

[26] See OECD. (2011). *Towards a better understanding of the informal economy by D. Andrews, A. Caldera Sánchez and A. Johansson* (OECD Economics Department Working Papers No. 873). OECD.

[27] See also Il fatto quotidiano, https://www.ilfattoquotidiano.it/2016/01/28/evasione-eurispes-fenomeno-di-massa-sommerso-per-540-miliardi-un-terzo-del-pil/2412877/ as accessed on October 23, 2020.

[28] See also Il fatto quotidiano, https://www.ilfattoquotidiano.it/2016/01/28/evasione-eurispes-fenomeno-di-massa-sommerso-per-540-miliardi-un-terzo-del-pil/2412877/ as accessed on October 23, 2020.

The sectors where the shadow economy is thriving the most are 'other activities and services' where it accounted for 32.9% of value added in 2013; trade, transport, hotels and catering (26.2%) and the construction sector (23.4%). In turn, the weight of non-declared value added was, especially relevant in professional services, with a share of 17.5% in 2013; in building and construction (14.2%) and in trade, transport, hotels and catering (13.9%). For the industrial sector, the percentage of undeclared value added was more evident in food processing and manufacturing of consumption goods (8.3%) and less important for investment goods (2.7%) (ISTAT, 2015).

The economic crisis of 2008/2009 and the following Eurozone crisis of 2011/12 seem to have exacerbated the phenomenon of irregular work in Italy. Measuring the irregularity rate of work through so-called "standard units" (ULA) of full time equivalent employment, these went from 11.8% in 2008 to around 15% of the labour force in 2013. In 2013, most irregular workers were dependent ones (2 million and 438 thousand ULAs) (ISTAT, 2015).

ISTAT (2015) indicates that the sector of personal and domestic services is one of the sectors with the highest degree of irregularity (45.0% in 2013), followed by the agricultural sector (17.6%), the sector of transport, hotels and catering (15.6%) and construction (15.4%) (ISTAT, 2015).

The value added deriving from irregular work is relevant in the sector of "Other personal services" (32.9% in 2013), trade, transport, hotels and catering (26.2% in 2013) and construction (23.4%) with similar rates as in 2011 (ISTAT, 2015).

By assessing the contribution to the total value added of the underground economy and comparing it to that of the regular economy by sector, it is possible to ascertain to what extent the underground economy supported the official one, especially during the Euro zone crisis (ISTAT, 2015).

In fact, in 2012, the growth of the value added in the underground economy significantly compensated the loss in the value added of the regular economy in some sectors. This was the case in food processing and manufacturing of consumption goods, where the underground economy contributed positively to the total value added by 1.2%, partially compensating for the loss of 5.9% of value added in the regular economy. Similarly, in the production of investment goods, the underground economy had a positive contribution of 1% while the regular economy decreased the total value added by 4.5%. Finally, in professional services, the underground economy increased the value added by 1.7%, while the regular economy decreased it by 3.1%. In other sectors, such as agriculture and fishery and in the production of intermediate goods, both the underground and the regular economy contributed positively to the total value added (ISTAT, 2015).

In 2013, it is possible to observe a similar dynamic, with the underground economy compensating the loss of the regular economy in trade, transport, hotels and catering. Moreover, it added 1.1% to the growth of the value added in agriculture (total +5.9%), while in construction it grew only modestly, by 0.1% and not enough to compensate for the loss of the official economy (total −4.8%) (ISTAT, 2015: 7).

Finally, and in line with the argument of this book, a good indicator of the relevance of the underground economy is given by the percentage of illegal immigrants in the labour market. Indeed, as already stated above, illegal migrants can only be employed as undeclared or irregular workers (Reyneri, 2003). On the basis of the OECD's published estimates of illegally employed migrants in Italy, the phenomenon is relevant as unauthorised migrants are estimated to account for 2% of total employment. This is second only to Greece and the US (OECD, 2011: 42).

Clearly, the regularisation of irregular migrants influences the dimensions and dynamics of irregular work. However, despite a number of regularisation programmes which resulted in 1,500,000 irregular migrants applying to become legal in the period between 1986 and 2002, already in 2004, it was estimated that two-thirds of documented migrants had spent a period as unauthorised residents (Blangiardo, 2005). Moreover, a few years after the regularisation waves, the number of irregular workers substantially increased again due to a constant increase of entries (Blangiardo, 2005).

Overall, it is easy to see how the availability of irregular or illegal work in the underground economy, where no documents are asked, favours the arrival of undocumented immigrants (Reyneri, 2003: 21). We can therefore talk of the existence of a vicious cycle between irregular migration and the underground economy: where the latter is more prominent, it promotes the expansion of the irregular entry of foreigners into the country.

This conclusion is substantiated by the fact, reported above, that the majority of migrants seem to be aware of similar opportunities in irregular labour markets. Moreover, as seen in the survey on the motivations of migration at the point of departure conducted by the author in Morocco, many do not consider it necessary to hold proper documents to work in EU countries (Chapter 3).

The next section will explore how host societies react to this attitude of migrants toward irregular or illegal activities with a special attention to the case of Italy.

The Attitude of Host Societies Towards Migration and Migrants' Deviant Behaviour

In this section, the attitude of Italians towards economic migrants, refugees and asylum seekers is assessed to ascertain whether such attitudes have been influenced by the dynamics of migrants' contributions to the underground economy. A survey by ISPI-RaiNews conducted by Ipsos in 2015 provides a wealth of information on Italian attitudes towards migration, with not all of them being reassuring (ISPI, 2015). From the survey, it emerges that 25% of the interviewees perceived migration as a threat even bigger than Islamic terrorism (Valtolina, 2015: 151). In turn, only 2% of the interviewees thought that migration contributed positively to Italy, while 67% considered it a threat. Finally, as reported in Table 6.3, just 28% were of the opinion that migratory flows were inevitable and therefore needed to be managed accordingly (ISPI, 2015).

Moreover, 39% of Italians interviewed thought the best way to address migration was to repatriate foreigners, perhaps using the army. Another 39% of the interviewees believed it would be best to try and prevent migrants from entering the country by negotiating with transit countries.

Table 6.3 Immigration as a threat for Italy (% values)

	Percentage
Yes, it is a threat and it is connected to terrorism	38
Yes, it is a threat but it is not connected to terrorism	29
No, it is not a threat it is a positive thing	2
No, it is not a threat and it is inevitable	28
Don't know	3

Source ISPI-RaiNews, "Gli italiani e le migrazioni"

On the other hand, only 16% of Italians interviewed considered it necessary to host refugees. This percentage changed after the Pope intervened asking Italian parishes to host a family of refugees (Tables 6.4 and 6.5).

In a survey taken by Demos and Pi in June 2015, it was reported that 42% of the Italians interviewed believed that migration was a security issue, the same percentage as 2007. Another 35% believed that immigrants threatened their employment (Valtolina, 2015: 156). Moreover, it emerged from a survey by SWG in April 2015 that 75% of Italians wanted migration to be reduced, with 57% wanting there to be a drastic reduction (Valtolina, 2015: 156) (Table 6.6).

Overall, it seems clear from the figures reported above that the attitude of Italians towards migration has been one of constant mistrust. On the one hand, Italians see third country nationals as a security threat while, on the other hand, they perceive migrants as competitors in the labour market. A part of the explanation for such a negative attitude can be traced back in the analysis referred to above concerning the involvement

Table 6.4 The best way to address immigration

	Percentages
Military intervention	17
Repatriation	22
Negotiations with transit countries	39
Host refugees	16
Do not know	6

Source ISPI-RaiNews, "Gli italiani e le migrazioni"

Table 6.5 Assessment of Pope Francis' request for each parish to host a family of refugees

	Percentages
Very positive, this could be an example for the government	25
Positive, the Catholic Church cannot be indifferent	45
Negative, it is not the role of the Catholic Church	9
Very negative, it is instrumental and does not solve the problem	15
Do not know	6

Source ISPI-RaiNews, "Gli italiani e le migrazioni"

Table 6.6 How should migrant flows be regulated in Italy?

	Percentages
Drastically reduced	57
Slightly reduced	18
Maintained as they are	10
Slightly increased	3
Drastically increased	2
Do not know	10

Source Swg (2015)

of irregular migrants in the underground economy. This dynamic exacerbates the marginalisation of migrants from the host society and the host society's related negative attitude towards migrants.

Based on Reyneri, it is possible to reconstruct the motivations behind a similar phenomenon (Reyneri, 2003). To start with, being inserted in the underground economy facilitates the perception of migrants as criminals and reduces the scope for considering them as positively contributing to the welfare of the host society. It does not help the fact that, as irregular workers, they do not fully pay taxes or social contributions but they do have access to public services, such as education, health care or even basic infrastructure such as streets and water. There is also a stigmatisation attached to migrant irregular workers, although this does not apply to Italian irregular workers. This is because in-group free riders are more accepted than out-group ones. However, there is also the tendency to consider immigrants involved in the underground economy more prone to committing crimes (Reyneri, 2003: 26).

Whether this is actually the case is a matter for discussion. Criminal statistics, as provided by the Italian Ministry for Justice, would appear to suggest that indeed immigrants are more likely to engage in deviant behaviours than the local population. However, it is necessary to introduce some considerations that make this statement more nuanced. First of all, in Italy, irregular migration is a crime in itself, according to law 94 of 2 July 2009 (Rosina, 2020). Second, it is imperative to look at how criminal statistics are constructed and at the proceedings of police activities and penal trials, to verify the extent to which migrants are the object of discrimination in the Italian criminal justice system (Palidda et al.,

1999).[29] Finally, it is indisputable that in Italy, especially after the introduction of the crime of illegal migration, a short circuit of securitisation prevailed which made the authorities privilege the adoption of repressive measures towards migrants instead of supportive ones.[30] Thus, although there is clearly a tendency in Italy for foreigners to be overrepresented in the jail population, giving rise to the phenomenon of 'ethnification' of jail populations, it is debatable whether this really reveals an 'ethnification' of crimes (Reyneri, 2003: 27).

Table 6.7 shows the extent to which the Italian jail population has been ethnified. Foreigners represent on average 33.6% of the total jail population in Italy, but in some regions, like Trentino Alto Adige, they are more than 70%.

This is not the place to attempt an evaluation of such a complex phenomenon and to try and identify why there is such a noticeable 'ethnification' of the jailed population in the country. It is, however, likely that, as undocumented migrants are by definition employed in the underground economy, this may contribute to the increase in deviant behaviour by foreigners. That is related also to the existence of a self-selection process amongst migrants linking irregular migration and employment in the underground economy with deviant behaviour. Indeed, it is possible to hypothesise that, if a migrant is willing to enter the country without proper documents and work in the shadow or illegal economy, they might also feel less bound to respect the rules. This might also mean that they could engage in illegal activities, considering that they can be more profitable. Even more likely is the possibility that migrants are obliged to engage in deviant or even criminal behaviour by those who helped them enter or smuggled them into the country. Criminal behaviour could also be necessary in order to repay the money that funded the journey. One final consideration is related to the fact that migrants might perceive Italy as a country where rules do not need to be respected because it offers easy unauthorised entry and easy irregular employment (Reyneri, 2003).

Concluding, it is possible to claim that the modalities of entry in Italy and access to the Italian informal or illegal economy produce a preselection of those foreigners that have less problems in undergoing an irregular migration process and in being involved in the prosperous Italian

[29] See Palidda et al., "Les Conduites Deviantes", 123–83.

[30] See Reyneri, "Immigration and Underground Economy", 26.

Table 6.7 Jailed population in Italy including foreigners, 30 April 2016

	N. Institutes	*Total Places*	*Total jailed*	*Women*	*Foreigners*	*% foreigners*
ABRUZZO	8	1,587	1,717	74	225	13.1
BASILICATA	3	470	485	7	112	23.1
CALABRIA	12	2,657	2,554	55	449	17.6
CAMPANIA	15	6,106	6,755	344	843	12.5
EMILIA ROMAGNA	11	2.8	3,094	131	1,510	48.8
FRIULI VENEZIA GIULIA	5	484	631	13	240	38.0
LAZIO	14	5.26	5,889	381	2,639	44.8
LIGURIA	6	1,109	1,374	77	726	52.8
LOMBARDY	18	6,125	8,077	388	3,757	46.5
MARCHE	7	853	867	19	283	32.6
MOLISE	3	263	310	0	56	18.1
PIEDMONT	13	3,842	3,665	129	1,607	43.8
APULIA	11	2,358	3,103	141	437	14.1
SARDENIA	10	2,632	2,051	55	425	20.7
SICILY	23	5.9	5,789	117	1,237	21.4
TUSCANY	18	3,406	3,320	121	1,558	46.9
TRENTINO ALTO ADIGE	2	506	450	15	331	73.6
UMBRIA	4	1,336	1,304	36	398	30.5
VALLE D'AOSTA	1	181	185	0	112	60.5
VENETO	9	1,704	2,105	110	1,129	53.6
Total Italy	193	49,579	53,725	2,213	18,074	33.6

Source Justice Ministry website: https://www.giustizia.it/giustizia/it/mg_1_14_1.wp?previsiousPage=mg_1_14&contentId=SST1232346

underground sector. On the other hand, it is precisely the presence of such a thriving non-emerged economy that attracts migrants in general and irregular ones in particular, representing an important pull factor for undocumented migration. On the other side, the easiness of access to a country's irregular labour markets can lead to migrants' marginalisation from the host society, incentivising their deviant tendencies (Andersson, 2014, 2016). The solution to this conundrum could be twofold: act to reduce the size of the Italian underground economy and increase the possibility for third country nationals to enter the country legally.

Conclusion

Concluding, the securitisation of migration policy overall increases insecurity. This is a consequence of the paradox of securitisation within globalisation. Indeed, as noticed above, migration is a structural component of globalisation and as such, it cannot be prevented through state policy. In a few words, the policy gap is real. It is an efficiency gap, as per Czaika and de Haas' classification (2013). Therefore, increasing securitisation increases irregular migration which, in turn, enhances insecurity through the involvement of organised crime in the process of migration, the marginalisation of migrants and the spiral of criminalisation. Paradoxically, allowing only irregular migrants to enter the country due to the lack of legal means of access and criminalising their entry and stay makes them more likely to engage in deviant activities. It also allows criminal organisations to be involved in their smuggling and exploitation. Therefore, increasing securitisation increases insecurity.

This is compounded by the paradox of irregular entry of protected refugees. Under the UN Convention of 1951 and the EU Dublin Convention (later regulation, 2003/2013), refugees have a right to ask for asylum in the first safe country of entry. However, they do not have a right to reach the first safe country regularly. They thus become irregular migrants up to the point to which they can ask for asylum. This dynamic, again, increases insecurity for both refugees and the local population. Indeed, migrants and refugees may die at sea, while the involvement of organised crime engaging in human trafficking, smuggling and exploitation of migrants makes host societies much less safe.

So, if the securitisation of migration in reality only increases insecurity for both migrants and locals, why is it so pervasive in the practice and discourse of managing migration around the globe? Who wins and who loses from such insecurity increasing securitisation policies?

In this chapter, we have shown how the losers of these practices are clearly both the migrants and the citizens of the receiving countries. In the next chapters we will see how, on the other hand, both employers and right-wing populist parties have something to gain from this situation.

References

Ambrosini, M. (2001). Immigrati e lavoro indipendente. In G. Zincone (Ed.), *Secondo rapporto sull'integrazione degli immigrati in Italia*. Il Mulino.

Ambrosini, M. (2013a). 'We are against a multi-ethnic society': Policies of exclusion at the Urban level in Italy. *Ethnic and Racial Studies, 36*(1), 136–155.

Ambrosini, M. (2013b). *Irregular migration and invisible welfare*. Palgrave Macmillan.

Ambrosini. M. (2015). Irregular but tolerated: Unauthorized immigration, elderly care recipients, and invisible welfare. *Migration Studies, 3*(2), 199–216.

Ambrosini, M. (2016). From 'illegality' to tolerance and beyond: Irregular immigration as a selective and dynamic process. *International Migration, 54*(2), 144–159.

Ambrosini, M. (2017). Why irregular migrants arrive and remain: The role of intermediaries. *Journal of Ethnic and Migration Studies, 43*(11), 1813–1830.

Anderson, B. (2008). *'Illegal immigrant': Victim or villain?* (COMPAS, Working Paper No. 64). University of Oxford (WP-08–64).

Andersson, R. (2014). *Illegality, Inc.: Clandestine migration and the business of bordering Europe*. University of California Press.

Andersson, R. (2016). Europe's failed 'fight' against irregular migration: Ethnographic notes on a counterproductive industry. *Journal of Ethnic and Migration Studies, 42*(7), 1055–1075. https://doi.org/10.1080/1369183X.2016.1139446

Bacon, D. (2008, September 1). *Illegal people: How globalization creates migration and criminalizes immigrants*. Boston: Beacon Press.

Baldwin-Edwards, M., & Arango, J. (1999). *Immigrants and the informal economy in southern Europe*. Frank Cass.

Blangiardo, G. C. (2005). I processi di immigrazione: dall'illegalità alla regolarizzazione. In M. Livi Bacci (a cura di), Giappichelli (Ed.), *L'incidenza economica dell'immigrazione* (pp. 41–56).

Borjas, G. J. (1990). *Friends or strangers*. New York: Basic Books.

Boswell, C., & Geddes, A. (2011). *Migration and mobility in the European Union*. London: Palgrave Macmillan.

Brettell, C., & Hollifield, J. F. (2008). *Migration theory: Talking across disciplines*. Front Cover. Routledge.

Bühn, A., Montenegro, C., & Schneider, F. (2010). New estimates for the shadow economies all over the world. *International Economic Journal, 24*(4), 443–461.

Cavounidis, J. (1998, November). *The immigrant labour force in the informal economy in Greece*. Paper presented at the Third International Metropolis Conference, Israel.

Czaika, M., & de Haas, H. (2011). *The effectiveness of immigration policies: A conceptual review of empirical evidence* (International Migration Institute Working Paper 33). IMI.

Czaika, M., & de Haas, H. (2013). The effectiveness of immigration policies. *Population and Development Review, 39*(3), 487–508.

De arcangelis, G., Di Porto, E., & Santoni, G. (2015, February). *Migration, Labor Tasks and Production structure, Centre for Studies in Economics and Finance* (Working Paper No. 390).

De Grauwe, P. (2012). *Interview*. http://aregan.wordpress.com/2012/03/20/interview-with-paul-de-grauw/. Accessed 18 October 2012.

de Haas, H. (2006). *Turning the tide? Why 'development instead of migration' policies are bound to fail* (IMI Working Paper).

de Haas, H. (2011). *The determinants of international migration: Conceptualising policy, origin and destination effects* (International Migration Institute Working Paper 32). IMI.

de Haas, H. (2015, January 7). *Borders Beyond Control?* Hein de Haas Blog, http://heindehaas.blogspot.co.uk/2015/01/borders-beyond-control.html. Accessed 25 August 2015.

De Haas, H., Castles, S. (Autore), & Miller, M. J. (2019). *The age of migration: International population movements in the modern world*.

EC. (1999). *Migrants' insertion in the informal economy, deviant behaviour and the impact on receiving societies* (CE/DGXII-Science 1999).

Esping-Anderson, G. (1999). *Social foundations of post-industrial economies*. Oxford University Press.

EURISPES. (2016). *Rapporto Italia 2016*. https://eurispes.eu/ricerca-rapporto/rapporto-italia-2016/. Accessed 23 October 2020.

EUROSTAT. (2015). *Migration and migrant Population statistics*. EUROSTAT web-site: http://ec.europa.eu/eurostat/statistics-explained/index.php/Migration_and_migrant_population_statistics. Accessed 25 July 2016.

EUROSTAT. (2017). *Migration and migrant Population statistics*. EUROSTAT web-site: http://ec.europa.eu/eurostat/web/population-demography-migration-projections/migration-and-citizenship-data. Accessed 7 March 2018.

FRONTEX. (2018). *The central Mediterranean rout*. http://frontex.europa.eu/trends-and-routes/central-mediterranean-route/

Freeman, G. (2008). *Migration and the political sustainability of welfare states*. http://cartss.colorado.edu/Freeman.doc. Accessed 23 June 2008.

Freeman, G. P., & Kessler, A. K. (2008). Political economy and migration policy. *Journal of Ethnic and Migration Studies, 34*(4), 655–678.

Geddes, A. (2000a). *Immigration and European integration: Towards the fortress Europe?* Manchester: Manchester University Press.

Geddes, A. (2000b). Lobbying for Migrant Inclusion in the European Union. *Journal of European Public Policy, 7*(4), 632–649.

Geddes, A. (2003). *The politics of migration and immigration in Europe*. London: Sage.

Geddes, A. (ed.). (2011). *International migration, sage library of international relations*. London: Sage.

Geddes, A. (2015). Governing migration from a distance: Interactions between climate, migration, and security in the South Mediterranean. *European Security, 24*(3), 473–490.

Guiraudon, V., & Lahav, G. (2006). *Immigration policy in Europe: The politics of control*. London: Routledge.

Hajer, M. (1993). *Discourse coalitions and the institutionalization of practice*.

Hollifield, J., Martin, P. L., & Orrenius, P. (eds.). (2014). *Controlling immigration: A global perspective* (3rd ed.).

Hopkins, T. K., & Wallerstein, I. (1986). Commodity chains in the world-economy prior to 1800. *Review (Fernand Braudel Center), 10*(1): 157–170. Accessed August 20, 2021. http://www.jstor.org/stable/40241052.

IOM. (2015). *World Migration Report*. IOM. http://www.iom.int/world-migration-report-2015. Accessed 25 July 2016.

ISMU. (2016). *Ventunesimo Rapporto sulle migrazioni 2015*. Franco Angelo web-site: http://www.ismu.org/volumi/. Accessed 7 April 2018.

ISPI, RaiNews, Ipsos. (2015). *Gli italiani e le migrazioni: percezione vs realtà*. http://www.ispionline.it/it/articoli/articolo/emergenzesviluppo-europa-italia-global-governance/gli-italiani-e-le-migrazioni-percezione-vs-realta-13562. Accessed 26 July 2016.

ISTAT. (2015). *L'Economia non osservata nei conti nazionali-Anni 2011–2013*. Rome ISTAT. https://www.istat.it/it/archivio/175791. Accessed 23 October 2020.

Jobert, B. (2003). Europe and the recomposition of national forums: The French case. *Journal of European Public Policy, 10*, 463–477.

Kosic, A., & Triandafyllidou, A. (2004). Albanian and Polish Migration to Italy: The Micro-Processes of Policy, Implementation and Immigrant Survival Strategies. *International Migration Review, 38*(4), 1413–1446, ISSN 0197-9183. https://doi.org/10.1111/j.1747-7379.2004.tb00242.x.

Magatti, M., & Quassoli, F. (2000, June). *The Italian case: socio-economic characteristics of immigrant businesses in Italy*. Second Conference of the Network 'Working on the Fringes: Immigrant Businesses, Economic Integration and Informal Practices', Jerusalem.

Massey, D. S. (1999). International migration at the dawn of the Twenty-First Century: The role of the state. *Population and Development Review, 25*(2): 303–322. http://www.jstor.org/stable/172427. Accessed August 20, 2021.

Martin, C. (2015). *Southern Welfare States: Configuration of the Welfare Balance between State and the Family. Campus. Southern Europe? Italy, Spain,*

Portugal, and Greece from the 1950s until the present day. https://halshs.archives-ouvertes.fr/halshs-01179541/document. Accessed 12 April 2016.

Mittelman, J. H. (2011, February 21). *Contesting global order development, global governance, and globalization* (304 Pages). Routledge. ISBN 9780415600965.

OECD. (2002). *Measuring non-observed economy*. https://www.oecd.org/sdd/na/1963116.pdf.

OECD. (2011). *Towards a better understanding of the informal economy, by D. Andrews, A. Caldera Sánchez and A. Johansson* (OECD Economics Department Working Papers No. 873). OECD.

OECD. (2015). *International Migration Outlook 2014*. OECD Publishing.

OECD. (2018). *Statistics*. http://stats.oecd.org/. Accessed 8 March 2018

Overbeek, H. W., & Léon, A. I. (2015). Neoliberal globalisation, transnational migration and global governance. In L. S. Talani & S. McMahon (Eds.), *Handbook of the international political economy of migration* (pp. 37–53). Edward Elgar. https://doi.org/10.4337/9781782549901.

Palidda, S., Frangoulis, M., & Et Papantoniou, A. (1999). *Les Conduites Deviantes Et La Criminalisation Des Immigrés*.

Pian, A. (2010). *Le cadre discursif du développement. Des discours et actions politiques concrètes, aux répertoires d'action des associations de refoulés* (Oxford Migration Institute, Working Paper 25).

Portes, A. (1994). The informal economy and its paradoxes. In N. J. Smelser & R. Swedberg (Eds.), *The handbook of economic sociology* (p. 08). Princeton.

Portes, A. (1995). Economic sociology and the sociology of immigration: A conceptual overview. In A. Portes (Ed.), *The economic sociology of immigration*. Russell Sage Foundation.

Reyneri, E. (1999a). *Immigration and the underground economy in new receiving South European Countries: Manifold negative effects, manifold deep-rooted causes*. Final Report Of The Project "Migrants' Insertion In The Informal Economy, Deviant Behaviour And The Impact On Receiving Societies"—CE/DGXII-Science, Research and Development, TSR Program, Contract No. Soe2-Ct95–3005, Mimeo.

Reyneri, E. (1999b). The mass legalisation of migrants in Italy: Permanent or temporary emergence from the underground economy? In M. Baldwin-Edwards & J. Arango, *Immigrants and the informal economy in southern Europe* (pp. 84–104). Frank Cass.

Reyneri, E. (2003, February 18–20), *Illegal immigration and the underground economy* (ANU National Europe Centre Paper No. 68). Paper presented to the conference The Challenges of Immigration and Integration in the European Union and Australia. University of Sydney. www.anu.edu.au/NEC/reyneri.pdf

Reyneri, E. (2007). Immigration in Italy: Trends and perspectives IOM. Argo (in print). http://www.portalecnel.it/portale/indlavdocumenti.nsf/0/466486C57FF3FF42C125737F0050A9EC/$FILE/Reyneri-percent20Immigrationpercent20inpercent20Italy.pdf. Accessed 20 May 2016.

Reyneri, E., & Baganha, M. (1999). *New migrants in South European Countries And Their Insertion In The Underground Economy*. Mimeo.

Rosina, M. (2020). *The Criminalization of migration in Italy and France* (PhD dissertation).

Rowlands, D. (1998, June 29–July 3). Poverty and environmental degradation as root causes of international migration: a critical assessment. In UN-IOM, *Technical symposium on international migration and development*.

Sassen, S. (1998). *Globalization and its discontents: Essays on the new mobility of people and money*. New York: New York Press.

Schneider. F., & Williams, C. (2013). *The shadow economy*. Institute of Economic Affairs. http://papers.ssrn.com/sol3/papers.cfm?abstract_id=2286334. Accessed 26 July 2016

Smith, M. P. (1994). Can you imagine? Transnational migration and the globalization of grassroots politics. *Social Text, 39*(15), 15–33.

Soysal, Y. (1998). Towards a post-national model of membership. In G. Shafir (Ed.), *The citizenship debates: A reader* (pp. 189–220). Minneapolis: University of Minnesota Press.

Soysal, Y. N. (2015). *Transnational trajectories in East Asia: Nation, citizenship, and region*. Routledge Taylor & Francis. 9781138819351.

SVIMEZ. (2015). *Rapporto 2015 sull'economia del Mezzogiorno*. Presented on 27 October 2015 at Camera dei Deputati. Rome. http://www.svimez.info/rapporto/2015. Accessed 7 April 2018.

Talani, L. S. (2017). *The political economy of Italy in the Euro*. Palgrave.

Talani, L. S., & Roccu, R. (2019). *The dark side of Globalisation*. Palgrave.

Thielemann, E. R. (ed.). (2003). European burden-sharing and forced migration. *Journal of Refugee Studies, 16*(3), 225–358.

Triandafyllidou, A. (Ed.). (2010). *Irregular migration in Europe: Myths and realities*. Ashgate.

Triandafyllidou, A. (Ed.). (2017). Beyond irregular migration governance: Zooming in on migrants' agency. *European Journal of Migration and Law, 19*, 1–11. Thielemann, E.R. (ed.) (2003) "European Burden-Sharing and Forced Migration". Journal of Refugee Studies 16:3, 225-358

Triandafyllidou, A., & Ambrosini, M. (2011). Irregular migration control in Italy and Greece: Strong fencing and weak gate-keeping strategies serving the labour market. *European Journal of Migration and Law, 13*, 251–273.

Triandafyllidou, A., & Dimitriadi, A. (2014). Deterrence and protection in the EU's migration policy. *The International Spectator, 49*, 142–162.

Triandafyllidou, A., & Maroukis, T. (2008). *The case of the Greek Islands: The challenge of migration at the EU's Southeastern Sea Borders* (Documentos CIDOB, Serie: Migraciones 17, pp. 63–82). CIDOB.

Triandafyllidou, A., & Maroukis, T. (2012). *Migrant smuggling: Irregular migration from Asia and Africa to Europe*. Palgrave Macmillan.

United Nations, World Population Monitoring. 1997.*Issue of International Migration and Development: Selected Aspects*.

Valtolina, G. G. (2015). Gli Italiani e l'Immigrazione: atteggiamenti ed orientamenti. In ISMU (2015), pp. 151.

Walters, W. (2010). Imagined migration world: The European Union's anti-illegal immigration discourse. In M. Geiger & A. Pécoud (Eds.), *The Politics of international migration management: Migration, minorities and citizenship* (pp. 73–95). Palgrave Macmillan.

Williams, C. C., & Windebank, J. (1995). Black market work in the European Community: Peripheral work for peripheral localities? *International Journal of Urban and Regional Research, 19*(1), 23–39.

Wilpert, C., & Laacher, S. (1999). New forms of immigration and the informal labour market in old receiving countries: France and Germany, mimeo.

Zanfrini, L. (2015a). Tra cibo e Terra. Il lavoro immigrate nella filiera alimentare, XXIV Rapporto immigrazione 2014. *Speciale Expo, Tan Editrice Todi (PG), 2015*, 329–358.

Zanfrini, L. (2015b). Il Lavoro. In ISMU, *Ventunesimo Rapporto sulle migrazioni 2015* (pp. 103–119). Franco Angelo.

CHAPTER 7

The Insertion of Migrants into the Labour Force of Receiving Countries: Competition or Complementarity?

This chapter addresses the insertion of migrants into the labour markets of receiving countries. The objective is to assess the political economy debate on whether migrant labour is competing with the local labour force for jobs or whether, in fact, migrant labour is complementing the local labour force. To this aim, the chapter will first analyse how neoliberal/institutionalist, realist and transnationalist approaches to international migration deal with the question of the impact of migration on employment and whether international migration should be liberalised. It will then refer to the existing data on the dynamics of the integration of migrants in the labour force of OECD countries. Conclusions will be drawn on the extent to which migrant labour can represent a valuable asset for receiving societies in terms of contributing to the welfare state and providing support for an ageing population.

The Political Economy Case for and Against International Migrant Workers

From the political economy viewpoint, the debate on job competition between migrant and local workers is long standing. It has generally been resolved differently according to the different theoretical perspectives.

L. S. Talani, *The International Political Economy of Migration in the Globalization Era*, International Political Economy Series,
https://doi.org/10.1007/978-3-030-79321-0_7

In the neoliberal economic tradition, there is a degree of consensus on the economic advantages of international migration.[1] For example, the work of Giovanni Peri analyses the long-run impact of immigration on employment, productivity and its skill bias. He concludes that a no crowding-out effect on the local population's employment rate can be identified.[2] This is a relevant contribution for analysing the insertion of regular migrants in the labour markets of receiving countries. In particular, Peri notices that:

> For economists, (...) international migration has the formidable ability of increasing total world income and productivity, generating huge global economic opportunities. (2012: 37)

Various studies by the World Bank (2005)[3] and others estimate massive increases in world GDP by increasing international migration, which has a much bigger effect than fully liberalising trade.[4] Some scholars from this tradition conclude that the total liberalisation of international migration

[1] Clemens, M. A. (2011). Economics and emigration: Trillion dollar bills on the sidewalk? *Journal of Economic Perspectives, 25*(3). A compelling argument that liberalizing migration rules would yield huge economic benefits. Clemens, M. A. (2015). Why it's time to drop the brain drain refrain. Center for Global Development blog. A counter to the common argument that emigration hurts developing countries Jaumotte, F., Koloskova, K., & Saxena, S. C. (2016). *Impact of migration on income levels in advanced economies*. Spillover Task Force, International Monetary Fund. Recent IMF research on whether immigration is good for growth and incomes. Sari Pekkala Kerr, J., & Kerr, W. R. (2011). *Economic impacts of immigration: A survey* (NBER Working Papers 16,736). A review of the research evidence on the economic impacts of immigration. Portes, J. (2018). *The economic impacts of immigration to the UK and New evidence on the economics of immigration to the UK*. VoxEU (April and October, respectively). News summaries of the research evidence on the economic impacts of migration to the UK on jobs, wages, productivity, and more.

[2] See Peri, G. (2013). The economic consequences of the proposed immigration reform. *Intereconomics, 48*(3), 191–192; Peri, G. (2012). The effect of immigration on productivity: Evidence from U.S. states. *Review of Economics and Statistics, 94*(1), 348–358; Peri, G. (2012, January). Immigration, labor markets, and prductivity. *Cato Journal 32*(1), 35–54.

[3] World Bank. (2005). *The potential gains from international migration*, World Bank web-site: https://documents1.worldbank.org/curated/en/507301468142196936/841401968_2005103190201050/additional/343200GEP02006.pdf.

[4] Pritchett (2006).

in OECD countries would produce an increase of world GDP by 150% over 50 years.[5]

The reasoning behind similar predictions are related to the increase in productivity guaranteed by international migration. The employment of migrants in countries where they can produce, on average, four to five times more value per hour of work than in their countries of origin makes human labour much more efficient (Clemens et al., 2009). The debate for liberals then moves to the issue of who will gain more from this large increase in productivity. Certainly, migrants will benefit most, but will they do so at the expense of local workers by reducing their wages and taking their jobs? Or is this increase in productivity a positive-sum game from which everyone benefits?

Peri demonstrates that more foreign workers do not necessarily displace native ones, i.e. that there is no competition between migrants and native workers for the same jobs (2012: 41). This is true in the medium to long term but also in the short term as the assumption that all workers are homogeneous and other variables are fixed is false.

Similarly, Portes notices that the distributional impacts of immigration, i.e. its alleged negative effect on the wages and employment of especially low skilled labour, is virtually non-existent (2019: 3–4).[6] Indeed, whereas it is true that immigration adds to the labour supply and that immigrants might even sometimes take the jobs of the locals, what is often overlooked is its effect in terms of enhanced demand for labour.

On the other hand, the effects on wages need to be assessed by looking at the data. A famous and often cited study on the consequences of an inflow of migrants on the wages of the local labour force is David Card's 1990 work of the Mariel boat lift. The arrival of a large number of Cuban refugees to Miami in 1980 represented a massive "supply shock" of low-skilled immigrants into Florida's labour market. This, however, did not have a great impact on the wages of native workers. The conclusions of this study were so controversial that Borjas disputed them in his 2016 monograph on the subject, as we will see below. However, Portes points to the existence of a wide consensus on the original research results by Card (Portes, 2019: 3–4). The case of recent immigration from central

[5] Klein and Ventura (2009).

[6] Portes, J. D. (2019). *The economics of migration*. Contexts. https://doi.org/10.1177/1536504219854712.

and eastern European countries to the UK confirms the relatively small impact of immigration on the labour market of natives. In fact, the inflow of foreign workers into the British labour market seems to have favoured an expansion of job opportunities for native workers (Portes, 2019). On the other hand, the withdrawal of the "bracero" programme in the US led to a contraction of employment in agriculture, rather than the creation of new positions for local workers.[7]

More specifically, categorising the way in which the economy and labour market respond to an inflow of migrant workers, Peri identifies four reactions (2012: 41).

Firstly, investments grow as firms have a larger labour supply and invest to keep the same ratio of capital per labour. This boosts the overall size of the economy, as well as its innovation content (Peri, 2012: 42). Portes (2019) notes that innovation benefits from the transfer of knowledge guaranteed by the free circulation of international workers through, for example, new patents, technology transfer, the development of multinational supply chains, and other similar effects.

Second, immigrant workers are not homogeneous, but they are notably divided between those with high levels of education (with tertiary education) and those with lower levels of education (without tertiary education) (Acemoglu, 2002; Goldin & Katz, 2008; Katz & Murphy, 1992). As they are inserted into different sections of the labour market, they affect different groups of native workers. Thus, if the labour supply of foreigners with lower levels of education is larger than that of natives, all less educated workers will see their wages decrease relative to highly educated workers. However, Peri shows that if the proportion is the same between foreigners and natives, then there is no effect on wages (2012: 43).

Moreover, even within low level or high level groupings of education there are various differences between natives and foreigners which makes it unlikely that there will be competition between them. Peri and Sparber (2009) demonstrate that immigrants, who have a limited knowledge of the local language, have a comparative advantage to specialise in manual jobs. This allows locals to upgrade their jobs, in all sections of the labour market, boosting not only overall productivity, but also the wages of native workers (Ottaviano & Peri, 2013; Peri, 2011). More generally, immigrants' skills and competencies are often complementary to those of

[7] 2018 American Economic Review paper, Clemens, Ethan G. Lewis, and Hannah M. Postel.

natives, thus improving the efficiency of the economy (Portes, 2019). The case of the US and Italy is relevant in this context. In both countries, the inflow of low-skilled female migrants working as carers has enhanced the ability of high-skilled native women to participate in the labour market (Portes, 2019; Chapter 6).[8]

Finally, empirical research demonstrates that immigrants are paid less than natives with the same skills and jobs. This helps reduce costs of production, increases productivity and creates more jobs (Ottaviano et al., 2010; Peri, 2012: 44). Portes (2018) demonstrates that in the case of the UK, the impact of immigration on productivity was positive. A study by the IMF in 2016 suggested that a 1% increase in the migrant share of the adult population resulted in around a 2% increase of GDP per capita and productivity gain (Portes, 2019).

In conclusion, the effects of more investment; complementarities between immigrants and local labour force; the upgrade of local workers and the enhancement in technology and innovation, guarantee more jobs for everyone, as well as a boost in GDP and an increase of productivity.

Moreover, Peri notes that immigration has helped relieve the shortage of labour in OECD countries resulting from an ageing population and the reduction in the number of younger workers (2012: 48). Thus, Peri advocates for a total liberalisation of immigration for highly educated workers. He also strongly advises for legal entry and residency for less educated workers. Indeed, Peri claims:

> Clearly the demand for immigrants in these kinds of jobs is one of the driving forces that have contributed to the problem of undocumented immigration. Without finding a legal way to satisfy that demand and the continued employment of those undocumented, there will not be a solution to the issue. (2012: 49)

Thus, for liberal economists, there is a very strong case for opening the borders to migrants. Similar views are shared by liberal institutionalists from a rights-based or libertarian perspective. Indeed, in recent years a number of academics from the liberal institutionalist perspective have started to address the question of the liberalisation of migration within the context of the establishment of a global migration governance regime

[8] 2011 articles by Patricia Cortés and José Tessada's in the American Economic Journal: Applied Economics and Guglielmo Baron and Sauro Mocetti in Labour Economics.

(Betts 2011a; Hansen et al., 2011; Koser, 2010; Koslowski, 2011; Kunz et al., 2011; Newland, 2010).[9]

Betts (2014)[10] specifies that the global governance of migration can take two forms. On the one hand, it can be represented by an institutional arrangement of some sort where states engage in collective action. However, on a more substantive level, global governance is given by the system of norms regulating migration at the international level. As far as migration is concerned, although no formal or coherent multilateral institutional framework regulating international migration has emerged, this does not mean that there is no global governance of the phenomenon (Betts, 2014: kindle ed. loc.:11,005). In fact, according to Betts (2014), there are various levels at which a global governance regime can be identified: multilateralism, embedded governance, regionalism, bilateralism and unilateralism with extraterritorial scope. At the multilateral level, a global governance regime for refugees, labour and international travel was established in the inter-war period (Koslowski, 2011). The rights of refugees are guaranteed by the 1951 Convention Relating to the Status of Refugees and by the role of the UNHCR. The rights of international travellers are guaranteed by the passport regime and the International Civil Aviation Organization (ICAO). Likewise, workers are protected by the International Labour Organization (ILO) treaties

[9] Betts, A. (Ed.). (2011a). *Global migration governance*. Oxford University Press; (2011b). The global governance of migration and the role of trans-regionalism. In R. Kunz, S. Lavenex, & M. Panizzon (Eds.), *Multi-layered migration governance*. Routledge. Cholewinski, R., MacDonald, E., & Perruchoud, R. (Ed.). (2007). *International migration law: Developing paradigms and key challenges*. Cambridge University Press. Hansen, R., Koehler, J., & Money, J. (Eds.). (2011). *Migration, the nation-state and international cooperation*. Routledge. Koser, K. (2010). International migration and global governance. *Global Governance, 16*(3), 301–316. Koslowski, R., (Ed.) (2011). *Global mobility regimes*. Palgrave Macmillan. Kunz, R., Lavenex, S., & Panizzon, M. (Eds.). (2011). *Multi-layered migration governance*. Routledge. Kuptsch, C., & Martin, P. (2011). Low-skilled labor migration. In A. Betts (Eds.), *Global migration governance* (pp. 34–59). Oxford University Press. Loescher, G. (2001). The UNHCR and world politics: State interests versus institutional autonomy. *International Migration Review, 35*(1), 33–56. Loescher, G., Betts, A., & Milner. J. (2008). *The United Nations high commissioner for refugees (UNHCR): The politics and practice of refugee protection into the twenty-first century*. Routledge. Newland, K. (2010). The governance of international migration: Mechanisms, processes and institutions. *Global Governance, 16*(3), 331–344. Salter, M. (2009). *The North Atlantic field of aviation security* (Working Paper).

[10] Betts, A. (2014). Commentary. In Hollifield et al. (Ed.), *Controlling immigration* (Kindle Edition). Stanford University Press.

(Kuptsch & Martin, 2011). Embedded governance refers to the rights attributed to international migrants by other areas of public international law. This is most notably done by international human rights law, international humanitarian law, World Trade Organization (WTO) law, maritime law and labour law as well as by UN and non-UN organisations not explicitly devoted to migration (Betts, 2014). The regional level of governance has been analysed in depth in Chapter 4 and is a staple of the liberal institutionalist approach to migration governance, given the thick level of institutionalisation of some formal regional economic communities (RECs) such as the EU and NAFTA (Hollifield, 2012). Bilateral agreements, such as those relating to visas, readmission agreements, border control and rescue operations, make up for the fourth and most substantial level of the global governance of migration. Finally, Betts also considers some aspects of unilateralism, if they entail an extraterritorial dimension, as part of global migration governance. All of the arrangements above have, according to liberal institutionalists, the ability to change the modalities in which states apply immigration control, most likely in favour of increased migration. Betts concludes that in the globalisation era, international institutions, when broadly defined, do significantly constrain states' ability to control immigration (2014). Migration, is in fact, propelled by rights and, as Hollifield puts it:

> As the world becomes more open, more democratic, and more liberal, people are freer to move than ever before. (Hollifield, 2012: 17)

On the other hand, realists, based on the primacy of the political consequences for states when allowing more migration, emphasise the need to limit their inflow. Without denying the economic benefits, they tend to underline the political costs of international migration. One such scholar is Rodrik (2012).[11] Starting from a typical, sceptical approach to globalisation, Rodrik notices how previous waves of globalisation have gone wrong as globalisation is inherently unstable. This occurs because global markets are not restrained by the same level of governance existing at the national level. This leads to the classical realist recipe of embedding the markets in a strong governance system that can only exist at the level of the nation state. In his words:

[11] Rodrik, D. (2012). *The globalisation paradox* . Oxford University Press.

> Markets work best not where states are weakest, but where they are strong. Second, capitalism does not come with a unique model. (...) Nations are likely to— and indeed are entitled to—make varying choices among these arrangements depending on their needs and values. (Rodrik, 2012 loc: 171)

Hence, the paradox of globalisation. It is impossible to simultaneously have economic globalisation, democracy and national determination. Moreover, as global democracy is unachievable, given the contrasting interests of the different national states, the only solution is to keep the nation state and national determination and give up deepening globalisation. This of course applies also to international migration, despite the fact that international migration could be beneficial to the state (Rodrik, 2012: 266). Therefore, from this viewpoint, the nation state needs to implement domestic policies aimed at reducing international migration, as well as other components of globalisation, because of the political pressures to do so. These political pressures, in turn, are the outcome of democratic politics and therefore need to be accommodated (Rodrik, 2019).[12] The backlash of hyper globalisation is indeed populism and this has taken the form, at least in Europe and the US, of a marked anti-migrant attitude:

> Then economic anxiety can be channeled into opposition to these groups. Immigrants and refugees can be presented as competing for jobs, making demands on public services, and reducing public resources available for natives. Indeed, a major source of support for far-right parties in Europe has been the fear that immigration will erode welfare state benefits, a fear that is heightened in countries experiencing austerity and recession. (Rodrik, 2018: 24)[13]

While Rodrik concedes that international migration is economically beneficial to the state but is politically unfeasible, Borjas (2016) takes this further by claiming that the economic case for allowing more migration should be questioned. Borjas concludes that:

[12] Rodrik, D. (2019). *What is driving populism*, https://www.socialeurope.eu/whats-driving-populism.

[13] Rodrik, D. (2018). Populism and the economics of globalization. *Journal of International Business Policy, 1*(1–2), 12–33. https://www.nber.org/papers/w23559.pdf

> It is not that immigrants do jobs that natives don't want to do. It is instead that immigrants do jobs that natives don't want to do at the going wage. (2016: 109)

As markets adapt to either an excess or a scarcity of migrants, firms are likely to pay more to native workers when there is not an excess of labour supply due to immigration. They do so to attract workers, as the law of demand and supply would suggest (Borjas, 2016). Indeed, in Borjas opinion, when researchers answer the question of the impact of immigration on the labour market of receiving countries, they are misled by a number of assumptions. Having worked on this puzzle for much of the '90s, with his Harvard colleagues Richard Freeman and Lawrence Katz, Borjas acknowledges that immigrants and natives do not have homogeneous skills, characteristics and levels of education. However, it is not easy, or even possible, for native workers to switch across skill groups to react to the arrival of immigrants in the labour market.

In his 2015 paper,[14] Borjas demonstrates that the laws of supply and demand apply also in the immigration context and that wages grow the least in the skill groups where more immigrants are concentrated. In fact, he shows how a 10% increase of migrants in a given skill group produces a reduction in the salary of that group by at least 3% (Borjas, 2015). While this contradicts the evidence from the Mariel case quoted above, Borjas claims that the Mariel study's conclusions were wrong. Borjas' calculations using the 1960–2000 decennial censuses and the pooled 2008–2012 American Community Surveys show that the wages of those within the skill groups most affected by the inflow of immigrants from Cuba dropped substantially after 1980, and did not recover until a decade later (Borjas, 2015).

Borjas then addresses the issue of complementarities between natives and immigrants, as stressed by liberal institutionalists. These, as noted above, underline how a supply shock in a specific skill group affects other skill groups, producing wage gains and increases of productivity due to the creation of complementarities. However, according to Borjas, similar outcomes are affected by a number of unsustainable assumptions, or at least, by an array of competing assumptions all leading to different

[14] Borjas, G. J. (2015, September). *The wage impact of the Marielitos: A reappraisa* (National Bureau of Economic Research Working Paper no. 21588).

answers (Borjas, 2016). In any case, even after accounting for all complementarities, Borjas claims that wages fall for the skill groups that received the most immigrants. Moreover, the drop-in income following an increase in migration is disproportionately borne by members of disadvantaged groups.[15] Hence, in his opinion, one cannot claim that immigration is not harmful to anyone, as the work by Ottaviano and Peri quoted above seems to imply. This outcome leads to social and political consequences that cannot be ignored by receiving countries. Therefore, receiving countries have to adopt policies to restrain immigration.

From a transnationalist point of view, the issue of whether immigrants are competition for domestic workers is addressed in the context of the labour structure of receiving countries. Some European countries are characterised by high levels of unemployment. However, such unemployment concerns mainly educated young people, living with their families and who can afford to wait for the right job proposal. They are therefore unlikely to compete with migrant workers for low skilled, often irregular jobs in the underground economy.[16] This phenomenon is actually considered in the literature as a powerful pull factor for irregular migration. This brings scholars to conclude that a demand for migrant workers does exist even in countries where the overall unemployment rate is high. However, the insertion of migrants into the informal sector strongly impacts their ability to contribute to states' welfare, producing instances of hostility towards irregular migrants and migrants in general and further contributing to their marginalisation (Ambrosini, 2018; Bacon 2008; Chapter 6).[17]

Scholars converge in noticing, as seen in previous chapters, that the securitisation of migration within globalisation has magnified the incidence of unauthorised migration (Ambrosini, 2018; Donato & Armenta, 2011).[18] The vast majority of unauthorised migrants are employed as

[15] Borjas, G. J. (2014). *Immigration economics* Harvard University Press.

[16] See Reyneri (1999: 7–14).

[17] See Reyneri (1999: 7–14).

[18] Donato, K. M., & Armenta, A. (2011). What we know about unauthorized migration. *Annual Review of Sociology, 37*, 529–543.

Ambrosini, M. *Irregular immigration in Southern Europe* (Migration, Diasporas and Citizenship) (Kindle Locations 108–110). Springer International Publishing. Kindle Edition.

vulnerable workers in the informal sector (Andersson, 2010[19]; Gomberg-Munoz & Nussbaum-Barberena, 2011: 373).[20]

In general terms it is possible to categorise entry or exit of migrants as follows:

1. Legal-legal: migrants who apply for immigrant status, gain legal entry and remain in the country as a legal migrant.
2. Illegal-legal: migrants who enter a country illegally, using false documents or by evading immigration restrictions, and who seek to change their status after their arrival (Triandafyllidou, 2010).[21]
3. Legal-illegal: this group is considered to encompass the largest percentage of the "illegal" migrant population. This happens, for example, when migrants exceed the terms of their visas (Baldwin-Edwards, 2008).[22]
4. Illegal-Illegal (independent): migrants who enter the country illegally and remain illegal but without the assistance of organised criminal groups.
5. Illegal-illegal (indentured): this is the most vulnerable category. They are undocumented migrants at the mercy of the criminals who procured their passage and employment. They have incurred large debts for their passage which require long periods of time to re-pay.
6. Legal-legal (indentured): those migrants who paid criminal organisations to gain legal access to and a legal stay in the host country. Legal status may improve job prospects, however, the amount of

[19] Anderson, B. (2008). *'Illegal immigrant': Victim or villain?* (COMPAS Working Paper No. 64 [WP-08-64]). University of Oxford (2010). Migration, immigration controls and the fashioning of precarious workers. *Work, Employment and Society, 24*(2), 300–317. Crossref.

[20] Gomberg-Munoz, R., & Nussbaum-Barberena, L. (2011). Is immigration policy labor policy? Immigration Enforcement, Undocumented Workers, and the State. *Human Organization, 70*(4): 366–375. Crossref.

[21] Triandafyllidou, A., (Ed.). (2010). *Irregular migration in Europe: Myths and realities*. Ashgate; ed. (2017). Beyond irregular migration governance: Zooming in on migrants' agency. *European Journal of Migration and Law, 19*, 1–11. Triandafyllidou, A., & Maroukis, T. (2012). *Migrant smuggling. Irregular migration from Asia and Africa to Europe*. Palgrave Macmillan.

[22] Baldwin-Edwards, M. (2008). Towards a theory of illegal migration: Historical and structural components. *Third World Quarterly, 29* (7): 1449–1459.

> money to pay back means that crime is often a necessary or attractive option.[23]

Hence, distinguishing between regular and irregular migration is not so clear-cut and most migrants are irregular at a certain point of their migratory process (Ambrosini, 2016, 2018).[24] Most importantly, however, unauthorised residence and irregular work very often go together.[25] As noticed by Ambrosini:

> The links among restrictive immigration policies, labour markets in search of flexibility, and migrant employment is stronger than commonly believed. (2018, Kindle Locations: 241–242)

According to the literature, migrants are generally unaware of the difference between authorised and unauthorised migration until they are very advanced in their migratory process (Triandafyllidou, 2017: 5). Only a few of the migrants interviewed by the author in her research on migration from Morocco to Europe admitted that they would enter Europe irregularly (Chapter 3).[26] This however would not stop them from entering a European country without a work permit, which reveals that they knew that work was easy to find in southern European countries, especially Spain and Italy (Chapter 3). As noticed in the scholarly debate, some migrants were motivated to enter Italy instead of, for example Greece, as they knew that the amount of money to be earned in the Italian underground economy was higher and it was easier to be involved in it (Baldwin-Edwards, 2008; King, 2002; Maroukis et al., 2011; Sassen, 2000; Triandafyllidou & Ambrosini, 2011; Triandafyllidou & Maroukis, 2008). Moreover, according to Reyneri (1999), the opportunity for irregular employment was cited as a reason for coming to Italy by many

[23] See Unodccp. (2001). *Global programme against smuggling of migrants*. Odccp Web Page, https://wwwOdccp.Org.

[24] Ambrosini, O. (2018). *Irregular immigration in Southern Europe* (Migration, Diasporas and Citizenship) (Kindle Locations, pp. 170–171). Springer International Publishing. Kindle Edition.

[25] See also Zincone, G. (1999). Illegality, enlightenment and ambiguity: A hot Italian recipe. In M. Baldwin-Edwards, & J. Arango (Eds.), *Immigrants and the informal economy in Southern Europe* (pp. 43–83). Frank Cass.

[26] See also Reyneri (1999: 24).

Moroccans, who often entering Europe through Spain head to Italy to find jobs in its thriving underground/illegal sector.[27]

Entering a country and being easily employed in its informal economy, as in the case of Italy, does not mean that the expectations of the migrant are met in terms of living standards and conditions. However, this does not stop migrants from deciding to migrate. The EU project discussed in Chapter 6 underlines that none of the migrants interviewed who found themselves in a difficult situation in the host country hinted at the possibility of returning to their country. They all said that their negative opinion would not stop their friends and relatives from undergoing their migratory process.[28] Moreover, migrants are very unlikely to return home without money, as this would signal that their migration experience was a failure. This makes it more likely that they will be involved in the non-legal economy of the host country and that they will prolong their stay even if their life is difficult and unlikely to become more comfortable. Finally, stricter rules on employment, entry and regularisation make it easier for migrants to be engaged in out-of-law activities (Ambrosini, 2018; Chapter 6).[29]

Migrants will often depict their life in the host country as better than it is, thus inducing their friends and relatives to follow their path. However, even if they tell the truth about their life abroad this would not serve as a deterrent (Rosina, 2020).[30] Indeed, quite apart from the fact that messages coming from returning migrants can be contradictory, their negative experiences would be discarded as irrelevant by the local population who are willing to flee the country. As already discussed at length in Chapter 6, migration is a self-fulfilling process. Migration leads to further migration, regardless of the information it transmits and of its individual success or failure for a migrant.

In fact, there is plenty of evidence that immigration in Italy is seldom a success story.[31] However, this does not stop other migrants from illegally entering the country and similarly being implicated in

[27] Ibid.

[28] Ibid.

[29] Ambrosini, M. (2018). *Irregulari immigration in Southern Europe* (Migration, Diasporas and Citizenship) (Kindle Location, 110). Springer International Publishing. Kindle Edition.

[30] Reyneri (1999: 25).

[31] See, for example, https://lacnews24.it/cronaca/migranti-sfruttati-sferlazza-politica-assente_107743/ as accessed on April, 10, 2020.

the informal or illegal economy, being criminalised and being further incentivised to adopt a deviant behaviour (Ambrosini, 2018). All this supports the hypothesis that the existence of the underground economy is a strong "pull" factor for migration (Baldwin-Edwards, 2008; King, 2002; Maroukis et al., 2011; Sassen, 2000; Triandafyllidou & Maroukis, 2008; Triandafyllidou & Ambrosini, 2011).[32]

See also https://lacnews24.it/cronaca/migranti-sfruttati-lavoro-nero-prostituzione-arresti_107737/ as accessed on April 10, 2020.

[32] Baldwin-Edwards, M. (2008). Towards a theory of illegal migration: Historical and structural components. *Third World Quarterly, 29* (2008), 1449–1459. Bakewell, O. (2008). Research beyond the categories. The importance of policy-irrelevant research into forced migration. *Journal of Refugee Studies, 12*, 432–453. Basch, L., Glick Schiller, N., & Szanton Blanc, C. (1994). Nations Unbound. *Transnational Projects, Postcolonial Predicaments and Deterritorialised Nation States* . Routledge. Betts, A. (Ed.). (2011). *Global migration governance.* Oxford University Press. Bigo, D. (2009). Security and immigration. Toward a critique of the governmentality of unease. *Alternatives: Social Transformation and Humane Governance, 27*, 63–92. Boccagni, P. (2009). Tracce transnazionali. Vite in Italia e proiezioni verso casa tra I migranti ecuadoriani. Franco Angeli. Cassarino, J.-P., Readmission Policy in the European Union, Brussels, European Parliament (2010), available online at http://cadmus.eui.eu/handle/1814/14957. Castles, S. (2004). Why migration policies fail. *Ethnic and Racial Studies, 27*, 205–227. Czaika, M., & de Haas, H. (2013). The effectiveness of immigration policies. *Population and Development Review, 39*, 487–508. Hinshaw, D. (2015, June 12). Allure of wealth drives deadly trek. *The Wall Street Journal*, available online at http://www.wsj.com/articles/young-men-in-senegal-joinmigrant-wave-despite-growing-prosperity-at-home-1434127244 (accessed 15 June 2015). Hollifield, J., Martin, P., & Orrenius, P. (Eds.). (2014). *Controlling immigration: A global perspective* (3rd ed.) Stanford University Press. King, R. (1995).Migration, globalisation and place. In D. Massey and P. Jess (Eds.), *A place in the world? Places, cultures and globalisation* (pp. 5–53). Open University and Oxford University Press; King, R. (2002). Towards a new map of European migration. *International Journal of Population Geography, 8*, 89–106. Kunz, R., Lavenex, S., & Panizzon, M. (Eds.) (2011). *Multilayered migration governance. The promise of partnership.* Routledge. Latour, B. (1999). On recalling ANT. *The Sociological Review, 47*, 15–25. Downloaded from Brill.com April 17, 2020 04:15:43PM via free access Editorial 11. *European Journal of Migration and Law 19*, 1–11 Levitt, P. (2001). *The transnational villagers.* University of California Press. Levitt, P. (2009). *God needs no passport.* New Press. Maroukis T., Iglicka, K., & Gmaj, K. (2011).'Irregular migration and informal economy in Southern and Central-Eastern Europe: Breaking the vicious cycle? *International Migration, 49*, 129–156. Papastergiadis, N. (2000). *The turbulence of migration. Globalization, deterritorialization and hybridity.* Cambridge: Polity Press. Sassen, S., Guests and Aliens (New York, NY: The New Press, 2000). Schneider, J. (2012). Practical measures for reducing irregular migration. In Federal Office for Migration and Refugees (BAMF) (Working Papers, No. 41), available online at http://www.bamf.de/SharedDocs/Anlagen/EN/Publikationen/EMN/NationaleStudien-WorkingPaper/emn-wp41-irregular-migration.html. Triandafyllidou, A., &

This is particularly true in the context of the paradox of securitisation as underlined by transnationalists, where borders are strictly closed. Those who are unable to obtain a work permit to enter regularly and work regularly in a country are cut off from the regular labour market. Despite this, thanks to the existence of the irregular economy they do not need to go back to their home country. Thus, the ready availability of employment in the underground economy, where no documents are required, promotes undocumented immigration. There is a feedback process between undocumented migration and work opportunities in the underground economy which can hardly be underestimated. It is the engine allowing for the continuous unhindered entry of irregular migrants (Ambrosini, 2018). Rather than being an effect of irregular immigration, the submerged economy would appear to be one of its causes. In a nutshell, if the underground economy did not offer the possibility of employment for undocumented migrants many of them would not enter receiving countries in the first place.

Thus, the issue of whether immigrants represent a competitive threat for domestic workers is to be addressed in the context of the labour structure of the receiving countries (Ambrosini, 2018).[33] For example, southern European countries are certainly characterised by high levels of unemployment. However, the demand for highly flexible and precarious workers, in a very segmented labour market, provides irregular workers with the possibility of finding jobs (Hollifield et al., 2014; for Europe: Triandafyllidou, 2010a).

Ambrosini, M. (2011). Irregular migration control in Italy and Greece: strong fencing and weak gate-keeping strategies serving the labour market. *European Journal of Migration and Law, 13*, 251–273. Triandafyllidou, A., & Dimitriadi, A. (2012). Deterrence and protection in the EU's migration policy. *The International Spectator, 49*, 142–162. Triandafyllidou, A., & Maroukis, T. (2012). *Migrant smuggling: Irregular migration from Asia and Africa to Europe*. Palgrave. Triandafyllidou, A., & Maroukis, T. (2008). The Case of the Greek Islands: The Challenge of Migration at the EU's Southeastern Sea Borders. *Documentos CIDOB, Serie: Migraciones, 17*, 63–82. CIDOB . Walters, W. (2010). Imagined migration world: The European Union's anti-illegal immigration discourse. In M. Geiger & A. Pécoud (Eds.), *The politics of international migration management. Migration, minorities and citizenship* (pp. 73–95). Palgrave Macmillan.

[33] Ambrosini, Maurizio. Irregular Immigration in Southern Europe (Migration, Diasporas and Citizenship) (Kindle Locations 1629–1631). Springer International Publishing. Kindle Edition.

As shown by the figures reported in the section below and confirmed in the literature (Castles et al., 2014),[34] most migrants find work in sectors such as care services, construction, agriculture, cleaning and catering. These are also sectors highly featured in the underground economy and providing a wealth of irregular jobs to both local workers and migrants, especially unauthorised ones (Talani, 2017). However, in southern Europe, given its strong segmentation of labour markets and vast underground economy, unauthorised and regular migrants tend to occupy job positions which are usually less appealing to the local unemployed population (Reyneri, 2013).[35]

Such unemployment concerns mainly educated young people, living with their families who can often afford to wait for the right job proposal. They are therefore unlikely to compete with migrant workers for low-skilled jobs (Ambrosini, 2018: Kindle Locations 1659–1660)[36] (Reyneri 1999, pp. 7–14).[37] Thus, scholars conclude that demand for migrant workers exists in southern European countries, although the overall unemployment rate is high (Ambrosini, 2018; Reyneri, 2013). Indeed, as noticed above, especially in southern Europe, nearly all migrants are employed with low level tasks, often below their qualifications and educational level. They are employed in personal and social services and in the industrial and agricultural sectors (see below this chapter). In southern European countries, migrants hold a wide range of jobs in the least skilled services, while the proportion of self-employed migrants is negligible (see below this chapter) (Reyneri & Fullin, 2011). Moreover, the availability of cheap, migrant labour in, for example, domestic services has allowed local

[34] Castles, Stephen, Hein de Haas, and Mark Miller. 2014.

Springer International Publishing. Kindle Edition. T*he age of migration. International population movements in the modern world.* (5th ed). Basingstoke: Palgrave Macmillan.

[35] Reyneri, Emilio. 2013. Irregular Immigration and the Underground Economy in Southern Europe: Breaking the Vicious Circle. In.

Europe's Immigration Challenge. Reconciling Work, Welfare and Mobility, ed. Grete Brochmann and Elena Jurado. London: I.B. Tauris.

Reyneri, Emilio, and Giovanna Fullin. 2011. Ethnic Penalties in the Transition to and From Unemployment: A West European Perspective.

International Journal of Comparative Sociology, 52(4), 247–263.

[36] Ambrosini, Maurizio. Irregular Immigration in Southern Europe (Migration, Diasporas and Citizenship) (Kindle Locations 1659–1660). Springer International Publishing. Kindle Edition.

[37] See Reyneri (1999, pp. 7–14).

women to upgrade their employment status or enter the labour market for the first time (Da Roit et al., 2013[38]; Tognetti Bordogna & Ornaghi 2012).[39] Finally, in the aftermath of the 2008–2009 crisis, the informal economy not only absorbed immigrant workers, but the underground economy also helped soften the impact of the recession by supporting the creation of value-added (Chapter 6).

Thus, there is scarce, if any, competition because in both the formal and the underground economy, most activities carried out by migrants are below the level accepted by domestic workers (Reyneri & Fullin, 2011: 250).[40].

For this reason, part of the literature refers to the existence of a "Mediterranean model of immigration" (Baldwin-Edwards & Arango, 1999; King & Black, 1997; Pugliese, 2002).[41]

Finally, the insertion of the vast majority of immigrants into the underground economy is a clear reason why southern European countries have quickly adopted negative attitudes towards immigration. This explains why it is politically convenient to adopt an anti-migration position (Ambrosini, 2013a; 2015). Working in the underground economy, as well as living without documents, contributes to both the marginalisation of migrants and the negative attitudes towards them of the local population (see Chapter 6).

Concluding, from the transnationalist perspective, the paradox of securitisation within globalisation makes competition between the local and immigrant labour force highly unlikely. Indeed, as the vast majority of especially low-skilled migrants are irregular at some point, if not for the whole of their migratory experience, often the only chance they have to support themselves is to engage with the underground economy. This is

[38] Da Roit, Barbara, Amparo González Ferrer, and Francisco J. Moreno-Fuentes. 2013. The Southern European Migrant Based Care Model: Long Term Care and Employment Trajectories in Italy and Spain.

European Societies 15 (4): 577–596.

[39] Tognetti Bordogna, Mara, and Annalisa Ornaghi. 2012. The 'Badanti' (Informal Carers) Phenomenon in Italy. *Journal of Intercultural Studies, 33*(1), 9–22.

[40] Reyneri, E., & Fulli, G. (2011). Ethnic Penalties in the transition to and from unemployment: A West European perspective. *International Journal*

[41] Pugliese, Enrico. 2002.

L'Italia tra migrazioni internazionali e migrazioni interne. Bologna: Il Mulino.

. 2012. Il lavoro agricolo immigrato nel Mezzogiorno e il caso di Rosarno. *Mondi Migranti, 6*(3), 7–28.

substantiated by an EU study conducted on the subject which reaches the following conclusions:

- New immigrants are mostly unauthorised people, but have various personal characteristics and migratory projects.
- In the receiving labour markets, there is more replacement and complementarity than competition, except for the self-employment field.
- Migrants are inserted in well-rooted and flourishing underground economies.
- The domestic underground economy promotes unauthorised immigration.
- The underground economy prevents immigrants from steadily settling in the host country.
- The underground economy facilitates migrants' marginalisation.
- The underground economy may contribute to increasing migrants' deviant behaviour.
- Medice cura te ipsum: southern European countries should realise that only by curbing the size and relevance of the underground economy for the domestic labour market it is possible to tackle the problems of unauthorised immigration.[42]

What kind of policy implications does this analysis suggest? The study above suggests interventions in a number of areas in relation to migration, especially in southern European countries. In particular, it stresses the extent to which undocumented migrants are creating wealth in their host countries, providing services and performing jobs that the labour force would not supply otherwise. Awareness of the relevance of migrant communities for the local economy could help integrate them into local civil societies, creating the necessary level of integration between migrants and local social capital to avoid violent outbursts of discontent (see Chapter 8).

[42] These Conclusions Were Contained in the Evaluation Report by the Author: Talani, L. S. (2002). Final evaluation report of the project migrants' insertion in the informl economy, deviant behaviour and the impact on receiving societies. Ce/Dgxii-Science, Research And Development, Tsr Program, Contract No. Soe2-Ct95-3005, Mimeo, p. 11.

In more general terms, trasnationalists advise to adopt measures aimed at the progressive eradication of the phenomenon of irregular migration. This could be obtained by, on the one hand, reversing the paradox of the securitisation of migration. On the other hand, it could be achieved by adopting more realistic migratory policies in terms of acceptance of migratory flows not only as an inevitable event, but also as a positive development for receiving countries.[43]

Moreover, these considerations help explain why migration has been an easy target for populist parties, as further addressed in Chapter 9. The hostility towards migrants is greatly enhanced by the factors facilitating the introduction of unauthorised migrants into the informal/illegal economies of receiving countries. These factors are reinforced by official migratory policies denying the right to entry on the basis of the existence of high unemployment levels in receiving countries. Moreover, this activates a vicious cycle between illegal entry, insertion into the informal/illegal sector, stigmatisation and increased likelihood to adopt deviant behaviours (Bacon, 2008). The interruption of such a cycle is not an easy task, but evidence shows that tightening border controls is not the solution (Rosina, 2020; Talani, 2002; Chapter 6).[44]

Overall, transnationalists would endorse a gradual opening to immigrant workers with an effective control of their insertion into the labour market of receiving countries (Leon & Overbeek, 2015). Indeed, from analysing the labour markets of EU countries, there does not seem to be competition between the domestic labour force and immigrant workers. This is despite the high level of unemployment recorded in some EU Member States.

To this we should add considerations about the difficulty of sustaining the welfare states of EU countries. This is especially so in the context of an ageing population, thus making the need to integrate immigrant workers in the social and political life of receiving countries more evident (Ambrosini, 2018).

[43] These Conclusions Were Contained In The Evaluation Report By The Author: Talani, L. S. (2002). Final evaluation report of the project migrants' insertion in the informl economy, deviant behaviour and the impact on receiving societies. Ce/Dgxii-Science, Research And Development, Tsr Program, Contract No. Soe2-Ct95-3005, Mimeo, p. 11.

[44] Ibid.

Finally, the legalisation of migration and the extension of social and political rights to foreign workers would contribute to elimination of social dumping against the local labour force ensuing from the wage depression effect that the use of unauthorised migrant workers in the black or even legal economy might have. It would also contribute to the eradication of the underground economy, since migrant workers with a legal status are undoubtedly more likely to denounce situations of exploitation and illegality than unauthorised migrants.[45]

In the next section, evidence is provided on the insertion of migrants into the labour markets of receiving countries.

The Integration of Migrant Workers into the Labour Markets: General Patterns

This section reports on the scarce data available relating to the integration of international migrants in the labour markets of receiving countries.

On the basis of the discussion above, it is interesting to find out which sectors of receiving countries' economies international migrants are usually employed in. However, data from the OECD on the sectors of employment of migrants stops at the census of the year 2000.[46] According to them, compared to the native-born, immigrants are over-represented in the low-skilled jobs of the agricultural and industrial sectors.[47]

Table 7.1 shows the situation of European OECD countries in 2000, highlighting the differences between foreign born and native workers in their patterns of employment in the different economic sectors. The table points at the existence of a clear gap between the percentage of foreign-born labour employed in the low skill sectors of agriculture and industry and the percentage of native-born labour working in the same sectors.

The gap was particularly clear in the case of Greece, where 24.3% more foreign men than natives work in the agricultural and industrial sectors. There was a similar pattern in Luxembourg (+14.1%), Italy (+10.7%) and Spain (+4.5%) (Table 7.1).

[45] See Talani (2002: 12/13).

[46] See OECD. (2008). *A profile of immigrant populations in the 21st century: Data from OECD countries*. OECD.

[47] Ibid., p. 150.

As far as foreign women are concerned, they tended to be overrepresented in the low skill personal and social services sectors, such as services for the elderly and for children, or services for restaurants and hotels. Although also native-born women were significantly involved in these activities, there was still a gap between the percentage of foreign-born and native-born women employed in these sectors. This phenomenon was more pronounced in southern European countries: Greece, where 16.3% more foreign women were involved in the personal and social services sector, Portugal (+7.9%), Italy (+7%) and Spain (+3.3%) (Tables 7.1 and 7.2).

Data from EUROSTAT allows us to compare how the situation evolved between 2008 and 2014.[48] Using the ISCO's classification of ten occupational categories, we can distinguish between the first three categories which require a high level of qualifications, the following four as medium skilled and including white collar, white collar/office and service jobs and blue collar/manual occupations, and the last two categories which represent unskilled workers.

From this data, in 2014, a tendency emerges of 'first-generation immigrants' working in less qualified jobs, especially elementary occupations and service workers.

Overall, still in 2014, more than 25% of 'first-generation immigrants' residing legally in the EU had an 'unskilled blue collar occupation' ('plant and machine operators and assemblers and elementary occupations') (see Fig. 7.1). These are the kinds of jobs that require a very low level of education, such as primary education. However, a substantial number of 'first-generation immigrants', almost 1 in 3, were employed in highly skilled office occupations, such as 'senior officials, managers', 'professionals' and 'technicians' which normally require a first or second stage of tertiary education. This is in contrast with more than 40% of 'natives with native background' working in these highly qualified occupations.

The crisis of 2008 seems to have had a greater impact on immigrants than on the native working force. While highly skilled migrants have been positively impacted by structural changes in the period between 2008 and 2014, with an increase of 3.7% in the number of 'first-generation immigrants' working in highly skilled non-manual jobs, the percentage

[48] See EUROSTAT https://ec.europa.eu/eurostat/statistics-explained/index.php?title=First_and_second-generation_immigrants_-_statistics_on_employment_conditions#Occupation as accessed on April 23, 2020.

Table 7.1 Sectors of activity of the native-born and the foreign-born by country of residence and gender percentage of the employed population aged 15 and above

			Native-born			Foreign-born			Total					
			Men	Women	Total	Men	Women	Total	Men	Women	Total			
	Country	Sector	(%)	(%)	(%)	(%)	(%)	(%)	(%)	(%)	(%)	Difference foreigners minus natives	Difference foreigners minus natives	Difference foreigners minus natives
AUT	Austria	Agriculture and industry	40.9	19.6	31.4	43.3	19.3	32.9	41.3	19.5	31.6	2.4	-0.3	1.5
		Producer services	11.7	13.0	12.3	11.9	16.4	13.8	11.7	13.5	12.5	0.2	3.4	1.5
		Distributive services	23.7	24.1	23.9	22.5	19.9	21.4	23.5	23.6	23.5	-1.1	-4.2	-2.5
		Personal and social services	23.7	43.3	32.4	22.3	44.4	31.9	23.5	43.4	32.4	-1.4	1.1	-0.6
BEL	Belgium	Agriculture and industry	39.5	14.1	28.8	39.3	10.9	28.7	39.5	13.9	28.8	-0.2	-3.3	-0.1
		Producer services	11.2	11.4	11.3	12.0	15.0	13.1	11.3	11.7	11.5	0.7	3.6	1.8
		Distributive services	23.4	20.4	22.1	22.3	19.2	21.2	23.3	20.3	22.0	-1.1	-1.2	-1.0
		Personal and social services	25.9	54.1	37.7	26.4	55.0	37.0	25.9	54.2	37.7	0.6	0.9	-0.7
CHE	Switzerland	Agriculture and industry	38.8	16.0	29.1	43.3	19.0	33.0	39.9	16.7	30.0	4.6	3.0	3.9
		Producer services	18.5	15.8	17.3	15.1	15.2	15.1	17.7	15.6	16.8	-3.3	-0.6	-2.2
		Distributive services	22.8	25.3	23.9	19.4	19.9	19.6	21.9	24.0	22.8	-3.4	-5.5	-4.3
		Personal and social services	20.0	42.9	29.8	22.1	46.0	32.3	20.5	43.7	30.4	2.1	3.1	2.5
CZE	Czech Republic	Agriculture and industry	55.3	32.6	45.1	50.8	35.8	44.2	55.1	32.7	45.0	-4.5	3.2	-0.8
		Producer services	6.9	9.4	8.0	7.4	7.8	7.6	6.9	9.3	8.0	0.5	-1.5	-0.4
		Distributive services	18.5	18.5	18.5	22.1	19.1	20.8	18.7	18.5	18.6	3.5	0.6	2.2
		Personal and social services	19.3	39.5	28.4	19.7	37.3	27.4	19.3	39.5	28.3	0.5	-2.2	-0.9
DNK	Denmark	Agriculture and industry	38.0	14.4	27.0	29.3	15.3	22.9	37.5	14.5	26.7	-8.7	0.8	-4.0
		Producer services	13.7	12.8	13.3	16.2	19.2	17.6	13.9	13.2	13.6	2.5	6.3	4.2
		Distributive services	25.3	17.2	21.5	25.2	16.1	21.1	25.3	17.2	21.5	-0.1	-1.1	-0.4
		Personal and social services	23.0	55.5	38.2	29.2	49.4	38.4	23.4	55.2	38.2	6.2	-6.1	0.2
ESP	Spain	Agriculture and industry	48.0	20.6	37.9	52.5	15.5	37.6	48.3	20.2	37.8	4.5	-5.1	-0.2
		Producer services	8.7	12.2	10.0	7.3	10.6	8.6	8.6	12.1	9.9	-1.4	-1.6	-1.4
		Distributive services	22.4	23.3	22.7	18.5	16.8	17.8	22.1	22.8	22.4	-3.8	-6.6	-4.9
		Personal and social services	21.0	43.9	29.5	21.7	57.2	35.9	21.0	44.9	29.9	0.7	13.3	6.5
FIN	Finland	Agriculture and industry	46.5	17.5	32.4	34.6	17.5	27.3	46.2	17.5	32.3	-11.9	0.0	-5.1
		Producer services	12.7	13.6	13.2	17.4	16.0	16.8	12.8	13.7	13.2	4.7	2.4	3.7
		Distributive services	22.5	16.8	19.7	19.5	19.5	19.5	22.4	16.8	19.7	-3.0	2.7	-0.2
		Personal and social services	18.3	52.1	34.7	28.4	47.0	36.3	18.5	52.0	34.7	10.1	-5.1	1.6
FRA	France	Agriculture and industry	38.6	15.9	28.3	38.6	13.4	28.3	38.6	15.7	28.3	0.0	-2.6	0.0
		Producer services	3.5	5.0	4.2	3.3	5.8	4.3	3.5	5.1	4.2	-0.3	0.8	0.1
		Distributive services	32.5	26.6	29.8	32.8	27.1	30.6	32.6	26.7	30.0	0.0	0.0	0.7
		Personal and social services	25.4	52.4	37.7	25.3	53.5	36.9	25.4	52.5	37.6	-0.1	1.1	-0.8
GBR	United Kingdom	Agriculture and industry	36.6	11.7	25.2	22.4	10.1	16.7	35.3	11.6	24.4	-14.2	-1.7	-8.4
		Producer services	16.9	17.1	17.0	21.6	20.1	20.9	17.3	17.4	17.4	4.7	2.9	3.9
		Distributive services	25.1	22.1	23.7	25.0	18.8	22.2	25.0	21.8	23.6	-0.1	-3.2	-1.5
		Personal and social services	21.4	49.1	34.1	31.1	51.0	40.2	22.3	49.2	34.7	9.6	1.9	6.1
GRC	Greece	Agriculture and industry	41.0	28.5	36.4	65.3	21.5	49.7	44.4	27.6	38.2	24.3	-7.1	13.3
		Producer services	7.7	12.0	9.3	4.0	9.8	6.1	7.2	11.7	8.9	-3.7	-2.2	-3.2
		Distributive services	25.6	20.2	23.6	16.7	13.1	15.4	24.4	19.2	22.5	-8.9	-7.0	-8.2
		Personal and social services	25.7	39.3	30.7	14.0	55.6	28.8	24.1	41.4	30.5	-11.7	16.3	-1.9
HUN	Hungary	Agriculture and industry	47.8	27.5	38.5	40.4	25.0	33.3	47.6	27.4	38.4	-7.4	-2.4	-5.2
		Producer services	9.2	9.7	9.4	10.5	10.2	10.3	9.3	9.7	9.5	1.2	0.5	0.9
		Distributive services	22.9	20.6	21.9	25.9	22.9	24.5	23.0	20.7	21.9	3.0	2.3	2.7
		Personal and social services	20.1	42.2	30.2	23.3	41.9	31.8	20.2	42.2	30.2	3.2	-0.4	1.6
IRL	Ireland	Agriculture and industry	47.2	14.9	33.8	37.9	14.6	28.0	46.1	14.9	33.1	-9.3	-0.4	-5.9
		Producer services	11.9	16.5	13.8	17.9	18.3	18.1	12.6	16.7	14.3	6.0	1.8	4.3
		Distributive services	21.6	20.0	21.0	18.3	17.4	17.9	21.3	19.7	20.6	-3.3	-2.6	-3.0
		Personal and social services	19.3	48.6	31.4	25.8	49.8	36.0	20.1	48.7	32.0	6.5	1.2	4.6
ITA	Italy	Agriculture and industry	46.6	26.1	38.6	57.3	26.2	45.2	47.2	26.1	39.0	10.7	0.2	6.5
		Producer services	9.3	11.0	10.0	5.1	7.7	6.1	9.1	10.8	9.8	-4.2	-3.3	-3.9
		Distributive services	20.3	17.8	19.4	16.4	14.0	15.5	20.1	17.6	19.2	-3.9	-3.8	-3.9
		Personal and social services	23.7	45.1	32.0	21.1	52.1	33.3	23.6	45.5	32.1	-2.6	7.0	1.2
LUX	Luxembourg	Agriculture and industry	25.4	15.6	21.3	39.5	8.0	26.5	31.3	12.5	23.5	14.1	-7.6	5.2
		Producer services	15.3	17.6	16.3	21.2	26.6	23.4	17.7	21.4	19.2	5.9	9.0	7.2
		Distributive services	24.2	19.2	22.1	17.2	15.7	16.6	21.3	17.8	19.8	-6.9	-3.5	-5.5
		Personal and social services	35.2	47.6	40.3	22.1	49.7	33.5	29.7	48.4	37.5	-13.1	2.1	-6.9
NLD	Netherlands	Agriculture and industry	35.6	12.0	25.9	35.6	13.2	26.2	35.6	12.1	25.9	0.1	1.2	0.4
		Producer services	15.5	14.6	15.1	16.4	21.3	18.5	15.5	15.2	15.4	1.0	6.7	3.4
		Distributive services	23.7	21.3	22.7	20.7	18.3	19.7	23.4	21.0	22.4	-3.0	-2.9	-3.0
		Personal and social services	25.3	52.1	36.3	27.2	47.1	35.6	25.4	51.7	36.3	1.9	-4.9	-0.7
NOR	Norway	Agriculture and industry	38.5	10.2	25.3	27.7	8.7	18.8	37.7	10.1	24.8	-10.9	-1.5	-6.5
		Producer services	13.2	10.4	11.9	15.8	12.3	14.2	13.4	10.5	12.0	2.6	1.9	2.3
		Distributive services	25.3	19.5	22.6	22.2	15.8	19.2	25.0	19.3	22.3	-3.1	-3.8	-3.4
		Personal and social services	23.0	59.9	40.3	34.3	63.2	47.8	23.8	60.1	40.8	11.3	3.3	7.6
POL	Poland	Agriculture and industry	55.9	33.7	45.7	44.7	31.9	39.4	55.8	33.7	45.6	-11.2	-1.8	-6.3
		Producer services	6.7	8.7	7.6	11.1	9.4	10.4	6.8	8.7	7.7	4.4	0.7	2.8
		Distributive services	21.2	20.5	20.9	19.6	17.7	18.8	21.2	20.4	20.9	-1.7	-2.8	-2.1
		Personal and social services	16.1	37.2	25.8	24.6	41.1	31.4	16.2	37.2	25.8	8.5	3.9	5.6
PRT	Portugal	Agriculture and industry	50.1	28.8	40.7	46.6	16.8	33.1	49.8	27.8	40.1	-3.5	-12.1	-7.7
		Producer services	6.9	8.4	7.5	9.8	13.0	11.3	7.1	8.8	7.9	3.0	4.7	3.8
		Distributive services	23.3	18.5	21.1	20.3	17.9	19.2	23.0	18.4	21.0	-2.9	-0.5	-1.9
		Personal and social services	19.8	44.3	30.6	23.3	52.3	36.4	20.1	45.0	31.1	3.5	7.9	5.9
SVK	Slovak Republic	Agriculture and industry	51.7	29.2	41.2	45.3	29.1	37.4	51.6	29.2	41.1	-6.5	-0.1	-3.8
		Producer services	7.5	7.9	7.7	8.6	7.8	8.2	7.5	7.9	7.7	1.1	-0.1	0.5
		Distributive services	20.1	19.3	19.8	20.7	18.5	19.6	20.1	19.3	19.7	0.6	-0.8	-0.1
		Personal and social services	20.7	43.5	31.4	25.4	44.5	34.8	20.8	43.5	31.5	4.8	1.0	3.4
SWE	Sweden	Agriculture and industry	40.7	11.6	26.8	35.2	14.4	25.0	40.2	11.9	26.6	-5.5	2.8	-1.8
		Producer services	15.6	12.3	14.0	14.7	13.9	14.3	15.5	12.4	14.0	-1.0	1.6	0.3
		Distributive services	22.5	15.4	19.1	22.3	12.6	17.5	22.5	15.1	18.9	-0.3	-2.8	-1.6
		Personal and social services	21.1	60.8	40.1	27.9	59.1	43.2	21.8	60.6	40.4	6.7	-1.6	3.1

Note Among the employeed men aged 15 and above born and living in Australia, 35.9% work in the agriculture and industry sector, 14 in the producer services sector, 28.2% in the Distributive services sector and 21.9% in the personal and social services sector

Source Database on imigiration in OECD countries (Dioc)

Table 7.2 Sectors of activity of the foreign-born population from the 5 main countries of origin percentage of the employed population aged 15 and above

	Country	Country of birth	‰ of total pop.	Agriculture and industry (%)	Producer services (%)	Distributive services (%)	Personal and social services (%)
AUT	Austria	Former Yugoslavia	53.8	39.5	13.4	20.5	26.6
		Turkey	18.6	45.9	12.2	20.7	21.1
		Germany	17.0	24.1	15.7	22.3	37.9
		Poland	6.5	30.7	14.2	21.8	33.2
		Romania	4.9	38.8	12.1	22.6	26.5
BEL	Belgium	France	12.6	23.2	11.9	22.6	42.3
		Italy	11.5	44.7	8.3	18.6	28.5
		Morocco	9.1	36.1	16.6	19.5	27.8
		Netherlands	8.6	32.0	16.2	23.8	28.1
		Germany	8.0	23.9	15.9	21.1	39.1
CHE	Switzerland	Former Yugoslavia	43.0	46.5	9.2	18.4	25.8
		Italy	36.2	42.6	13.2	22.0	22.3
		Germany	26.8	23.8	20.4	17.9	37.9
		Portugal	20.5	41.1	9.6	17.8	31.5
		France	14.7	20.7	18.5	20.5	40.4
CZE	Czech Republic	Slovak Republic	24.7	46.4	7.6	15.5	30.5
		Former USSR	4.9	52.3	8.4	15.7	23.5
		Vietnam	2.5	3.5	1.4	92.4	2.7
		Poland	2.2	62.3	3.9	12.1	21.7
		Former Yugoslavia	0.8	32.8	9.5	32.7	24.9
DNK	Denmark	Turkey	5.4	21.9	19.5	21.8	36.8
		Germany	4.7	26.0	14.5	19.1	40.4
		Former Yugoslavia	4.7	34.9	18.4	21.0	25.7
		Sweden	4.0	19.2	16.7	20.2	44.0
		Norway	2.9	17.7	16.2	18.4	47.7
ESP	Spain	Morocco	10.1	55.5	5.2	15.9	23.4
		Ecuador	8.7	42.8	5.3	12.8	39.1
		Colombia	5.3	31.3	8.6	14.2	45.9
		France	5.2	34.1	10.8	23.6	31.5
		Germany	3.6	29.5	12.6	25.3	32.5
FIN	Finland	Sweden	6.2	32.8	13.0	22.0	32.2
		Former USSR	6.0	27.1	16.2	24.8	31.9
		Germany	0.7	25.5	20.5	16.7	37.3
		United Kingdom	0.7	20.7	20.3	11.0	47.9
		Vietnam	0.5	52.6	6.5	13.8	27.1
FRA	France	Algeria	21.6	22.2	4.4	30.7	42.6
		Portugal	13.7	44.4	5.9	23.9	25.9
		Morocco	9.4	29.5	3.6	31.8	35.2
		Tunisia	5.0	24.7	4.8	33.4	37.1
		Spain	4.2	34.1	5.4	26.0	34.4
GBR	United Kingdom	India	8.6	24.2	16.9	28.8	30.1
		Ireland	8.6	23.9	17.7	17.8	40.6
		Germany	5.1	18.6	20.4	21.8	39.1
		Pakistan	4.3	22.6	12.8	38.7	25.9
		Kenya	3.2	17.4	22.4	31.8	28.4
GRC	Greece	Albania	54.1	67.6	3.2	9.1	20.0
		Former USSR	22.3	44.6	5.9	14.7	34.7
		Germany	12.9	25.0	10.3	27.0	37.8
		Bulgaria	5.9	52.0	3.6	6.4	38.0
		Romania	4.1	62.4	4.0	8.8	24.7
HUN	Hungary	Romania	16.1	38.4	9.6	20.9	31.1
		Former USSR	3.4	25.6	10.8	25.5	38.0
		Former Yugoslavia	2.6	33.7	10.5	28.5	27.3
		Slovak Republic	1.9	34.4	10.9	16.1	38.7
		Germany	1.2	34.1	13.5	22.2	30.2
IRL	Ireland	United Kingdom	76.8	30.2	16.6	19.3	33.8
		United States	4.8	23.5	21.3	13.6	41.6
		Former USSR	3.9	47.6	10.7	17.2	24.4
		France	2.8	22.4	26.9	15.8	35.0
		Germany	2.8	28.4	24.5	17.7	29.3
ITA	Italy	Switzerland	5.0	42.1	8.4	18.6	30.9
		Morocco	3.9	65.3	2.1	17.4	15.1
		Albania	3.6	70.5	2.1	9.5	18.0
		Germany	3.4	41.0	7.6	18.4	33.0
		Former Yugoslavia	3.3	54.6	4.7	16.2	24.5
LUX	Luxembourg	Portugal	150.3	44.7	12.0	17.0	26.4
		France	57.6	14.1	31.3	17.9	36.6
		Belgium	45.2	13.9	39.1	17.1	29.9
		Italy	30.4	25.4	19.6	17.7	37.3
		Germany	29.6	14.8	30.3	17.2	37.7
NLD	Netherlands	Turkey	10.2	44.6	13.7	18.5	23.2
		Indonesia	9.8	22.3	18.3	14.2	45.2
		Morocco	8.2	34.3	15.4	21.4	28.9
		Germany	7.2	22.2	14.0	27.9	35.9
		United Kingdom	3.8	23.2	28.4	20.4	28.0
NOR	Norway	Sweden	7.3	18.8	14.5	21.5	45.2
		Denmark	5.0	23.5	12.3	19.0	45.3
		Former Yugoslavia	3.3	24.8	13.6	20.0	41.5
		United Kingdom	3.3	24.5	19.7	16.1	39.7
		Germany	2.4	20.0	12.8	15.3	52.0
POL	Poland	Former USSR	5.9	40.0	10.3	17.5	32.2
		Germany	1.2	44.5	10.3	18.8	26.5
		France	0.3	45.6	11.5	17.4	25.5
		Czech Republic	0.1	37.2	10.0	19.1	33.7
		Former Yugoslavia	0.1	51.7	7.1	23.5	17.7
PRT	Portugal	Angola	25.9	26.2	12.5	19.8	41.4
		Mozambique	12.0	19.7	15.4	23.1	41.7
		France	11.4	38.2	8.3	23.3	30.2
		Brazil	6.5	35.2	9.5	18.2	37.1
		Cape Verde	6.1	47.6	11.9	9.4	31.1
SVK	Slovak Republic	Czech Republic	17.9	36.8	8.6	19.4	35.3
		Former USSR	2.0	37.6	7.0	18.7	36.7
		Hungary	1.1	44.4	5.7	17.1	32.8
		Poland	0.6	44.0	6.2	17.0	32.9
		Romania	0.5	61.4	4.2	14.9	19.6
SWE	Sweden	Finland	23.1	31.6	12.8	14.7	41.0
		Former Yugoslavia	14.4	38.4	12.2	17.6	31.8
		Iran	5.7	15.7	11.5	23.4	49.4
		Poland	4.9	21.5	15.2	16.4	46.9
		Norway	4.1	23.5	14.3	19.4	42.8

Note Among the employeed men aged 15 and above born and living in Australia, 24.8% work in the agriculture and industry sector, 17.9% in the producer services sector, 22.6% in the Distributive services sector and 34.7% in the personal and social services sector
Source Database on imigiration in OECD countries (Dioc)

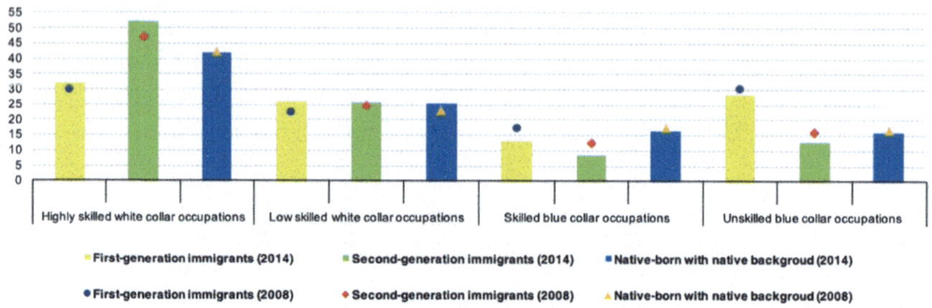

Fig. 7.1 Occupation of employees by migration status and year, EU, 2008 and 2014 (%) (*Source* Eurostat, EU LFS AHM2014/2008)

of 'first-generation immigrants' working in 'unskilled blue collar occupations' was lower in 2014 than in 2008 (–1.5%). Similarly, the percentage of 'first-generation immigrants' working in skilled manual jobs decreased by 4.3% between 2008 and 2014.

Looking at the differences between 'first-generation immigrants' from the EU and from outside the EU, in 2014, foreign-born workers of 'EU origins' were more likely to be employed in 'skilled occupations' than those of 'non-EU origins'. Authorised non-EU migrants were instead mostly employed either as unskilled manual workers (29.4%) or as 'highly skilled professionals' (30.6%) (Fig. 7.2).

Table 7.3 details the three main job occupations of 'first-generation

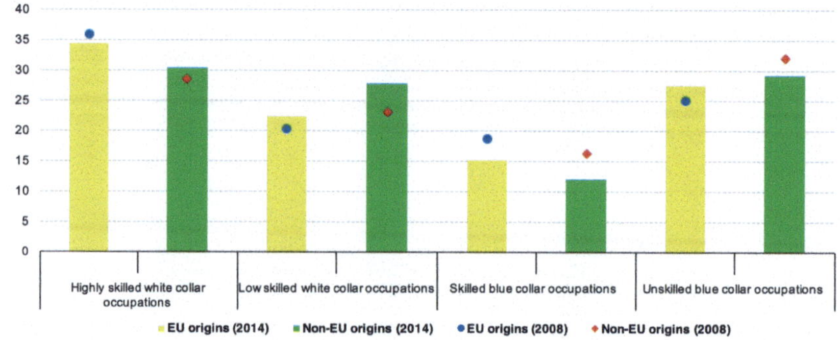

Fig. 7.2 Occupation of first-generation immigrant employees, by origin, EU, 2008 and 2014 (%) (*Source* Eurostat, EU LFS AHM2014/2008)

Table 7.3 Top three occupational groups of first-generation immigrant employees, 2014

	First	Second	Third
EU-28	Elementary occupations	Service workers and shop and market sales workers	Professionals
Belgium	Elementary occupations	Professionals	Service workers and shop and market sales workers
Bulgaria	:	:	:
Czech Republic	Service workers and shop and market sales workers	Craft and related trades workers	Professionals
Denmark	:	:	:
Germany	:	:	:
Estonia	Plant and machine operators and assemblers	Craft and related trades workers	Service workers and shop and market sales workers
Ireland	:	:	:
Greece	Elementary occupations	Service workers and shop and market sales workers	Craft and related trades workers
Spain	Elementary occupations	Service workers and shop and market sales workers	Craft and related trades workers
France	Service workers and shop and market sales workers	Professionals	Elementary occupations
Croatia	Service workers and shop and market sales workers	Craft and related trades workers	Elementary occupations
Italy	Elementary occupations	Service workers and shop and market sales workers	Craft and related trades workers
Cyprus	Elementary occupations	Service workers and shop and market sales workers	Professionals
Latvia	Elementary occupations	Technicians and associate professionals	Legislators senior officials and managers
Lithuania	Craft and related trades workers	Professionals	Service workers and shop and market sales workers
Luxembourg	Professionals	Technicians and associate professionals	Service workers and shop and market sales workers
Hungary	Professionals	Service workers and shop and market sales workers	Craft and related trades workers
Malta	Legislators senior officials and managers	Professionals	Service workers and shop and market sales workers
Netherlands	:	:	:
Austria	Service workers and shop and market sales workers	Elementary occupations	Craft and related trades workers
Poland	Professionals	Service workers and shop and market sales workers	Legislators senior officials and managers
Portugal	Professionals	Service workers and shop and market sales workers	Elementary occupations
Romania	:	:	:
Slovenia	Craft and related trades workers	Elementary occupations	Plant and machine operators and assemblers
Slovakia	Plant and machine operators and assemblers	:	:
Finland	Professionals	Service workers and shop and market sales workers	Elementary occupations
Sweden	Professionals	Service workers and shop and market sales workers	Technicians and associate professionals
United Kingdom	Professionals	Service workers and shop and market sales workers	Elementary occupations

Note Population aged in 23–54
Source Eurostat, EU LFS AHM2014/2008

immigrants' in EU Member States in 2014. In countries like Italy and Spain, as well as four other EU Member States, regular immigrants were mostly employed in 'elementary occupations'. Also, overall in the EU, 'elementary occupations' ranked first among the occupation of 'first-generation immigrants'.

In 2014, EU countries able to attract the most highly skilled 'first-generation immigrants' were Luxembourg, Hungary, Poland, Portugal, Finland, Sweden and the UK.

In 2017, across the OECD, the percentage of immigrants holding low-skilled jobs, or elementary occupations was 18% as opposed to 11% of the native workers. In the EU this difference was more accentuated, with figures of 20% and 8%, respectively. Overall, immigrant workers are more concentrated in lower skilled employment in most countries. This is especially true in southern Europe, where the percentage of immigrants holding low-skilled jobs or elementary occupations is at least 30% with the exception of Portugal. This figure was three times higher than the number of native workers employed in those sectors (OECD, 2018). As Fig. 7.3 below shows, in Greece, immigrants are employed in elementary occupations six times more than the local labour force (33.1% vs 5.2%) and four times more than in Nordic countries. In the EU and in the US, more than 20% of low-skilled jobs are held by immigrants. However, this level exceeds 40% in Austria, Germany, Sweden and Norway, and 60% in Switzerland and Luxembourg. Moreover, migrants from outside the EU were far more likely to have unskilled jobs in 2017, except in the UK, Ireland, and Hungary.

Usually, the percentage of immigrants working in highly skilled occupations is lower than that of the native-born in all OECD and EU countries except in Australia, New Zealand, Portugal, Malta, Turkey and some Central European countries (such as Bulgaria and the Slovak Republic and Poland) (see Fig. 7.3). In the EU in 2017, 11% less migrants were employed in highly skilled positions than native ones. This gap has actually increased between 2007 and 2017 (OECD, 2018) especially in Nordic countries (with the exception of Sweden).

Looking at the precise sector of employment of migrants, Eurostat uses the statistical classification of economic activities in the European Community (NACE). Table 7.4 presents the main sectors of activity by migration status.

In 2014, most 'first-generation immigrants' were employed in 'manufacturing' together with 'wholesale and retail trade; repair of motor

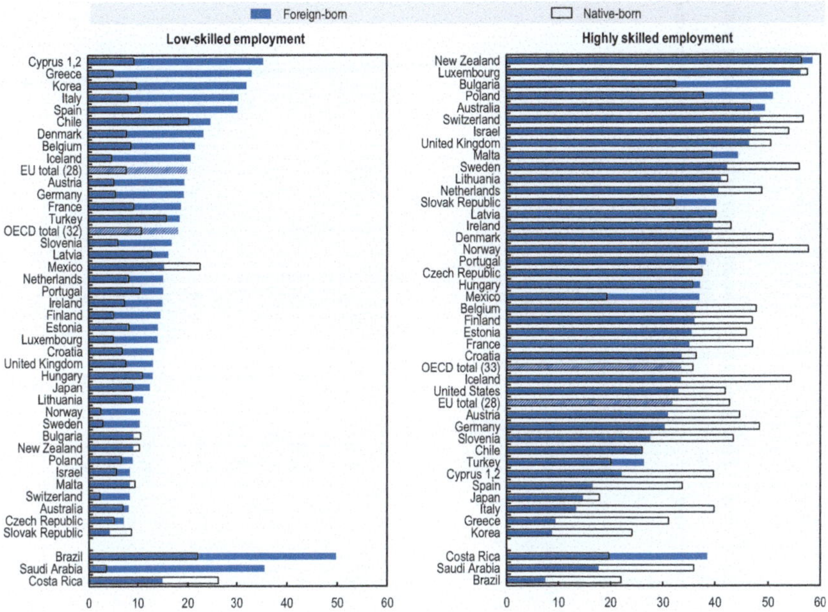

Fig. 7.3 Low- and highly educated Percentages of 15- to 64-year-olds not in education, 2017 (*Source* OECD [2018], https://www.intlnursemigration.org/wp-content/uploads/2019/08/OECD-2018-Immigrant-Integration.pdf)

vehicles and motorcycles'. These sectors were also those employing most workers generally, and almost a third (30.2%) of the 'native-born with native background' (16.0% and 14.2%, respectively). However, pointing at some form of complementarity between the existing and incoming labour force, it should be noted that there were many activities where the percentages of 'native-born with native background' and 'second-generation immigrants' were similar while those of 'first-generation immigrants' were higher. As it may be expected, these were: 'construction', 'accommodation and food service activities', 'administrative and support services' and 'activities of households as employers'.

Especially in the case of the latter, 'first-generation immigrants' were 7.4% as compared to less than 1% for the other two groups. Similarly, in the hospitality sector, they were almost double than the other two categories. On the other hand, they were less represented in the categories

Table 7.4 Main activity sector of immigrants, by migration status and year, EU, 2014 and 2008 (%)

Activity	*2014*			*2008*		
	First-generation migrants	*Second-generation migrants*	*Native with native back-ground*	*First-generation migrants*	*Second-generation migrants*	*Native with native back-ground*
A-Agriculture	2.7	1.1	5.3	1.7	1.3	4.8
B-Mining and quarrying	0.3	0.2	0.5	0.3	0.3	0.5
C-Construction	8.4	6.0	7.0	11.1	7.0	8.1
C-Manufacturing	12.2	11.5	15.0	16.6	14.9	18.1
D-Electricity	0.3	0.8	0.8	0.3	0.7	0.8
E-Water supply; sewerage	0.6	0.7	0.9	0.6	0.5	0.8
G-Wholesale and retail trade; repair of motor vehicles and motorcycles	12.0	12.6	14.2	12.0	13.2	13.8
H-Transportation and storage	5.6	5.9	5.4	5.3	6.6	5.4
I-Accommodation and food service activities	9.5	4.1	3.7	8.8	3.2	3.2
I-Accommodation and food service activities	9.5	4.1	3.7	8.8	3.2	3.2
J-Information and communication	3.1	4.7	3.1	2.6	4.2	3.0
K-Financial and insurance activities	2.3	3.9	3.1	2.2	3.6	3.3
L-Real estate activities	0.8	1.3	0.8	0.7	1.0	0.7
M-Professional	5.0	7.8	5.5	3.9	6.5	4.9
N-Administrative and support service activities	6.2	4.7	3.5	6.9	4.0	3.2

Activity	*2014*			*2008*		
	First-generation migrants	*Second-generation migrants*	*Native with native back-ground*	*First-generation migrants*	*Second-generation migrants*	*Native with native back-ground*
O-Public administration and defence; compulsory social security	2.8	7.8	7.7	3.4	8.2	7.9
P-Education	5.6	8.6	8.0	4.6	7.4	7.4
Q-Human health and social work activities	9.9	12.2	9.8	9.0	10.9	9.8
R-Arts	1.5	2.3	1.6	1.5	2.0	1.4
S-Other service activities	2.8	2.6	2.3	2.6	2.8	2.2
T-Activities of households as employers;undifferentiated goods-and services-producing activities of household for own use	7.4	0.6	0.6	5.1	0.7	0.6
U-Activities of extraterritorial organisations and bodies	0.6	0.1	0.0	0.3	0.2	0.0
No answer	0.5	0.5	0.2	0.5	0.7	0.3

Source Eurostat, EU LFS AHM2014/2008

of 'education' and 'public administration and defence' (5.6% compared to 8.6% and 8%) and 'compulsory social security' (2.9% compared with 7.7% and 7.8%).

One of the consequences of the global financial crisis was the decline, between 2008 and 2014, in the number of migrants in general working in 'manufacturing' and 'construction'. While, during the same period, 'first-generation immigrants's' activities increased substantially (+2.3%) in the category of 'activities of households as employers'.

Looking at the differences between EU born and non-EU born 'first-generation immigrants' the biggest differences were in the categories of 'construction' and 'manufacturing'. In those activities, 'EU immigrants' were more represented by 3% (see Fig. 7.5). These two sectors were decreasing from 2008 to 2014 while still remaining two of the largest. Also, in 2014, more 'first-generation immigrants' of 'EU' than of 'non-EU origins' were employed in the 'professionals' sector (5.9% as opposed to 4.5%). Whereas, in 2014, the number of non-EU migrants was around 2% higher in the sectors of 'human health and social work', 'accommodation and food service', 'wholesale and retail' and 'activities of households as employers'.

In 2014, the most common sector of employment for first-generation immigrants in the EU 28 and in the Czech Republic, Estonia, Croatia, Italy, Latvia, Lithuania, Poland, Portugal, Slovenia and Slovakia was 'manufacturing' (Table 7.6).

The second sector employing the most 'first-generation immigrants' in the EU was wholesale and retail, which however ranked first only in Hungary while for Italy, Latvia, Lithuania and Luxembourg it did not feature in their top three activity sectors.

Human health and social work activities ranked third as a sector of activity for 'first-generation immigrants' in the EU, with France, Sweden and the UK having it ranked first.

Very important for Spain, Cyprus and Italy are immigrants working in the 'activities of household as employers' while Greece and Malta had the highest percentage of immigrants employed in the tourist sector.

Unique was the situation of Luxembourg where the three top activities for migrants were: 'financial and insurance activities', 'activities of extraterritorial organisations and bodies' and 'professional' activities.

Table 7.5 Main activity sectors of first-generation immigrant employees by origin, EU, 2014 and 2008 (%)

Activity	*2014*		*2008*	
	EU origins	*Non-EU origins*	*EU origins*	*Non-EU origins*
A-Agriculture	3.1	2.5	2.3	2.0
B-Mining and quarrying	0.3	0.3	0.3	0.2
C-Construction	10.5	7.3	14.1	12.0
C-Manufacturing	14.1	11.1	15.1	13.4
D-Electricity	0.3	0.3	0.3	0.2
E-Water supply; sewerage	0.7	0.6	0.5	0.4
G-Wholesale and retail trade; repair of motor vehicles and motorcycles	10.7	12.8	10.7	12.4
H-Transportation and storage	5.4	5.7	5.0	5.5
I-Accommodation and food service activities	8.0	10.1	7.2	9.5
J-Information and communication	2.9	3.2	2.8	2.9
K-Financial and insurance activities	2.6	2.1	2.9	2.1
L-Real estate activities	0.8	0.7	1.0	0.7
M-Professional	5.9	4.5	5.2	3.5
N-Administrative and support service activities	6.0	6.3	5.2	6.6
O-Public administration and defence; compulsory social security	2.6	3.0	3.4	3.3
P-Education	6.2	5.3	6.0	4.2
Q-Human health and social work activities	8.1	10.8	8.0	9.0
R-Arts	1.9	1.4	1.9	1.3

(continued)

Table 7.5 (continued)

Activity	*2014*		*2008*	
	EU origins	*Non-EU origins*	*EU origins*	*Non-EU origins*
S-Other service activities	2.6	2.8	2.2	2.5
T-Activities of households as employers; undifferentiated goods-and services-producing activities of household for own use	6.4	8.0	4.7	7.4
U-Activities of extraterritorial organisations and bodies	0.8	0.5	0.6	0.2
No answer	0.1	0.6	0.5	0.7

Source Eurostat, EU LFS AHM2014/2008

In the EU, self-employment seems to be mostly confined to the native population.[49] In 2014, 18.9% of those who were self-employed were natives. In relation to 'second-generation immigrants' only 14.6% with an 'EU origin' were self-employed, with the rate being 11.5% for 'non-EU origin' migrants. Moreover, the self-employment percentage of 'first-generation immigrants' was 16.7% for EU migrants and 15.5% for non-EU ones. These figures have increased for non-EU 'first-generation immigrants' since 2008 (Fig. 7.4).

In countries like Greece, Italy and Cyprus, the percentage of immigrants in self-employment was lower than that of natives (EU, 2015).

In 2017, across the OECD and the EU, the percentage of immigrants in self-employment was around 12% of the total immigrant population, the same rate as for natives. However, great variation exists among countries. For example, there were more immigrants in self-employment than

[49] Self-employment Definition: The self-employed are people who work in their own firms or create their own businesses, sometimes hiring employees. Self-employment includes entrepreneurs, liberal professions, artisans, traders and many other freelance activities. OECD 2018.

Table 7.6 Top three activities of first-generation immigrant employees, 2014

	First	Second	Third
EU-28	Manufacturing	Wholesale and retail trade	Human health and social work activities
Belgium	Administrative and support service activities	Wholesale and retail trade	Manufacturing
Bulgaria	:	:	:
Czech Republic	Manufacturing	Wholesale and retail trade	Construction
Denmark	:	:	:
Germany	:	:	:
Estonia	Manufacturing	Wholesale and retail trade	Transportation and storage
Ireland	:	:	:
Greece	Accomodation and food service activities	Wholesale and retail trade	Construction
Spain	Activities of households as employers	Accomodation and food service activities	Wholesale and retail trade
France	Human health and social work activities	Wholesale and retail trade	Construction
Croatia	Manufacturing	Construction	Wholesale and retail trade
Italy	Manufacturing	Activities of households as employers	Construction
Cyprus	Activities of households as employers	Wholesale and retail trade	Accomodation and food service activities
Latvia	Manufacturing	Transportation and storage	Education
Lithuania	Manufacturing	:	:
Luxembourg	Financial and insurance activities	Activities of extraterritorial organisations and bodies	Professional
Hungary	Wholesale and retail trade	Manufacturing	Accomodation and food service activities
Malta	Accomodation and food service activities	Wholesale and retail trade	:
Netherlands	:	:	:
Austria	:	:	:
Poland	Manufacturing	Wholesale and retail trade	:
Portugal	Manufacturing	Wholesale and retail trade	Human health and social work activities
Romania	:	:	:
Slovenia	Manufacturing	Construction	Wholesale and retail trade
Slovakia	Manufacturing	:	:
Finland	Manufacturing	Human health and social work activities	Wholesale and retail trade
Sweden	Human health and social work activities	Education	Wholesale and retail trade
United Kingdom	Human health and social work activities	Wholesale and retail trade	Manufacturing

Note Population aged in 23–54. 'wholesale and retail' trade also includes repair of motors vehicle and motorcycles
Source Eurostat, EU LFS AHM2014/2008

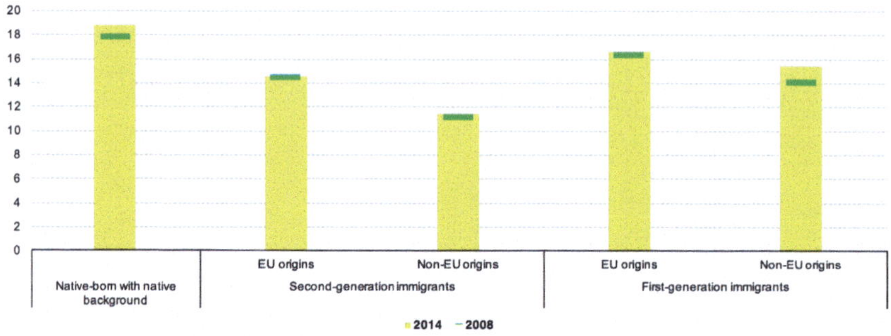

Fig. 7.4 Self-employment by migration status and origin, EU, 2008 and 2014 (%) (*Note* Population aged 25–54. *Source* Eurostat, EU LFS AHM2014/2008)

native workers in Central and Eastern Europe, especially in Poland, with a percentage twice as that of locals. On the other hand, in southern Europe, Japan, Korea and the Latin American OECD countries, immigrants were underrepresented among the self-employed. For example, in Greece, Italy and Iceland the number of self-employed among the natives was twice that of the migrants, and in Korea four times (OECD, 2018).[50] Immigrant businesses in OECD countries are however smaller than the business of native-born individuals, apart from in Australia, New Zealand, Central Europe and the Baltic countries (OECD, 2018). This is especially true in the EU, where up to 75% of immigrant businesses have no employees, in comparison with seven native businesses in ten. Moreover, in 2017 in the EU, central and eastern Europe, more native-owned businesses than immigrant-owned businesses had over 10 employees, especially in Luxembourg, Denmark and Switzerland.

All immigrants from outside the EU were more likely than those from the EU to hold temporary employment contracts.[51] In 2014, while only

[50] https://www.intlnursemigration.org/wp-content/uploads/2019/08/OECD-2018-Immigrant-Integration.pdf.

[51] Definition: In most countries, temporary work denotes any kind of wage-earning employment governed by a fixed-term contract, including apprenticeships, temporary employment agency work, and remunerated training courses. In Australia, temporary work is defined as work without paid leave. No such definition of temporary work exists in the United States (OECD, 2018).

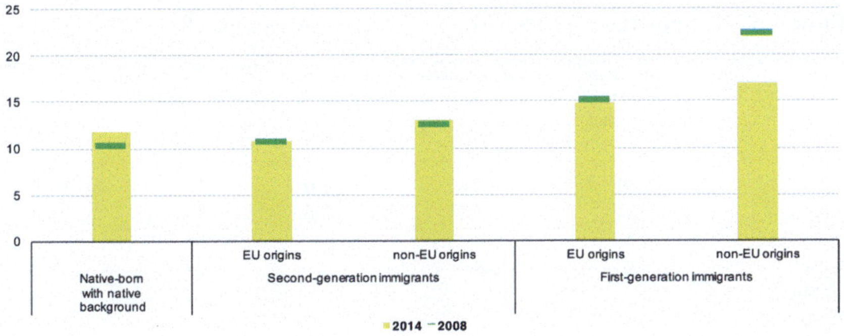

Fig. 7.5 Temporary contract by migration status and origin, 2008 and 2014 (%) (*Note* Population aged 25–54. *Source* Eurostat, EU LFS AHM2014/2008)

11.9% of natives had a temporary contract, the rate was 17.0% for 'first-generation immigrants' from outside the EU (Fig. 7.5).

This trend was true for all countries within the EU, where 'first-generation immigrants' have always had higher rates of temporary contracts than natives, pointing to a higher degree of precariousness of immigrants, and of non-EU immigrants in particular, as opposed to natives.

In the majority of EU Member States, those with the highest number of temporary contracts were immigrants from outside the EU. In France, the difference between non-EU and EU immigrants was 8.9%, in Belgium 6.8%, in Sweden 6.7% and in Luxembourg 6.2%. While in other EU countries the variation was less than 5% and in Italy, the Czech Republic, Austria, Spain and Portugal, EU migrants had more temporary contracts than non-EU ones. It must be noted however that in southern European countries, given the relevance of irregular jobs, it is fairly likely that many non-EU migrants have no contracts at all.

Table 7.7 shows that overall in the EU and in most Member States, apart from France and Hungary, 'first-generation immigrant' employees of 'EU origins' recorded a higher percentage of temporary contracts in 2014 than 'native-born with native background'. In Spain, Belgium, Greece, Portugal, Italy and Cyprus the variation between EU 'first-generation immigrants' and natives was significant (e.g. 12% in the case of Spain).

Table 7.7 Employees with temporary contracts, by origin, 2014 (%)

	Native-born with native background	*Second-generation immigrants*		*First-generation immigrants*	
		EU	*Non-EU*	*EU*	*Non-EU*
-28	11.9	10.8	13.0	14.9	17.0
Belgium	5.4	7.9	12.6	13.5	20.3
Bulgaria	4.8	–	–	–	–
Czech Republic	7.0	8.7	:	11.7	9.2
Denmark	–	–	–	–	–
Germany	–	–	–	–	–
Estonia	2.2	–	–	0.0	–
Ireland	–	–	–	–	–
Greece	10.3	11.6	26.6	18.6	18.8
Spain	21.8	30.6	26.4	33.5	31.8
France	12.1	12.4	21.1	11.8	20.6
Croatia	14.6	18.7	20.1	–	19.3
	12.3	17.7	–	19.2	14.6
Cyprus	10.1	16.9	–	16.0	–
Latvia	3.3	–	–	–	–
Lithuania	2.4	–	–	–	–
Luxembourg	5.4	–	–	7.6	13.7
Hungary	10.0	–	–	5.5	–
Malta	6.0	0.0	–	–	–
Netherlands	–	–	–	–	–
Austria	4.3	6.1	–	7.8	5.9
Poland	26.1	18.3	–	–	–
Portugal	18.9	34.9	–	25.1	23.5
Romania	1.2	–	–	–	–
Slovenia	12.3	15.9	28.6	–	22.4
Slovakia	7.1	–	–	–	–
Finland	12.5	12.4	–	18.0	20.3
Sweden	8.9	10.9	–	11.8	18.5
UK	3.2	3.4	4.0	7.0	7.3

Note Population aged 25–54
Source Eurostat, EU LFS AHM2014/2008

In 2015–2016, across the OECD and the EU, 15% of foreign workers were on temporary contracts, as opposed to 16% and 12% among native workers, respectively. In the EU, non-EU migrants have the highest share of temporary contracts, with 18%. That was not the case, however, in most central and eastern European countries, the UK, Austria and Italy

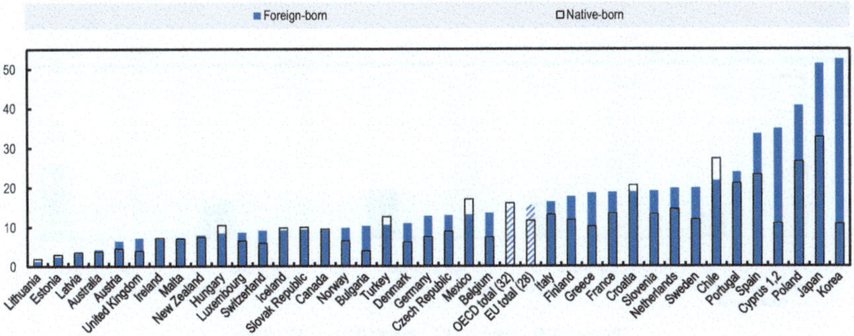

Fig. 7.6 Workers on temporary contracts, percentage of all wage-earners, 15- to 64-year-olds, 2015–2016 (*Source* OECD 2018, https://www.intlnursemigration.org/wp-content/uploads/2019/08/OECD-2018-Immigrant-Integration.pdf)

where EU-born migrants had slightly more temporary contracts (OECD, 2018).

As shown in Fig. 7.6, in around half of the EU countries the share of temporary contract workers is at least 5% higher among migrants. This gap is especially wide in Spain, Greece and Poland. However, temporary work of migrants is uncommon in most central and eastern European countries, where it accounts for less than 10% of immigrant employment. Over the period 2005–2006 and 2015–2016 there does not seem to have been a significant change in temporary contracts as a share of employment arrangements (OECD, 2018).

Another indicator of the higher precariousness for immigrants workers in comparison with the local labour force, which may also hint at a lack of competition between the two, is the fact that in 2014 'first-generation immigrants' of both 'EU origins' and 'non-EU origins' had the highest percentages of part-time employment.[52]

Figure 7.7 below indicates the prevalence of part-time work among both first and second-generation female immigrants, with a gap of up to 18% with the male population. Again, the rate of part-time jobs was higher for 'first-generation immigrants', both from the EU and from

[52] Part-time employees are persons whose usual working hours are less than the normal working hours. It can be either voluntary (e.g. for family reasons) or involuntary (when the person would like to work more hours but cannot find a suitable contract). This analysis does not distinguish between the two because of sample size limitations.

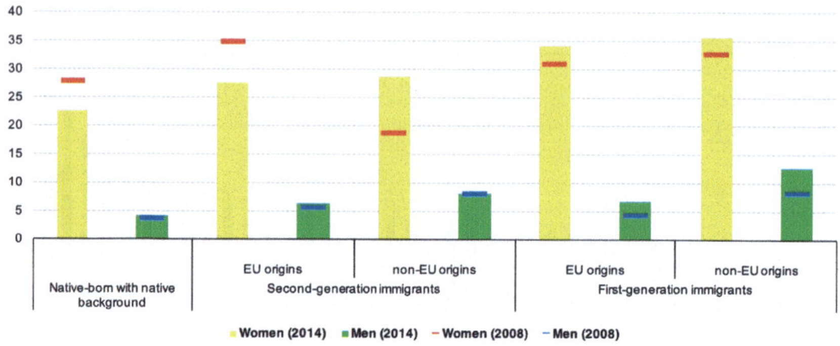

Fig. 7.7 Part-time employment by migration status, sex and origin, EU, 2008 and 2014 (%) (*Note* Population aged 25–54. *Source* Eurostat, EU LFS AHM2014/2008)

outside the EU. This rate represented more than one-third of female 'first-generation immigrants' and 6.7% ('EU origins') and 12.7% ('non-EU origins') of male ones in 2014. For the male employees, those with the highest rate of part-time positions, more than three times that of native men, were 'first-generation immigrants' from outside the EU, testifying their higher degree of vulnerability. As may be expected overall, part-time positions were higher among both male and female 'first-generation immigrants' with 'non-EU origins'.

In relation to 2008, in 2014 part-time work for 'first-generation immigrants' of both genders and from either the EU and outside the EU increased. Although the data does not include whether such part-time contracts were voluntary or involuntary, this could well point to a worsening of 'first-generation immigrants' position in the labour markets of receiving countries after the global economic crisis.

At the member state level, the situation was quite variegated. However, in all Member States, apart from Cyprus, Luxembourg and Austria, 'first-generation immigrants' from outside the EU had a higher rate of part-time jobs than any other category, with the highest gaps being in the UK (7.4%) and Italy (4.3%).

The difference between part-time positions in the native labour force and in the 'non- EU first-generation immigrant' category was significant in Greece (13.6%), Spain (10.7%) and Italy (12%). In general, 20.4% of 'first-generation immigrants' from the EU and 23.3% from outside the EU had a part-time job in the EU 28 in 2014. This percentage was around 25% or higher in Belgium, Italy, Spain and Austria for both EU and non-EU 'first-generation immigrants' (see Table 7.8). Only in

Table 7.8 Part-time employment, by origin, 2014 (%)

	Native-born with native background	*Second-generation immigrants*		*First-generation immigrants*	
		EU	*Non-EU*	*EU*	*Non-EU*
EU-28	13.2	16.7	18.0	20.4	23.3
Belgium	24.0	24.8	20.4	24.7	27.5
Bulgaria	1.4	–	–	–	0.0
Czech Republic	4.1	5.4	–	6.7	5.2
Denmark	–	–	–	–	–
Germany	–	–	–	–	–
Estonia	6.5	6.0	–	–	–
Ireland	–	–	–	–	–
Greece	7.8	13.7	–	22.8	21.4
Spain	14.9	19.6	25.6	26.0	25.6
France	16.6	16.1	21.9	21.4	20.8
Croatia	2.0	–	–	–	7.8
Italy	18.0	19.6	–	25.7	30.0
Cyprus	9.2	13.3	–	10.5	5.2
Latvia	4.1	–	–	–	–
Lithuania	5.6	–	–	–	–
Luxembourg	21.8	18.5	–	14.2	17.4
Hungary	5.2	10.7	–	–	22.3
Malta	13.4	–	–	–	–
Netherlands	–	–	–	–	–
Austria	28.2	27.3	31.7	31.0	25.5
Poland	4.7	6.6	–	–	–
Portugal	5.6	–	–	–	11.1
Romania	0.7	–	–	–	–
Slovenia	5.7	5.4	8.2	–	8.2
Slovakia	4.6	–	–	–	–
Finland	8.4	–	–	11.0	14.9
Sweden	19.5	21.0	–	21.5	20.9
UK	21.5	20.5	18.6	16.1	23.5

Note Population aged 25–54
Source Eurostat, EU LFS AHM2014/2008

Cyprus, Slovenia, Croatia and the Czech Republic did the percentage of non-EU 'first-generation immigrants' with part-time jobs fall below 10%.

In general, in the majority of OECD countries more immigrants, especially women, are employed in part-time jobs (Fig. 7.8).

Finally, another indicator of the lack of competition with the native labour force is the share of atypical jobs, defined as jobs with atypical working hours, such as working evenings, nights or the weekend. Here again, the rate for immigrant statuses was higher than native populations. Men were also more likely than women to have atypical working hours. As it could be expected, given their higher degree of vulnerability, the highest rate of atypical jobs was 54.4% for 'first-generation immigrants' of 'non-EU origins'. Even 'first-generation immigrant' women from outside the EU recorded a rate of 49.5% in 2014 (see Fig. 7.9). As in the case of part-time jobs, atypical work also generally increased between 2008 and 2014.

Looking at the differences between Member States, the highest shares (over 70%) among EU 'first-generation immigrants' of 'EU origins' were recorded in Croatia (87.2%). For 'second-generation immigrants' of 'non-EU origin', Greece had the highest shares (76.0%) (see Table 7.9).

Overall in the EU 28, the highest rates of atypical work were registered again among 'first-generation immigrants' of 'non-EU origin' (52.1%).

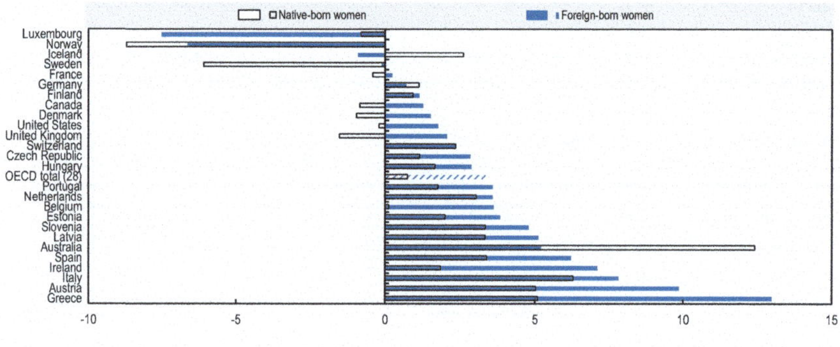

Fig. 7.8 Change in the share of women working part-time between 2007 and 2016 Percentage points, 15–64 (*Sources* OECD/EU (2018), Settling In 2018: Indicators of Immigrant Integration, OECD Publishing, Paris/EU, Brussels, https://doi.org/10.1787/9789264307216-en.OECD/EU 2019)

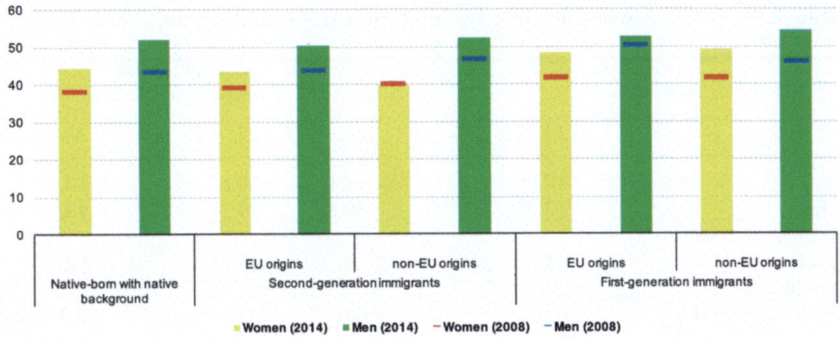

Fig. 7.9 Atypical working time by migration status, sex and origin, EU, 2008 and 2014 (%) (*Note* Population aged 25–54. *Source* Eurostat, EU LFS AHM2014/2008)

Although this was not the case for the UK, Croatia and Poland, where immigrants from the EU had higher shares.

In 2015–2016, the percentage of foreign-born workers likely to be employed in jobs generating physical health risks (46% average) was higher than for native workers (35% on average) in all European countries. The gap was at least 20% in Germany, Slovenia, Estonia and Sweden (OECD, 2018). Only in Norway and Denmark did immigrants have an occupational risk similar to that of native-born workers. Such employment is typically low skilled, and in fact, around three in five low educated immigrants had physically risky jobs in 2015, as opposed to only half of their native-born peers. However, also for highly skilled jobs, the percentage of migrants in occupational risk jobs is higher than that of native workers (Fig. 7.10).

Moreover, migrant workers are overrepresented among the unemployed in all OECD countries except in the Slovak Republic, Mexico, Poland, Hungary and Greece.[53] In the Nordic countries, and in Switzerland and Belgium, the unemployment rate for those who are foreign-born is more than twice that of the native-born. In France, it is 7.6% points higher and in Germany 9.4% (Fig. 7.11 and Table 7.10).

The economic recession of 2008–2009 seems to have heavily impacted the unemployment rate of immigrants who, in the period between 2009

[53] Ibid., p. 114.

Table 7.9 Atypical working time by migration status and origin, 2014 (%)

	Native-born with native background	*Second-generation immigrants*		*First-generation immigrants*	
		EU	*Non-EU*	*EU*	*Non-EU*
EU-28	48.4	47.0	45.5	50.6	52.1
Belgium	42.5	43.7	50.8	39.1	43.1
Bulgaria	39.1	–	–	–	–
Czech Republic	43.9	53.8		45.3	53.3
Denmark	0.0	0.0	0.0	0.0	0.0
Germany	0.0	0.0	0.0	0.0	0.0
Estonia	42.3	48.0	55.5	–	55.4
Ireland	0.0	0.0	0.0	0.0	0.0
Greece	56.0	73.7	76.0	64.5	69.8
Spain	47.0	45.8	43.4	45.0	53.8
France	44.4	45.0	45.7	40.3	43.4
Croatia	69.2	62.4	59.8	87.2	75.7
Italy	42.7	41.3	75.8	48.8	52.3
Cyprus	36.1	38.3	–	57.6	76.5
Latvia	37.4	44.9	34.2	–	40.2
Lithuania	41.0	35.8	54.3	–	37.1
Luxembourg	46.8	45.9	–	45.1	48.8
Hungary	47.1	44.0	–	48.6	44.9
Malta	47.9	–	–	55.0	56.2
Netherlands	0.0	0.0	0.0	0.0	0.0
Austria	48.0	50.0	50.6	55.9	56.1
Poland	61.2	62.8	49.1	74.9	57.5
Portugal	47.8	48.7	64.9	50.6	52.1
Romania	48.2	–	–	–	:
Slovenia	61.8	57.9	66.0	69.1	79.8
Slovakia	56.2	57.0	–	58.8	–
Finland	54.0	65.8	–	52.3	55.2
Sweden	34.1	34.3	–	33.8	41.4
UK	51.8	51.4	42.6	60.4	56.6

Note Population aged 25–54
Source Eurostat, EU LFS AHM2014/2008

and 2014, had an average unemployment rate that was 5% higher than native workers, at 15% in OECD countries (OECD, 2019). Particularly difficult is the situation in southern European countries (except Portugal) as well as in Sweden, Finland and France where more than 13% of

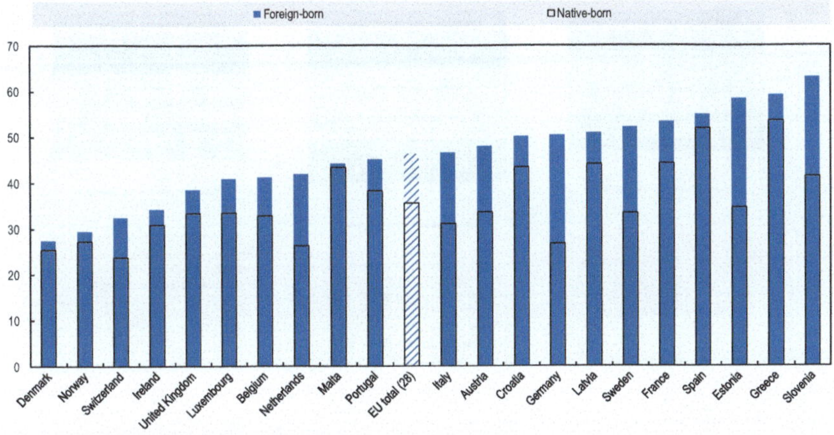

Fig. 7.10 Shares of the foreign- and native-born in occupations that put their physical health at risk Percentages of 15- to 64-year-olds in employment, 2015–2016 (*Sources* OECD/EU (2018), Settling In 2018: Indicators of Immigrant Integration, OECD Publishing, Paris/EU, Brussels, https://doi.org/10.1787/9789264307216-en 2019)

migrants were unemployed in 2018. Smaller gaps were recorded in the US, Australia and Canada (see Fig. 7.12).

Moreover, according to an analysis by the OECD (2018),[54] immigrants are more at risk of labour market exclusion.[55] Indeed, in the OECD, 2.2 million unemployed immigrants, roughly more than 30% of the total, could not find a job for at least one year. In 2015–2016 in the EU, 50% of unemployed non-EU workers, and 44% of EU ones, were long-term ones (almost 2 million). This is in stark contrast with 2005–2006 when immigrants were less likely, than native-born workers, to experience long-term unemployment across the OECD and the EU.

[54] https://www.intlnursemigration.org/wp-content/uploads/2019/08/OECD-2018-Immigrant-Integration.pdf.

[55] Definition: The long-term unemployment rate is the share of job seekers who have been without a job for at least 12 months among all the unemployed. Involuntarily inactive people are those who are not seeking work but are willing to take up work. They include, among others, discouraged workers, who are not seeking work because they believe no suitable jobs are available. OCED 2018.

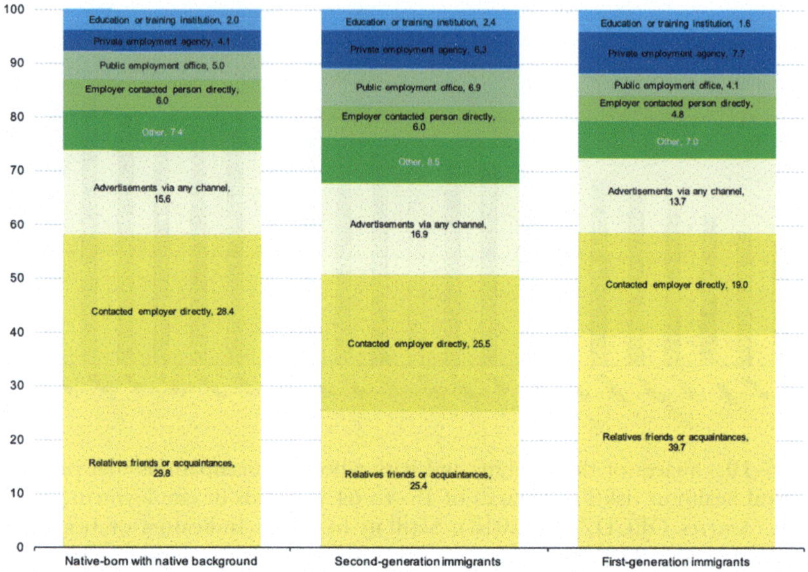

Fig. 7.11 Method to find a job by migration status, 2014 (%) (*Note* Population aged 25–54. *Source* Eurostat, EU LFS AHM2014/2008)

This is particularly true in countries hardly hit by the economic crisis, such as Ireland, Latvia, Greece and Spain. In Denmark, Sweden, Switzerland and Lithuania, the rates of foreign-born workers in long-term unemployment are more than 10 percentage points higher than those for natives. Moreover, inactivity is involuntary in one-quarter of those who are immigrants in the EU against one-sixth of the inactive native-born. This is less the case in the US, where involuntary inactivity is less than 10% among both native and foreign workers (Table 7.11 and Fig. 7.13).

Apart from unemployment and employment rates, the OECD records the number of younger immigrants who are neither in employment, education or training (NEET). In the EU in the period between 2013 and 2018, more than 18% of immigrants aged 15–24 were neither in employment, education or training. This is compared to 11% for their native counterparts in 2018. This share was lower in non-European OECD countries (Fig. 7.14) where this gap was smaller, except in New Zealand and Mexico

Table 7.10 Immigrants'labour market outcomes in OECD countries in 2018 (OECD, 2019 https://www.oecd-ilibrary.org/sites/c3e35eec-en/1/2/2/index.html?itemId=/content/publication/c3e35eec-en&_csp_=5484c834d3b947b42e43a8aee995b48b&itemIGO=oecd&itemContentType=book)

	2018		*Annual change*		*Gap with the native-born in 2018*	
	Percentages		*Percentage points*			
	Unemployment rate	*Employment rate*	*Unemployment rate*	*Employment rate*	*Unemployment rate*	*Employment rate*
Australia	5.5	72	−0.4	1.2	0.1	−3
Austria	9.4	68	−1.3	2.5	5.7	−6.4
Belgium	11.5	58.3	−1.8	1.8	6.8	−7.7
Canada	6.4	72.3	−0.4	−0.4	0.6	−2
Chile	7.5	76.9	1.7	3.1	−0.8	16.7
Czech Republic	2.5	79.4	−0.5	1.9	0.2	4.8
Denmark	9.8	66.4	-0.8	1.4	5.5	−10.6
Estonia	7.9	70.1	1.4	−1.6	2.7	−5.2
Finland	14.1	62.2	−1.7	1.9	7	−10.6
France	14.6	58.5	−0.8	1.9	6.3	−7.9
Germany	6	69.5	−0.4	1.4	3.1	−8.1
Greece	28.6	52.8	−1.3	0.1	10	−2.3
Hungary	4.6	71.7	1.2	−2	0.9	2.5
Iceland	5.1	82.5	2.4	−5.6	2.6	−2.8
Ireland	7.2	70.7	−1	1.7	1.8	2.7
Israel	3.5	78.8	−0.2	−0.1	−0.8	12.1
Italy	13.7	60.9	−0.5	1	3.4	2.8
Korea	4.6	70.9	–	–	–	–

(continued)

Table 7.10 (continued)

	2018		*Annual change*		*Gap with the native-born in 2018*	
	Percentages		*Percentage points*			
	Unemployment rate	*Employment rate*	*Unemployment rate*	*Employment rate*	*Unemployment rate*	*Employment rate*
Latvia	7.7	69	−0.3	2.4	0.1	−3
Lithuania	7.4	71.1	0.9	1	1.2	−1.4
Luxembourg	6.4	71.4	−1.5	0.6	2	9.6
Mexico	4.1	51.8	−0.1	−0.4	0.7	−9.7
Netherlands	7	64.9	−1.9	1.9	3.6	−14.3
New Zealand	4.1	77.2	−0.5	0.8	−0.5	−0.5
Norway	7.9	69.7	−1.2	0.3	5	−6.8
Poland	4.7	73	−3.6	3.1	0.8	5.7
Portugal	8.5	75.1	−1.5	0.8	1.4	6
Slovak Republic	n.r.	73.3	n.r.	3.5	n.r.	5.7
Slovenia	6.5	67	−2	0.7	1.4	−4.6
Spain	20.7	61.6	−2.7	1.9	6.5	−1
Sweden	15.7	66.7	0.2	0.4	11.8	−14.1
Switzerland	7.9	76.6	−0.1	0.7	4.5	−5.3
Turkey	12.1	47.4	−3.3	1.2	1	−4.7
UK	4.7	73.7	−0.5	1.2	0.7	−1.2
US	3.5	71.6	−0.5	0.7	−0.6	2.4
OECD average	8.7	68.3	−0.7	0.8	2.9	−2.4
OECD Total	7.1	69	−0.7	1	1.4	1.9
EU28	10.6	66	−1	1.6	4.1	−2.9

Source OECD (2019) https://www.oecd-ilibrary.org/sites/c3e35eec-en/1/2/2/index.html?itemId=/content/publication/c3e35eec-en&_csp_=5484c834d3b947b42e43a8aee995b48b&itemIGO=oecd&itemContentType=book

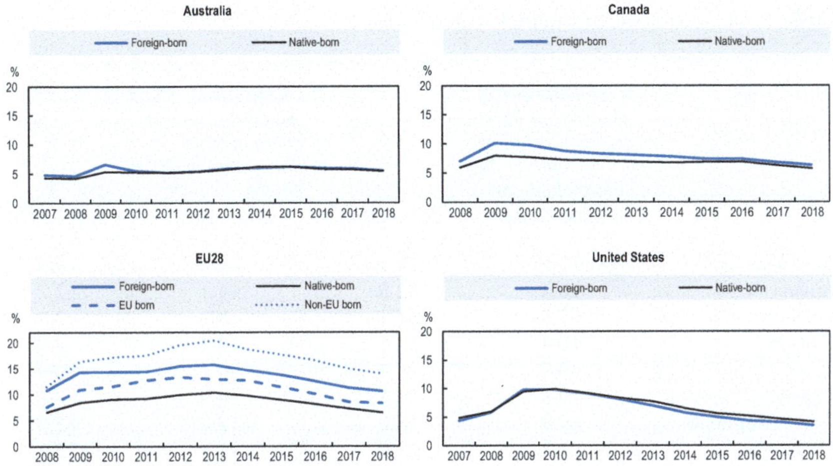

Fig. 7.12 Unemployment rates by place of birth, 2007–2018. Percentages of the active population aged 15–64 (*Note* The data for the EU28 refers to the first three quarters for the year 2018. The series on non-EU born and EU born exclude Germany. *Source* European countries: Labour Force Surveys (Eurostat); Australia, Canada: Labour Force Surveys; United States: Current Population Surveys)

Another more general example of discrimination in the integration of foreign-born workers in western labour markets is the fact that the majority of foreign-born workers are employed in jobs below their qualifications, or, in other words, they are overqualified for their job placements. Indeed, still in 2018, the share of immigrant workers with tertiary education in low- and medium-skilled jobs remained disproportionally high across OECD countries (OECD 2019). This is detailed in the figure below, which shows how, apart from the case of Switzerland, the rate of over-qualification is systematically higher among highly educated immigrant workers than their native peers. Looking at OECD countries, in 2018 the average gap between foreign and native-born workers increased by 1.5 percentage points to 12% from 2007. This gap is higher in the UK, Germany, Austria and Denmark while it decreased substantially only in Greece and Spain (Fig. 7.15).

Huge differences also exist in the labour market outcomes of immigrants by regions of origin. Most notably in the EU in the period between 2013 and 2018, migrants from the Middle East appear to be

Table 7.11 Long-term unemployment rate

	Long-term unemployment of the foreign-born population (% of total unemployment)		*Differences with the native-born (% points) +: higher than natives −: Lower than natives*	
	2006–2007	*2015–2016*	*2006–2007*	*2015–2016*
Australia	17.9	24.3	+1.4	−0.3
Austria	30.4	32.5	+4.1	+2.7
Belgium	57.2	57.1	+8.5	+8.3
Canada	10.4	13.5	+3.2	+2.8
Croatia	60.4	58.4	+0.3	+0.9
Cyprus1,2	19.6	39.2	+0.9	−7.7
Czech Republic	69.9	48.8	+17.0	+7.8
Denmark	20.1	33.8	+1.8	+12.2
Estonia	58.8	38.3	+12.1	+3.9
Finland	32.0	28.1	+10.1	+1.8
France	45.7	49.6	+7.1	+7.7
Germany	56.7	57.7	−0.1	−0.3
Greece	44.5	71.1	−8.2	−1.6
Hungary	41.9	53.8	−4.2	+8.1
Iceland	..	13.1	−0.1	+3.5
Ireland	24.5	52.3	−9.6	−5.5
Israel	..	12.5	..	−0.7
Italy	41.2	55.2	−8.3	−4.1
Korea	–	2.1	..	+0.8
Latvia	28.2	50.5	−2.8	+7.8
Lithuania	..	54.3	+2.8	+14.3
Luxembourg	29.8	30.6	+1.7	+2.1
Malta	–	48.1	−7.7	+6.4
Netherlands	50.2	50.3	+10.8	+9.9
New Zealand	10.4	9.5	−0.8	−2.4
Norway	31.1	34.4	+13.1	+7.9
Portugal	42.2	51.9	−7.2	−5.0
Slovenia	54.8	57.7	+7.9	+5.7
Spain	11.9	48.2	−11.1	−0.3
Sweden	18.7	27.6	+6.6	+13.0
Switzerland	46.3	43.6	+16.2	+14.3
Turkey	–	21.9	+0.0	−2.6
UK	24.0	24.1	+1.0	−5.0
US	6.6	11.8	+0.2	+0.4

(continued)

Table 7.11 (continued)

	Long-term unemployment of the foreign-born population (% of total unemployment)		*Differences with the native-born (% points) +: higher than natives −: Lower than natives*	
	2006–2007	*2015–2016*	*2006–2007*	*2015–2016*
OECD total (29)	29.2	37.3	−2.1	+4.6
EU total (28)	41.3	48.4	−3.7	+0.1

Note 2012–2013 data for Turkey. Turkey is not included in the OECD total
Source https://www.intlnursemigration.org/wp-content/uploads/2019/08/OECD-2018-Immigrant-Integration.pdf

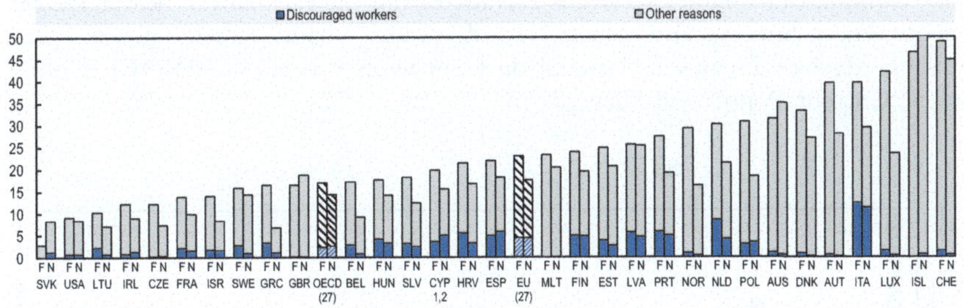

Fig. 7.13 Involuntary inactivity due to discouragement or other reasons Percentages among the inactive foreign-born (F) and native-born (N), 15- to 64-year-olds, 2015–16 (*Source* https://www.intlnursemigration.org/wp-content/uploads/2019/08/OECD-2018-Immigrant-Integration.pdf)

less integrated than migrants from other countries of origin. In 2018, the unemployment level of migrants from the Middle East was still as high as in 2013 at around 22%.

Similarly, migrants from the Northern African region in the EU registered unemployment levels above 21% in 2018, much higher than that of sub-Saharan migrants. Also in Australia in 2018, the migrants presenting the highest rates of unemployment were those from North Africa and the Middle East. They recorded an unemployment rate twice the average of foreign-born migrants at nearly 11%. While in Canada, migrants from the MENA area and sub-Saharan Africa topped the list.

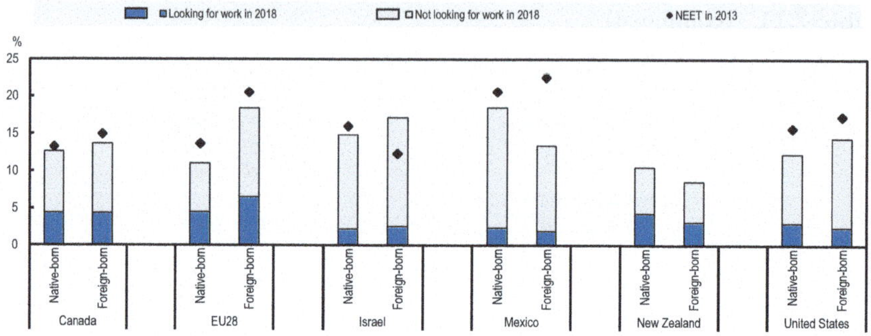

Fig. 7.14 NEET rates by place of birth in selected OECD countries, 2013 and 2018 Share of the population aged 15–24 that is not in employment, education or training (*Note* The data for European countries refers to the first three quarters only. Compulsory military service is excluded from the calculation. *Source* EU28: Labour Force Surveys (Eurostat); New Zealand, Canada, Israel: Labour Force surveys; Mexico: Encuesta Nacional de Ocupación y Empleo (ENOE); United States: Current Population Surveys)

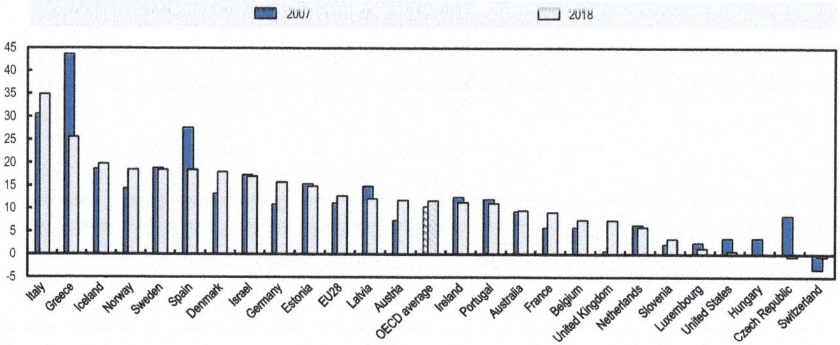

Fig. 7.15 Differences in over-qualification rates between foreign- and native-born workers, 2007 and 2018 Percentage points (*Note* The reference population are persons with a high education level aged 15–64 who are not in education, except in Israel where the calculation includes persons in education. The data for European countries and Turkey refers to the first three quarters only in 2018. The data for Australia refers to the years 2014 and 2017. *Source* European countries: Labour Force Survey (Eurostat); United States: Current Population Survey; Israel: Labour Force Survey. OECD, 2019)

In the US in 2018, the level of unemployment for Mexican and South American migrants increased by 5 percentage points from 2013 (see Table 7.12).

Another indicator of the poor performance of migrants in labour markets in the wake of the 2008–2009 economic crisis is the increase in the share of immigrant workers living below the poverty threshold. This occurred in all EU countries and at a higher rate than for native workers.

In 2017, the European Union recorded a percentage of around 18% of immigrant workers aged 18- to 64-years-old living below the poverty line as opposed to only 8% of the local workforce (Fig. 7.16). The gap between the two populations actually increased from 6 to 10% in the last ten years. This rise has been especially marked in Spain and Italy, where about 30% of foreign-born workers were poor in 2017–2018. Similarly, in Denmark, Germany and the Netherlands, poverty rates increased at a fast

Table 7.12 Employment and unemployment rates by region of origin in selected OECD countries in 2013 and 2018

Percentages

	Region of birth	*2013*	*2018*	*2013*	*2018*
	Other Oceania	75.7	77.2	6.2	5.7
	Europe	73.9	77.9	4.5	4
	North Africa and the Middle East	47.7	50.9	12.1	10.7
	Sub-Saharan Africa	74.1	75.6	6.1	6.9
Australia	Asia	66.7	69.4	6.4	5.7
	Americas	73.7	79.1	5.3	5
	Foreign-born (total)	69.7	72	5.9	5.2
	Native-born	73.3	74.9	5.8	5.4
	Sub-Saharan Africa	65.3	69.9	12.3	9.5
	North Africa	60.8	69.5	14.8	8
	Middle East	59.1	61.4	12	9.7
	Asia	69.4	72.6	8.1	5.7
	Europe	74.3	77.3	5.8	5.4
Canada	Oceania	79.2	76.4	5.8	3.4
	Central and South America and Caribbean	71.8	73.2	8.7	7
	Other North America	70.8	69.5	6.4	5.1
	Foreign-born (total)	69.9	73.8	8.2	5.9

(continued)

Table 7.12 (continued)

Percentages					
	Region of birth	*2013*	*2018*	*2013*	*2018*
	Native-born	73.2	74.4	6.9	5.7
	EU28 and EFTA	66.2	72	13.5	8.3
	Other European countries	54.9	62.2	19.7	13
	North Africa	45.6	50.3	28.9	21.1
	Sub-Saharan Africa	58.7	64.9	21.1	14
EU28 countries	Middle East	50.6	50.2	22	22
	North America	69.1	70.8	6.4	6.9
	Central and South America and Caribbean	56.8	64.7	27.2	16.1
	Asia	64.3	66.1	10.4	6.9
	Other regions	62.6	66.2	11.4	11.2
	Foreign-born (total)	60.9	65	17.1	12
	Native-born	64.3	67.4	10.3	7.2
		66.2	70.9	7.7	3.7
	Other Central American countries	73.6	74.9	6.5	3.5
	South America and Caribbean	69	73.6	8.7	4.1
	Canada	73.2	71.3	6.2	2.7
US	Europe	70.6	75.4	6.2	3
	Africa	66.9	71.4	9.4	4.5
	Asia and the Middle East	68.1	69.2	5.3	3
	Other regions	63.6	68.8	7.8	4.6
	Foreign-born (total)	68.4	71.6	7	3.5
	Native-born	65.7	69.2	7.7	4.1

Source OECD (2019)

pace in the last ten years, although not as much as in southern Europe (Fig. 7.16).

Behind the poverty levels of immigrant workers are precisely the phenomena described above. In particular, such a percentage of immigrant workers are subjected to poverty due to their large concentration in low-skilled jobs, their working conditions being below standard in terms of hours worked, their contract type and in general their more vulnerable and precarious position in the labour markets of receiving countries.

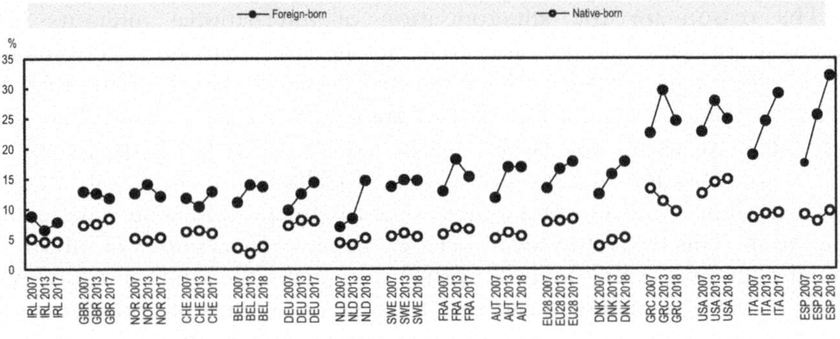

Fig. 7.16 Poverty rates of workers by place of birth in selected OECD countries, 2007, 2013 and 2017–2018 (*Note* The poverty rate used here is the share of workers living below the poverty threshold as defined by Eurostat (60% of the median equivalised disposable household income in each country). *Sources*: European countries: Eurostat dataset (population aged 18–64) [ilc_iw16] extracted on 10 July 2019; United States: Current Population Survey (population aged 15–64)

Concluding, the data above clearly demonstrates that foreign-born populations in general, and Muslim migrants in particular, endure labour market conditions far below those of their native counterparts. On average they experience higher unemployment than the native-born population for each level of education. Furthermore, those with North African origins are the most overrepresented in terms of unemployment rates. Moreover, more than locally born workers, foreign-born labourers are on average employed in the low-skilled agricultural, care and industrial sectors, and they are often overqualified for their job placements. It seems therefore plausible that their positioning in the labour markets of receiving countries is complementary with that of the native workers.

Conclusion

The discussion above clarifies that international migrant workers can be beneficial to the economies of receiving countries as they cover unfilled positions in the labour market of host countries or complement the skills of the local labour force, thus increasing the overall productivity and efficiency of the economy.

The reason for the antagonisation of international migrants by receiving societies then has more to do with politics than with economics. On the one hand, as we will analyse in Chapter 9, governments might want to capitalise on the fact that migrants do not vote, using them as scapegoats for their citizens' discontent. On the other hand, the paradox of the securitisation of migration within globalisation makes legal migration very limited, promoting a process of irregularisation of international migration. This feeds into local societies' negative perceptions of international migrants, further adding to their marginalisation and vulnerability. This, however, does not mean that they do not contribute positively to the economic life of their receiving countries. This happens despite the fact that they are often underemployed, are relatively less employed than local population and have to accept working conditions below standards, especially if they are irregular migrants, as is often the case for unskilled workers.

In the next chapter, the question of integration of migrants in host societies will be analysed more in depth by looking at the role of social capital in favouring integration and at the relationship between social capital and Islamophobia.

References

Acemoglu, D. (2002, March). Technical change, inequality, and the labor market. *Journal of Economic Literature, 40*(1), 7–72.

Ambrosini, M. (2016). From "illegality" to tolerance and beyond: Irregular immigration as a selective and dynamic process. *International Migration, 54*, 144–159. https://doi.org/10.1111/imig.12214.

Ambrosini, M. (2018). *Irregular immigration in Southern Europe (migration, diasporas and citizenship)* (Kindle Locations 170–171). Springer International Publishing. Kindle Edition.

Anderson, B. (2010). Migration, immigration controls and the fashioning of precarious workers. *Work, Employment and Society, 24*(2), 300–317.

Anderson, B. (2008). *llegal immigrant': Victim or Villain?* (COMPAS, Working Paper No. 64 [WP-08-64]). University of Oxford.

Bacon, D. (2008, September 1). *Illegal people: How globalization creates migration and criminalizes Immigrants*. Boston: Beacon Press.

Bakewell, O. (2008). Research beyond the categories. The importance of policy-irrelevant research into forced migration. *Journal of Refugee Studies, 12*, 432–453.

Baldwin-Edwards, M., & Arango, J. (Eds.). (1999). *Immigrants and the Informal Economy in Southern Europe*. Routledge.

Baldwin-Edwards, M. (2008). Towards a theory of illegal migration: Historical and structural components. *Third World Quarterly, 29*(7), 1449–1459.

Basch, L., N., Schiller, G., & Szanton Blanc, C. (1994). *Nations unbound: Transnational projects, postcolonial predicaments and deterritorialised nation States*. London: Routledge.

Betts, A., (2014). Commentary. In Hollifield et al. (Ed.), *Controlling immigration* (Kindle Edition). Stanford University Press.

Betts, A. (Ed.). (2011a). *Global migration governance*. Oxford University Press.

Betts, A. (2011b). The global governance of migration and the role of trans-regionalism. In R. Kunz, S. Lavenex, & M. Panizzon (Eds.), *Multi-layered migration governance*. Routledge.

Bigo, D. (2002). Security and Iimmigration. Toward a critique of the governmentality of unease. *Alternatives: Social Transformation and Humane Governance, 27* , 63–92.

Boccagni, P. (2009). *Tracce transnazionali. Vite in Italia e proiezioni verso casa tra I migranti ecuadoriani*. Franco Angeli.

Borjas, G. J. (2014). *Immigration economics*. Harvard University Press.

Borjas, G. J. (2015). Immigration and globalization: A review article. *Journal of Economic Literature, 53*(4), 961–974. http://dx.doi.org/10.1257/jel.53.4.961. Accessed 13 September 2019.

Borjas, G. J. (2016). *We wanted workers: Unraveling the immigration narrative*. W. W. Norton & Company.

Cassarino, J.-P. (2010). *Readmission policy in the European Union, Brussels, European Parliament*. Available online at http://cadmus.eui.eu/handle/1814/14957.

Castles, S. (2004). Why migration policies fail. *Ethnic and Racial Studies, 27*, 205–227.

Castles, S., de Haas, H., & Miller, M. (2014). *The age of migration: International population movements in the modern world* (Kindle & 5th ed.). Palgrave Macmillan and Springer.

Cholewinski, R., MacDonald, E., & Perruchoud, R. (Eds.). (2007). *International migration law: Developing paradigms and key challenges*. Cambridge University Press.

Clemens, M. A. (2011). Economics and emigration: Trillion dollar bills on the sidewalk? *Journal of Economic Perspectives, 25*(3). A compelling argument that liberalizing migration rules would yield huge economic benefits.

Clemens, M. A. (2015). *Why it's time to drop the brain drain refrain*. Center for Global Development blog. A counter to the common argument that emigration hurts developing countries.

Czaika, M., & de Haas, H. (2013). The effectiveness of immigration policies. *Population and Development Review, 39* 487–508.

Roit, Da., Barbara, A. G., & Ferrer, and Francisco J. Moreno-Fuentes. . (2013). The Southern European migrant based care model: Long term care and employment trajectories in Italy and Spain. *European Societies, 15*(4), 577–596.

Donato, K. M., & Armenta, A. (2011). What we know about unauthorized migration. *Annual Review of Sociology, 37*, 529–543.

Goldin, C., & Katz, L. (2008). *The race between education and technology*. Belknap Press for Harvard University Press.

Gomberg-Munoz, R., & Nussbaum-Barberena, L. (2011). Is immigration policy labor policy? Immigration enforcement, undocumented workers, and the state. *Human Organization, 70*(4), 366–375.

Hansen, R., Koehler, J., & Money, J. (Eds.). (2011). *Migration, the Nation-State and International Cooperation*. Routledge.

Hinshaw, D., Allure of Wealth Drives Deadly Trek, The Wall Street Journal. (2015, June 12). Available online at http://www.wsj.com/articles/young-men-in-senegal-joinmigrant-wave-despite-growing-prosperity-at-home-1434127244. Accessed 15 June 2015.

Hollifield, J. F. (2012). Governing migration. In K. R. Khory (Eds.), *Global migration*. New York: Palgrave Macmillan. https://doi.org/10.1057/9781137007124_9.

Hollifield, J., Martin, P., & Orrenius, P. (Eds.). (2014). *Controlling immigration: A global perspective* (3rd ed.). Stanford University Press.

Katz, L. F., & Murphy, K. M. (1992, February). Source: The quarterly. *Journal of Economics, 107*(1), 35–78.

King, R. (1995). Migration, globalisation and place. In D. Massey & P. Jess (Eds.), *A Place in the World?* (pp. 5–53). Open University and Oxford University Press.

Jaumotte, F., Koloskova, K., & Saxena, S. C. (2016). *Impact of migration on income levels in advanced economies*. Spillover Task Force, International Monetary Fund

King, R. (2002). Towards a new map of European migration. *International Journal of Population Geography, 8*, 89–106.

King, R., & Black, R. (1997). *Southern Europe and the new immigration*. Sussex Academic Press.

Klein, P., & Ventura, G. (2009). Productivity differences and the dynamic effects of labor movements. *Journal of Monetary Economics, 56*, 1059–1073.

Koser, K. (2010). International migration and global governance. *Global Governance, 16*(3), 301–316.

Koslowski, R. (Ed.). (2011). *Global mobility regimes*. Palgrave Macmillan.

Kunz, R., Lavenex, S., & Panizzon, M. (Eds.). (2011). *Multi-layered migration governance*. Routledge.

Kuptsch, C., & Martin, P. (2011). Low-Skilled Labor Migration. In A. Betts (Ed.), *Global migration governance* (pp. 34–59). Oxford University Press.

Latour, B. (2017). On recalling ANT. *The Sociological Review, 47* (1999) 15–25. Downloaded from Brill.com April 17, 2020 04:15:43PM via free access Editorial 11. *European Journal of Migration and Law 19,* 1–11

Lavenex, S., & Panizzon, M. (Eds.). (2011). *Multilayered migration governance. The promise of partnership*. Routledge.

Levitt, P. (2009). *God needs no passport*. New Press.

Levitt, P. (2001). *The transnational villagers*. University of California Press.

Loescher, G. (2001). The UNHCR and World Politics: State Interests versus Institutional Autonomy. *International Migration Review, 35*(1), 33–56.

Loescher, G., Betts, A., & Milner, J. (2008). *The United Nations High Commissioner for Refugees (UNHCR): The politics and practice of refugee protection into the twenty-first century*. Routledge.

Maroukis T., Iglicka, K., & Gmaj, K. (2011). Irregular migration and informal economy in Southern and Central-Eastern Europe: breaking the vicious cycle? *International Migration, 49*, 129–156.

Newland, K. (2010). The governance of international migration: Mechanisms, processes and institutions. *Global Governance, 16*(3), 331–344.

OECD. (2008). *A profile of immigrant populations in the 21st century: Data from OECD countries*. OECD.

Ottaviano, G. I. P., Peri, G., & Wright, G. C. (2010). *NBER working paper series Immigration, offshoring and American jobs* (Working Paper 16439). http://www.nber.org/papers/w16439.

Papastergiadis, N. (2000). *The turbulence of migration*. Polity Press.

Pekkala Kerr, S., & Kerr, W. R. (2011). *Economic impacts of immigration: A survey* (NBER Working Papers 16736). A review of the research evidence on the economic impacts of immigration.

Peri, G. (2012, January). Immigration, labor markets, and productivity. *Cato Journall, 32*(1), 35–54.

Peri, G. (2012). The effect of immigration on productivity: Evidence from U.S. states. *Review of Economics and Statistics, 94*(1), 348–358

Peri, G., (2013). The economic consequences of the proposed immigration reform. *Intereconomics, 48*(3), 191–192.

Portes, J. D. (2019). The economics of mIgration. *Contexts*. https://doi.org/10.1177/1536504219854712.

Portes, J. (2018). *The Economic Impacts of Immigration to the UK" and "New Evidence on the Economics of Immigration to the UK*. VoxEU (April and October, respectively).

Pritchett, L. (2006). *Let their people Come Breaking the Gridlock on International Labor Mobility* (Working Paper). Washington, DC: Center for Global Development.

Pugliese, E. (2012). Il lavoro agricolo immigrato nel Mezzogiorno e il caso di Rosarno. *Mondi Migranti, 6*(3), 7–28.

Pugliese, E. (2002). *L'Italia tra migrazioni internazionali e migrazioni interne.* Il Mulino.

Reyneri, E., & Fullin, G. (2011). Ethnic penalties in the transition to and from unemployment: A West European perspective. *International Journal of Comparative Sociology, 52*(4), 247–263.

Reyneri, E. (1999). *Immigration and the underground economy in new receiving South European Countries: Manifold negative effects, manifold deep-rooted causes. Final Report of the project "Migrants' Insertion in the Informal Economy, Deviant Behaviour and the Impact on Receiving Societies".* CE/DGXII-Science, Research and Development, TSR Program, Contract No. Soe2-Ct95–3005, Mimeo.

Reyneri, E. (2013). Irregular immigration and the underground economy in Southern Europe: Breaking the vicious circle. In G. Brochmann and E. Jurado (Eds.), *Europe's Iimigration challenge: Reconciling work, welfare and mobility* . I.B. Tauris.

Rodrik, D. (2012). *The globalisation paradox.* OXford University Press.

Rodrik, D. (2018). Populism and the economics of globalization. *Journal of International Business Policy, 1*(1–2), 12–33.

Rodrik,D. (2019). *What is driving populism.* https://www.socialeurope.eu/whats-driving-populism.

Rosina, M., (2020). *Deterrence and international migration: The criminalisation of irregular entry and stay in Italy and France* (PhD dissertation, mimeo).

Sassen, S. (2000). *Guests and Aliens.* New Press.

Schneider, J. (2012). *Practical measures for reducing irregular migration* (Federal Office for Migration and Refugees [BAMF] Working Papers, No. 41). Available online at http://www.bamf.de/SharedDocs/Anlagen/EN/Publikationen/EMN/NationaleStudien-WorkingPaper/emn-wp41-irregular-migration.html.

Talani, L. S. (2002). *Final evaluation report of the project migrants' insertion in the informal economy, deviant behaviour and the impact on receiving societies.* Ce/Dgxii-Science, Research And Development, Tsr Program, Contract No. Soe2-Ct95–3005, Mimeo (p. 11).

Tognetti Bordogna, M., & Ornaghi, A. (2012). The 'Badanti' (Informal Carers) Phenomenon in Italy. *Journal of Intercultural Studies, 33*(1), 9–22.

Triandafyllidou, A., & Dimitriadi, A. (2014). Deterrence and protection in the EU's migration policy. *The International Spectator, 49*, 142–162.

Triandafyllidou, A., & Ambrosini M. (2011). Irregular migration control in Italy and Greece: Strong fencing and weak gate-keeping strategies serving the labour market. *European Journal of Migration and Law, 13*, 251–273.

Triandafyllidou, A., & Maroukis, T. *The case of the Greek Islands: The challenge of migration at the EU's Southeastern sea borders.*

Triandafyllidou, A. (2017). Beyond irregular migration governance: Zooming in on migrants' agency. *European Journal of Migration and Law, 19*, 1–11.

Triandafyllidou, A. (Ed.). (2010). *Irregular migration in Europe: Myths and realities*. Ashgate.

Triandafyllidou, A., & Maroukis, T. (2012). *Migrant smuggling: Irregular migration from Asia and Africa to Europe*. Palgrave Macmillan.

Unodccp. (2001). *Global programme against smuggling of migrants*. Odccp Web Page. https://Www.Odccp.Org.

Walters, W. (2010). Imagined migration world: The European Union's anti-illegal immigration discourse. In M. Geiger & A. Pécoud (Eds.), *The politics of international migration management: Migration, minorities and citizenship* (pp. 73–95). Palgrave Macmillan.

World Bank. (2005). *The potential gains from international migration*. World Bank Web-Site: https://documents1.worldbank.org/curated/en/507301468142196936/841401968_200510319020105O/additional/343200GEP02006.pdf.

Zincone, G. (1999). Illegality, enlightenment and ambiguity: A Hot Italian Recipe. In M. Baldwin-Edwards & J. Arango (Eds.), *Immigrants and the informal economy in southern Europe* (pp. 43–83). Frank Cass.

CHAPTER 8

Migration and Social Capital: From Islamophobia to Social Unrest and Terrorism

Social capital has increasingly become an essential notion for studying migration. An indicator of this is the fact that measures of social capital among migrants are now used by the British government to understand their level of integration in receiving societies.[1]

Can the notion of social capital between and within different ethnic communities be used to understand instances of hostility against Muslim communities or "Islamophobia"? Can diversity and the lack of integration of immigrant communities lead to phenomena of social unrest, violence or even terrorism? To answer these questions it is important to understand what social capital is.

This chapter will review the use of the notion of social capital in relation to the question of social unrest, terrorism and Islamophobia to ascertain to what extent it can be useful in understanding these phenomena.

To this aim, the chapter will first address the debate about the definition of social capital and the question of how to measure it. The chapter will then explore the extent to which social capital may help, or hinder the integration of Muslim migrants in receiving societies. Finally,

[1] See Home Office. (2019). *2019 Indicators of itegration framework*. https://assets.publishing.service.gov.uk/government/uploads/system/uploads/attachment_data/file/835573/home-office-indicators-of-integration-framework-2019-horr109.pdf.

L. S. Talani, *The International Political Economy of Migration in the Globalization Era*, International Political Economy Series,
https://doi.org/10.1007/978-3-030-79321-0_8

the relationship between social integration, or lack thereof, social unrest and Islamophobia will be addressed to verify whether this relationship provides fertile ground for radicalisation and terrorist tendencies among migrant communities.[2]

Social Capital and Migration[3]

The notion of social capital was devised in 1916 thanks to the seminal work of L. Judson Hanifan.[4] The study of social capital was expanded by Bourdieu, who defined it as resources embedded in social networks which are accessed and used by actors for action (Bourdieu, 1980).[5] This differs from the broader approach adopted by Coleman who emphasised the impact of social capital at the community level (Coleman, 1988).[6] Thanks to the contribution of Robert Putnam, Bourdieu's definition has been extensively applied in political science (Putnam, 2000, 2002).[7] At the centre of the notion of social capital is the intuition that a person's life is deeply affected by their social assets, including family, friends and associates. The existence of a thriving system of social connections allows a person to more efficiently resolve various crises and to obtain economic and other advantages. This does not only apply to the individual, but also to communities including states. Indeed, the density of social capital

[2] See Body-Gendrot, S. (2007a). Urban "riots" in France: Anything new? In P. Ponsaert (Ed.), *Local security policy in the Netherlands and Belgium*. Boom Juridische Utig; Body-Gendrot, S. (2007b). Order and disorder in the urban landscape. In R. Burdett & D. Sudjik (Eds.), *The urban age book*; Bourdieu, P. (1980). "Le capital social: notes provisoires". *Actes de la Recherche in Sciences Sociales, 31*, 2–3; Brehm J., & Rahn W. (1997). Individual-level evidence for the causes and Consequences of Social capital. *American Journal of Political Science, 41*(3), 999–1023; Coleman, J. (1988). *Social capital in the creation of human capital. American Journal of Sociology, 94*, 95–120.

[3] Kindler, M., Ratcheva, V., & Piechowska, M. (2015). *Social networks, social capital and migrant integrationat local level*. European literature review (IRiS Working Paper Series, No. 6/2015). Institute for Research into Superdiversity.

[4] See Putnam, R. (2000). *Bowling alone: The collapse and revival of American community* (p. 19). Simon & Schuster.

[5] See Bourdieu, P. (1980). Le capital social: Notes provisoires. *Actes de la Recherche in Sciences Sociales, 31*, 2–3.

[6] See Coleman, J. (1988). Social capital in the creation of human capital. *American Journal of Sociology, 94*, 95–120.

[7] See Putnam, R. (2002). *Democracies in flux* (p. 3). Oxford Scholarship online. www.oxfordscholarship.com. As accessed on February 19, 2007.

in different communities can determine their resilience to crises. For example, a lack of social capital could exacerbate situations of difficulty and instability. Thus, social capital becomes a relevant heuristic variable in social sciences, or, as Putnam puts it:

> civil society affects the health of our democracies, our communities and ourselves. (Putnam, 2002: 6)[8]

The existence of social capital in a given community or state allows for social policy to become more effective as, besides the social provisions established by law, people can rely on informal mutual welfare delivered by social networks, associations, friends and/or family. This adds to the quality and quantity of social policy, making it more effective and less expensive. Hence, social capital is a communal asset based on interpersonal networks and relations. It is facilitated by trust and reciprocity but also by solidarity norms and values and, in the case of family relations, by emotional ties. There are three types of social capital identified in the literature: bonding social capital, bridging social capital and linking social capital (Putnam, 2000, 2002).

Bonding social capital refers to relationships founded on identity and sense of belonging. These are the closest personal relationships, involving family and close friends, characterised by various degrees of emotional involvement. Bridging social capital are those social relations connecting different families and communities, tying them to larger social groups. Bridging social capital allows larger civic communities to pursue common goods and objectives. Finally, linking social capital allows for the connections between civic communities and political institutions, including state ones (Putnam, 2000, 2002). Looking at the case of migrant communities, the existence of bridging social capital would allow their connection with other migrant and non-migrant communities. It would also allow their integration into larger civic communities. Linking social capital, on the other hand, would allow their representation at the institutional level as well as bringing together instances of informal welfare and institutional welfare provisions in a sustainable way (Putnam, 2000, 2002).

In migration studies, one of the most persistent debates is which kind of social capital leads to inclusion or exclusion, to disadvantage or advantage (Kindler et al., 2015). In particular, the role of bonding social

[8] Ibid., p. 6.

capital is discussed and whether it poses barriers or facilitates bridging social capital, which as previously noticed, is necessary for integration. Based on Putnam's classification above, many scholars have tried to verify whether, to what extent and when ethnic diversity hinders the creation of social capital in Europe (Eijk, 2012; Gijsberts et al., 2012; Lancee & Dronkers, 2011; Laurence, 2011, 2013; Laurence & Heath, 2008).[9] Amin (2005), instead of putting the emphasis on how a variety of cohesive ethnic communities limits the creation of bridging social capital, focuses on their economic and social conditions.

Besides, the amount of social capital in a community is not given; it varies across time and space, making it imperative to study this notion in different societies and different historical periods.

Social capital can be defined as a private good because of its "internal value" for the people belonging to a specific social group or association. It provides psychological and social comfort, but also plain economic returns to members of the group or association. One classical example is the case of employment opportunities created by the intervention of friends, relatives or "social connections". In the literature there is a tendency to believe that migrant networks' social capital helps in finding jobs. However, there is also evidence that such jobs can be low standard ones and do not necessarily facilitate economic integration (Drever & Hoffmeister, 2008; Engbersen et al., 2006; Van Meeteren et al., 2009).

Social capital also has an "external value" which configures it as a public good, not only a private one. This refers to what economists call the positive externalities arising from the existence of social capital (Knack & Keefer, 1997). This means that the existence of social capital does not only provide benefits for those who belong to the social networks in question but also to those who do not take part in them. A typical example is a neighbourhood watch against crime which allows also those who do not belong to it to enjoy a safer environment. Looking at social capital from this perspective it acquires the value of a public good. It also allows for the resolution of collective action problems by relying on reciprocity and trust. In fact, social interactions based on trust and reciprocity help people act for the achievement of a common good or objective without coercion or immediate reciprocation, thus again, configuring social capital as a public good.

[9] Lancee and Dronkers (2011), Gijsberts et al. (2012), Laurence and Heath (2008), Eijk (2012), Laurence (2011, 2013).

From the operative point of view, we can have a narrow or a broader definition of social capital (Worms, 2002).[10]

A narrow conception of social capital only includes groups or associations joined by individuals on a voluntary basis which involves some form of face-to-face interaction. This is the case for sports clubs or cultural associations.

A broader definition of social capital includes all non-voluntary associations, to which people belong by chance, birth or constraint. For example, these are the family of origin, a social class or a neighbourhood.

The widest definition of social capital refers to sporadic and casual interactions, like those characterising groups of commuters, participants in a public demonstration or those in a queue.

Finally, the degree of social cohesion of a society can be measured either within each group and communities or between different groups and communities.

In a multi-ethnic and multi-cultural environment, the distinction between the existence of social capital within or between communities is particularly important as one can have a high degree of social cohesion within ethnic communities but not between them. This, as seen above, according to some scholars hinders integration (Eijk, 2012; Gijsberts et al., 2012; Lancee & Dronkers, 2011; Laurence, 2011, 2013; Laurence & Heath, 2008). Also, the distinction between a narrow and wider definition of social capital acquires relevance in multi-ethnic societies. Narrow forms of social capital such as voluntary associations and grassroots social groups are quicker in acquiring a multicultural dimension while wider forms, such as neighbourhoods, are slower and may require structural intervention to integrate different ethnicities and cultures.

It is indeed highly debated in the literature whether structural, public intervention can help create social capital or not (Knack & Keefer, 1997; Worms, 2002). The solution to this debate is not an easy one, as the relations between state and society are complex and can be studied from a variety of different ideological as well as theoretical approaches. Overall, and without any ambition of being exhaustive, one can hypothesise a dialectic relationship between state and society, with the state shaping society through policies and legislation and society shaping the

[10] See Worms, J. (2002). France: Old and new civic and social ties in France. In R. D. Putnam (Ed.), *Democracies in Flux: The evolution of social capital in contemporary society* (pp. 137–188, p. 138). Oxford University Press.

state through its culture and ideology. In a Gramscian perspective, both civil society and the state proper, i.e. political institutions, would reflect the power relations formed in the economic structure (Talani, 2010). Looking at Putnam's conceptualisation, institutional integration policies are related to the notion of linking social-capital, based on the positive relationship existing between participation in civil society and good governance (Putnam, 2000, 2002).

Another ongoing debate in the social sciences relates to how to measure social capital. In particular, the discussion focuses on whether the level of participation in voluntary associations and, in a related way, the density of associations, can be considered a good measure of social capital. The discussion also focuses on whether an increased number of associations necessarily allows for the increase of trust in host communities (Knack & Keefer, 1997).

There are a number of empirical studies considering trust as the best measure of social capital. In turn, the existence of social capital is related to the increase in economic growth and a more efficient provision of public goods (Knack & Keefer, 1997; Tabellini, 2005).[11] The availability of trust in society allows it to cope with market imperfections and moral hazards in uncertain environments. This is especially true in cases with no strong formal institutions able to enforce contracts or in economies where contracts are not sufficiently reinforced by formal institutions (Knack & Keefer, 1997; Tabellini, 2005).[12]

On the other hand, scholars tend to associate social capital and trust to membership of voluntary associations (Glaeser et al., 2002).[13] In their reasoning, belonging to the same association or club would favour the creation of relations based on trust. In turn, membership of voluntary

[11] See Knack, S., & Keefer, P. (1997). Does social capital have a economic payoff, A cross-country comparison. *Quarterly Economic Journal* (112), 1251–1288; Tabellini, G. (2005). Culture and institutions: Economic development in the regions of Europe (IGIER working paper n°292).

[12] See Knack, S., & Keefer, P. (1997). Does social capital have a economic payoff, A cross-country comparison. *Quarterly Economic Journal* (112), 1251–1288; Tabellini, G. (2005). Culture and institutions: Economic development in the regions of Europe (IGIER working paper n°292).

[13] See Glaeser, E. L., Laibson, D., & Sacerdote, B. (2002). An economic approach to social capital. *Economic Journal, 112*(483).

associations is recognised as a factor predisposing to participation in politics and public life. Finally, it also means adhering to the same set of values and following the same social norms.

However, there are scholars who dispute the relevance of participation in voluntary associations and groups in order to detect the presence of social capital. They underline how the promotion of the interests of particularistic groups could even result in less societal trust overall in the community (Knack & Keefer, 1997).[14]

Thus, the opinion of scholars is divided. First are those who measure the stock of social capital in a society by looking at level of participation to voluntary associations (Glaeser et al., 2002).[15] The second are those who relate it to the degree of trust and civic cooperation in a society, regardless of the density of associations (Knack & Keefer, 1997).[16]

In support of the first interpretation, in a study based on data from the US General Survey, Brehm and Rahn (1997)[17] prove a positive relation between interpersonal trust and participation in voluntary associations. This observation underlines how participation in voluntary associations increases trust. In particular, they found out that being members of the same group increases trust in the other members, whereas the relation between trust in other people and willingness to take part in a group is less marked (Brehm & Rahn, 1997).[18]

This position is critiqued by authors who do not believe that participation in close associations always increases civic trust and creates a positive attitude towards public good issues. There are cases in which the adoption of very selective and restrictive rules to enter associations can create highly exclusive groups leading to negative, instead of positive, social capital. For that matter, it is possible to identify associations which are based on values and rules contrary to trust and civic behaviour. They therefore themselves constitute negative social capital, as for example,

[14] See Knack and Keefer (1997).

[15] See Glaeser et al. (2002).

[16] See Knack and Keefer (1997).

[17] See Brehm J., & Rahn, W. (1997). Individual-level evidence for the causes and consequences of social capital. *American Journal of Political Science, 41*(3): 999–1023.

[18] See Brehm, J., & Rahn, W. (1997). Individual-level evidence for the causes and consequences of social capital. *American Journal of Political Science, 41*(3), 999–1023.

Mafia organisations or terrorist groups (see section below).[19] Knack and Keefer[20] make an empirical investigation over the relation between social capital, measured as trust and respect of civic norms, and membership of voluntary associations. Data was taken from World Values Surveys. An indicator of trust at the country level was the percentage of respondents who believed that others would adopt co-operative behaviour in a prisoner's dilemma type of situation. Respect for civic norms was measured on the basis of positive responses given to questions relating to the willingness of people to co-operate with strangers in relation to various issues involving public goods (like evading taxes). They concluded that there is no relation between trust and participation in voluntary associations.

When applied to the case of political Islam or, more generally, the existence of a multitude of Muslim voluntary associations, we will see that membership of those associations does not represent a good measure of social capital. Indeed, it can create diversity and reduce trust among host communities, rather than increasing it. For example, this can help explain why, despite a growing number of Muslim associations, the level of hostility towards Muslim communities has not decreased. And, vice versa, the likelihood of violence and social unrest by young Muslims has increased.

In the next section, attention is turned to whether the lack of social capital and integration can explain instances of violence and unrest in immigrant Muslim communities.

[19] Berman, S. (1997). Civil society and the collapse of the Wiemar Republic. *World Politics, 49*, 401–29; Raab, J., & Briton Milward, H. (2003). Dark networks as problems. *Journal of Public Administration Research and Theory, 13*, 413–439; Paxton, P. (2000). Social capital and democracy: An interdependent relationship. *American Sociological Review, 67*, 254–277; Portes, A. (1998). Social capital: Its origins and applications in modern sociology. *Annual Review of Sociology, 22*, 1–24; Woolcock, M. (1998). Social capital and economic development: Toward a theoretical synthesis and policy framework. *Theory and Society, 27*, 151–208. 7. Coleman, J. S. (1988). Social capital and the creation of human capital. *American Journal of Sociology, 94S*, S95–S120; Putnam, R. D. (2000). *Bowling alone: The collapse and revival of American Community*. Simon & Schuster; Sageman, M. (2004). *Understanding terrorist networks*. University of Pennsylvania Press.

[20] See Knack and Keefer (1997).

Between Social Unrest and Terrorism: The Consequences of the Lack of Social Capital in Immigrant Muslim Communities in Europe

Social capital-based explanations of social unrest and terrorism are becoming more and more common, especially with reference to the experience of Muslim migrants in Europe. For example, Sophie Body-Gendrot has explored in various occasions the role of social capital in her analysis of social unrest in inner cities (Sophie Body-Gendrot et al., 2014).[21] Young Muslims of immigrant origin, usually second generation youths, are often linked in the media with urban violence, riots, Islamic fundamentalism and terrorism (Body-Gendrot et al., 2014: 44).[22]

One of the most serious waves of social unrest involving young second generation migrants in France was the events spreading from Paris to many French provinces in October/November 2005, involving the disruption of public and private property by young Muslim migrants. This event has been related by Sophie Body-Gendrot to the relationship between migration, Muslim migration, social unrest and social capital. In her opinion:

> It is not every death of a youth which causes civil unrest and it is not every banlieue which is a springboard for violent reactions. Every locality is differentiated by accumulated/or lack of (social) capital.[23]

In particular, Body-Gendrot notices how in 2005 young Muslim unrest did not have political or religious motivations, but was a spontaneous manifestation of social disease. This is why she prefers to talk about "social unrests" and not "riots", whose meaning is usually connected to political violence. Looking at the specific case of the French unrests of 2005, it is possible to identify the social rationales behind them. From the social capital perspective, the motives are to be found in weak social networks

[21] Policing the Inner City in France, Britain, and the US, S. Body-Gendrot, C. de Wenden, Catherine Wihtol de Wenden, November 20, 2014, Springer; Globalization, fear and insecurity: The challenges for cities North and South, S. Body-Gendrot May 29, 2012, Springer.

[22] See Body-Gendrot, S. (2007a). Urban "riots" in France: Anything new? In P. Ponsaert (Ed.), *Local security policy in the Netherlands and Belgium*. Boom Juridische Utig.

[23] Ibid., p. 9.

and social fragmentation at the local level. This happened notwithstanding the density of Muslim associations at the national level. This position is shared by other scholars, such as Worms (2002) and Putnam (2002).

In fact, the regions experiencing less disorders during this period, such as Marseille and Saint-Denis, were characterised by strong local ties, social control by families, local community organisations and/or the existence of social networks. There thus seems to be a great correlation between the lack of social capital, in all three dimensions of bonding, bridging and linking social capital, and the likelihood of social unrest among the marginalised migrant youth.

Moreover, in the recent theoretical debate (Helfstein, 2014: 363)[24] social factors, most notably social capital, seem to have acquired a much more relevant role in explaining the dynamics of terrorism. While it is true that many analyses of terrorism, especially of a quantitative nature, explain the use of violence against civilians by reverting to economic and political variables, it is increasingly underlined in the literature that social variables do play a role (Abadie, 2006; Chenoweth, 2010; Li, 2005; Piazza, 2006; Piazza & Walsh, 2009).

The definition of terrorism, as found in the literature, is "the use or threatened use of violence against civilians to advance political aims" (Helfstein, 2014: 364). Terrorism differs from generic criminal behaviour as it often contains some misjudged altruistic motivations. It is therefore a social phenomenon in that strong in-group ties may make it easier to justify violent acts on its behalf. If one feels strong social ties to a community, it may be easier to organise and justify violent acts on their behalf (Bowles & Gintis, 2002). It is thus believed that the notion of social capital can help explain why people engage in terrorist activities. However, in the literature it is less clear whether social capital favours terrorist activities or restrains them, as Sophie Body-Gendrot (2012) and Sophie Body-Gendrot et al. (2014) believe is the case with social unrest.

Helfstein notes that social capital could foster the formation of terrorist groups by helping individuals overcoming collective action problems (2014: 364). This happens whichever definition of social capital is adopted, whether it is considered by definition a public good, like

[24] Social capital and terrorism Scott Helfstein To cite this article: Helfstein, S. (2014). Social capital and terrorism. *Defence and Peace Economics, 25*, 4, 363–380, https://doi.org/10.1080/10242694.2013.763505. To link to this article: https://doi.org/10.1080/10242694.2013.763505.

in Putnam (1994), or in more neutral terms (Bowles & Gintis, 2002; Coleman, 1988; Agrawal & Ostrom, 2001). On the other hand, a high density of social capital encourages adhering to social norms that consider violence against civilians to be unacceptable. Consequently, in Helfstein's opinion, while a high density of social capital makes the formation of terrorist cells easier, especially among friends and next to kin (Magouirk et al., 2008), it also constraints their behaviour making them less violent.

It is therefore hypothesised that a higher density of social capital should produce more terrorist organisations of a less violent nature.[25]

[25] References Abadie, A. (2006). Poverty, political freedom, and the roots of terrorism. *The American Economic Review, 96*(2), 50–56. Anderson, L., Mellor, J. M., & Milyo, J. (2004) Social capital and contributions in a public-goods experiment. *The American Economic Review, 94*(2), 373–376. Asal, V., & Rethemeyer, R. K. (2008). *Journal of Politics, 70*, 437–449. Axelrod, R. (1984). *Evolution of cooperation*. Basic Books. Axelrod, R. (1986). An evolutionary approach to norms. *American Political Science Review, 80*(4), 1095–1111. Basuchoudhary, A., & Shughart, W. F. II. (2010). On ethnic conflict and the origins of transnational terrorism. *Defence and Peace Economics, 21*(1), 65–87. Baum, F. (1999). Social capital: Is it good for your health? Issues for a public health agenda. *Journal of Epidemiology and Community Health, 53*, 195–196. Blomberg, S. B., Hessa, G. D., & Weerapana, A. (2004). Economic conditions and terrorism. *European Journal of Political Economy, 20*, 463–478. Blomberg, S. B., Engel, R. C., & Sawyer, R. (2008). *The life-cycle of a terrorist organization*. Paper Presented at the Terrorism and Policy Conference at the University of Texas at Dallas School of Economic, Political and, Policy Sciences, May 21–22. Bourdieu, P. (1985). The social space and the genesis of groups. *Theory and Society, 14*, 723–744. Bowles, S., & Gintis, H. (2002). Social capital and community governance. *The Economic Journal, 112*(483), F419–F436. Brehm, J., & Rahn, W. (1997). Individual-level evidence for the causes and consequences of social capital. *American Journal of Political Science, 41*, 999–1023. Chenoweth, E. (2010). Democratic competition and terrorist activity. *Journal of Politics, 72*(1), 16–30. Coleman, J. S. (1988). Social capital in the creation of human capital. *American Journal of Sociology, 94*(Supplement: Organizations and Institutions: Sociological and Economic Approaches to the Analysis of Social, Structure), S95–S120. Collier, P., & Hoeffler, A. (2004). Greed and grievance in civil war. *Oxford Economic Papers, 56*(4), 563–595. Dess, G. G., & Shaw, J. D. (2001). Voluntary turnover, social capital, and organizational performance. *Academy of Management Review, 26*(3), 446–456. Eubank, W. L., & Weinberg, L. (1994). Does democracy encourage terrorism? *Terrorism and Political Violence, 6*(4), 417–435. Fearon, J., & Laitin, D. (2003). Ethnicity, insurgency, and civil war. *American Political Science Review, 97*(1), 75–90. Finkel, S. E., Muller, E. N., & Opp, K. (1989). Personal influence, collective rationality, and mass political action. *American Political Science Review, 83*(3), 885–903. Freedom House Index. (2011). http://www.freedomhouse.org. Glaeser, E. L., Laibson, D., & Sacerdote, B. (2002). An economic approach to social capital. *The Economic Journal, 112*(483), F437–F458. Gleditsch, N. P., Wallensteen, P., Eriksson, M., & Strand, H. (2002). Uppsala Conflict Data Program (UCDP) at the Department of

Asal et al. are bolder in hypothesising that social factors, such as joint membership of associations, ideological and/or cultural homogeneity

Peace and Conflict Research, Uppsala University and Centre for the Study of Civil War at the International Peace Research Institute, Oslo: PRIO, http://www.prio.no/CSCW/Datasets/Armed-Conflict/UCDP-PRIO/. Hegre, H. (2001). Toward a democratic civil peace? democracy, political change, and Civil War, 1816–1992. *American Political Science Review, 95*, 33–48. Helliwell, J. F., & Putnam, R. D. (2007). Education and social capital. *Eastern Economic Journal, 33*(1), 1–19. Knack, S. (2002). Social capital and the quality of government: Evidence from the states. *American Journal of Political Science, 46*(4), 772–785. Knack, S., & Keefer, P. (1997). Does social capital have an economic payoff? A cross-country investigation. *Quarterly Journal of Economics, 112*, 1251–1288. KOF Index of Globalization. (2011). globalization.kof.ethz.ch. Krueger, A. B., & Malečková, J. (2003). Education, poverty and terrorism: Is there a causal connection? *Journal of Economic Perspectives, 17*(4), 119–144. Kydd, A. H., & Walter, B. F. (2006). The strategies of terrorism. *International Security, 31*(1), 49–80. LaPorta, R., Lopez-de-Salanes, F., Shleifer, A., & Vishny, R. (1997). Trust in large organizations. *American Economic Review Paper and Proceedings, 87*, 333–338. Lederman, D., Loayza, N., & Menendez, A. M. (2002). Violent crime: Does social capital matter? *Economic Development and Cultural Change, 50*(3), 509–539. Li, Q. (2005). Does democracy promote or reduce transnational terrorist incidents? *Journal of Conflict Resolution, 49*(2), 278–297. Li, Q., & Schaub, D. (2004). Economic globalization and transnational terrorism. *Journal of Conflict Resolution, 48*(2), 230–258. Magouirk, J., Atran, S., & Sageman, M. (2008). Connecting terrorist networks. *Studies in Conflict and Terrorism, 31*(1), 1–16. McBride, M., & Richardson, G. (2012). Stopping suicide attacks: Optimal strategies and unintended consequences. *Defence and Peace Economics, 23*, 413–429. Montalvo, J. G., & Reynal-Querol, M. (2005). Ethnic polarization, potential conflict, and civil wars. *American Economic Review, 95*(3), 796–816. Moore, W. H. (1995). Rational rebels: Overcoming the free-rider problem. *Political Research Quarterly, 48*(2), 417–454. Ostrom, E. (1990). Governing the commons: The evolution of institutions for collective action. Cambridge University Press. Piazza, J. A. (2006). Rooted in poverty? Terrorism, poor economic development, and social cleavages. *Terrorism and Political Violence, 18*(1), 159–177. Piazza, J. A. (2008). Do democracy and free markets protect us from terrorism? *International Politics, 45*, 72–91. Piazza, J. A., & Walsh, J. I. (2009). Transnational terror and human rights. *International Studies Quarterly, 53*, 125–148. SOCIAL CAPITAL AND TERRORISM 379. Pittel, K., & Rübbelke, D. T. G. (2012). Decision processes of a suicide bomber—The economics and psychology of attacking and defecting. *Defence and Peace Economics, 23*(3), 251–272. Putnam, R. D. (1994). *Making democracy work: Civic traditions in modern Italy*. Princeton University Press. Putnam, R. D. (2001). *Bowling alone: The collapse and revival of american community*. Simon and Schuster. Sandler, T. (1995). On the relationship between democracy and terrorism.*Terrorism and Political Violence, 7*(4), 1–9. Schmid, A. P. (1992). Terrorism and democracy.*Terrorism and Political Violence, 4*(4), 14–25. Stewart, F. (2000). Crisis prevention: Tackling horizontal inequalities. *Oxford Development Studies, 28*(3), 245–262. Terrorist Organization Profiles (TOPS). (2008).*University of Maryland: National Consortium for the Study of Terrorism and Responses to Terrorism*. http://www.start.umd.edu/start/data/tops/. Victoroff, J. (2005). The mind of the terrorist: A review and critique of psychological approaches. *Journal of Conflict Resolution, 49*, 3–42. World Bank Data and Statistics. (2011). http://web.worldbank.org.

and kinship and/or geographical proximity, accentuate trust and make belonging to the underground, including terrorist groups, easier (2014: 403).[26] In practice, they found that members of groups used their social capital to achieve a common terrorist objective. This substantiates the literature mentioned above highlighting the "dark side" of social capital[27] and identifying the potentially negative impact of social capital in terms of favouring political extremism and terrorism. One of those authors is Coleman, who already in 1988 noticed how South Korean student radical activists had socialised in high schools or through churches. They then proceeded to form study circles which represented the nucleus of political opposition (Coleman, 1988).[28] Fukuyama also noticed how the Mafia and the Ku Klux Klan had shared norms leading to co-operative practices, which implied the existence of social capital, although this was very negative for the rest of society (Fukuyama, 2001).[29]

Robert Putnam himself had recognised, in Bowling Alone, that social capital was not always positive and indeed could be connected to terrorism, as in the case of the terrorist attack in Oklahoma City (Putnam, 2000).[30] Also Sageman identified a clear connection between social

[26] To cite this article: Asal, V. H., Nagar, N., & Karl Rethemeyer, R. (2014). Building terrorism from social ties: The dark side of social capital. *Civil Wars, 16*, 4, 402–424, https://doi.org/10.1080/13698249.2014.981942. To link to this article: https://doi.org/10.1080/13698249.2014.981942.

[27] Berman, S. (1997). Civil society and the collapse of the Wiemar Republic. *World Politics 49*, 401–29; Raab, J., & Briton Milward, H. (2003). Dark networks as problems. *Journal of Public Administration Research and Theory, 13*, 413–39; Paxton, P. (2000). Social capital and democracy: An interdependent relationship. *American Sociological Review, 67*, 254–277; Portes, A. (1998). Social capital: Its origins and applications in modern sociology. *Annual Review of Sociology, 22*, 1–24; Woolcock, M. (1998). Social capital and economic development: Toward a theoretical synthesis and policy framework. *Theory and Society,27*, 151–208. 7. Coleman, J. S. (1988). Social capital and the creation of human capital. *American Journal of Sociology, 94S*, S95–S120; Putnam, R. D. (2000). *Bowling alone: The collapse and revival of American community*. Simon & Schuster; Sageman, M. (2004). *Understanding terrorist networks*. University of Pennsylvania Press.

[28] Coleman, J. S. (1988). Social capital and the creation of human capital. *American Journal of Sociology, 94S*, S95–S120.

[29] Fukuyama, F. (2001). Social capital, civil society and development. *Third World Quarterly, 22*, 7–20, 8.

[30] Putnam, R. D. (2000). *Bowling alone: The collapse and revival of American community*. Simon & Schuster.

capital and terrorism, pointing to the existence of social bonds facilitating recruitment for Al-Qaeda (Sageman, 2004).[31]

Asal et al., building on previous work, distinguish between structural and cognitive social capital, arguing that both dimensions concur to facilitate the formation of underground groups (2014: 404). While structural social capital refers to the forms of social organisations, especially networks, cognitive social capital is represented by shared norms, values, attitudes and beliefs, leading to trust and obligations. Thus, terrorism has a structural social component, given by participation in political/ideological associations or networks, albeit one with negative consequences. It also clearly has a cognitive element, as the members of terrorist associations develop trust and obligations of the basis of shared values, albeit of a terrorist nature.

The distinction between bonding and bridging capital also becomes relevant in this context (Asal et al., 2014: 406). As already noticed, bonding social capital refers to strong in-group relations, of an affective, ideological or even geographical nature (neighbours). On the other hand, bridging social capital refers to social ties across different groups in society. In the literature, as already seen above, bridging social capital is related to positive outcomes, such as the integration of migrants, whereas bonding social capital can lead to negative consequences, including terrorism.[32]

The solution to the negative spillovers of social capital could be represented by what Putnam calls linking social capital (Putnam, 2000). This leads to the question of how to create good social capital and whether this can be facilitated by political institutions and public policies. According to Sophie Body-Gendrot, another dimension needs to be added to this analysis of the role of social capital in limiting the social unrest of young, second generation migrants: space. Some territorial and urban elements, such as poorly designed urban planning, insufficient and costly public transportation, scarce public services, are indeed useful to explain why

[31] Sageman, M. (2004). *Understanding terrorist networks*. University of Pennsylvania Press.

[32] Harpham, T. (2002). Emma Grant and Elizabeth Thomas, 'Measuring social capital within health surveys: Key issues'. *Health Policy and Planning, 17*, 106–111; Rothstein, B., & Stolle, D. (2003). Introduction: Social capital in Scandinavia. *Scandinavian Political Studies, 26*, 1–26. 50. Leana, C. R., & Van Buren, H. J. (1999). Organizational social capital and employment practices. *Academy of Management Review, 24*, 538–555; Portes (note 6). 51. Portes (note 6). 52. Brewer, M. B. (1999). The psychology of prejudice: Ingroup love or outgroup hate? *Journal of Social Issues, 55*, 429–44.

there are less dense social relations in the Banlieues of Paris, Lyon and Marseille (Sophie Bodie Gendrot, 2007b).[33]

Regarding this, there seems to be evidence that spatial intervention and the redefinition of urban spaces can help increase the amount of social capital in society. Urban differences mirror social imbalances and inequalities. There is a tendency for social unrest to happen in marginalised locations, in spaces where social interactions resemble social struggles. It is a fact that social problems are intensified by spatial constraints, like in the case of the French unrest of 2005. Thus, social marginalisation and spatial marginalisation go together, with people secluded in the banlieues, alienated from the city centre by difficult public transport and unable to enjoy public spaces, to interact in pleasant environments, to fully take part in the life of the city. Similar considerations lead scholars such as Saskia Sassen, Sophie Body-Gendrot, Richard Sennet and Bauman to advocate for urban solutions to social problems, redesigning marginal neighbourhoods to integrate both migrant and local marginalised communities (Body-Gendrot, 2007b).

There remains the question of whether urban planners, local authorities and architects can actually facilitate integration by redesigning living spaces and social centres in the neighbourhoods or this task can only be performed by public policies and state intervention. Sophie Body-Gendrot believes that, "[o]ne of the great tasks of urban design lies in creating spaces which do not foreground fear" (Body-Gendrot, 2007b: 12). She asks, "[h]ow can a sense of safety emanate from public space? What mechanisms maintain social inclusion, mutual trust and civic involvement?" (Body-Gendrot, 2012: 173). One possibility could be to enlarge the green spaces shared by the population to allow people to enjoy a serene leisure environment without distinctions of class or ethnicity (Body-Gendrot, 2012).

Saskia Sassen advocates for more intervention by local communities to modify their surrounding spaces with urban planners intervening only to a limited extend by establishing common market places or social centres (Body-Gendrot, 2007b: 15). Cities are the geographical locus of new political claims raising from below to political institutions. Urban spaces allow for the materialisation of political activism by citizens that

[33] See Body-Gendrot, S. (2007b). Order and disorder in the urban landscape. In R. Burdett, & D. Sudjik (Eds.), *The urban age book*.

transcends established forms of political participation (Sassen in Body Gendrot, 2012: xvii–xviii).

On the other hand, for Richard Sennet new urban settings should be the result of spontaneous behaviours and not of the planned activity of policy-makers (Sennet, 1996).[34] Therefore, he declares himself against predictable planning as the way to facilitate social and ethnic integration. Solutions should come from below and not be imposed from above by urban planners or policy-makers. Feelings of trust and belonging can only ensue from a community of people sharing the same problems and working together to find suitable common solutions to them. That is what social capital is about.

Also for the co-existence of different ethnic or religious communities, including Islam, there is a need to develop relations of trust between them, something which is not necessarily facilitated by the multiplication of Muslim political associations. In the next section, we will elaborate on political Islam with the aim of identifying the relation between its increase and Islamophobia.

Islamophobia and Political Islam

Plenty of research shows that in European countries forms of discrimination hitting specifically Muslim communities exist and predate the events of September the 11th (Allen, 2010: 3; Glavanis, 1999[35]; Runnymead Trust, 1997; Wolferys, 2018[36]). The ensemble of discriminatory practices, ranging from education to political representation and from general legislation to gender, has been referred to as "Islamophobia" (Allen

[34] Sennet, R. (1996). *Uses of disorder: Personal identity and City life*. Faber and Faber.

[35] See Glavanis, M. P. (1999). Muslim voices" in the European Union: The stranger within-community, identity & employment project no. ERB-SOE2-CT96-3024. CCSR/University of Manchester, Mimeo.

[36] Wolfreys, J. (2018). *Republic of Islamophobia.The Rise of Respectable Racism in France*. Oxford University Press.

2010, 2020; Glavanis, 1999[37]; Halliday, 1999[38]; Kumar, 2012[39]; Sheehi, 2011[40]; Wolfreys, 2018; Runnymede Trust 1997[41]).

The debate about the definition, or even the existence of Islamophobia, is ongoing (Allen, 2020). In line with the prevailing view, here Islamophobia is considered an eminently contemporary phenomenon, although its historical roots are not denied (Allen, 2010, 2020; Halliday, 1999: 179; Sheehi, 2011). The academic discussion of Islamophobia distinguishes between a narrow definition, according to which Islamophobia is really 'Muslim-phobia' (e.g. Halliday, 1999[42]) and a broad one, where Islamophobia is about culture not just people (e.g. Modood, 2005[43]).[44]

There is a tendency in the current scholarly debate to include the definition of Islamophobia more within the second category (Allen, 2020: Ch.3). More precisely, it seems appropriate in the literature to equate Islamophobia to a kind of cultural racism (Allen, 2020; Carr, 2016; Jackson, 2018). To this categorisation critics often respond that "Islam" is a religion, not a race, and its members are such by choice, not by biology. However, those who consider Islamophobia racism or akin to it, stress the process of racialisation aimed at 'othering' Muslims, more than the existence of a distinct Muslim race (Allen, 2020; Carr, 2016).

Underlying the 'racist' nature of "Islamophobia" makes it possible to present it as a discriminatory phenomenon which, in turns, makes it a very "real" phenomenon, not only an academic one. Thus, Allen stresses how

[37] See Glavanis, M. P. (1999). "Muslim voices" in the European Union: The stranger within-community, identity & employment project no. ERB-SOE2-CT96-3024. CCSR/University of Manchester, Mimeo.

[38] Halliday, F. (1999). *Islam and the mythof confrontation: Religion and politics in the Middle East*. I.B. Tauris.

[39] Kumar, D. (2012). *Islamophobiaand the politics of empire*. Haymarket Books.

[40] Sheehi, S. (2011). *Islamophobia: The ideological campaign against Muslims*. SCB Distributors.

[41] Islamophobia: A challenge for us all, Report of the Runnymede Trust Commission on British Muslims and Islamophobia, was published in 1997. http://www.runnymede trust.org/projects/commissionOnBritishMuslims.html. As accessed on October 12, 2008.

[42] Halliday, F. (1999). *Islam and the mythof confrontation: Religion and politics in the Middle East*. I.B. Tauris.

[43] *Multicultural Politics: Racism, Ethnicity and Muslims in Britain* by Tariq Modood. Edinburgh University Press.

[44] A very detailed account of the scholarly debate about Islamophobia is contained in Allen (2020: Chapter 3).

any definition of Islamophobia needs to be appropriately 'real' to allow for the operationalisation and measurement of the phenomenon (Allen, 2020).

Looking at the use of the notion of Islamophobia in more operative terms, one of the most accurate and authoritative approaches to the question is the one proposed in a report commissioned by the Runnymede Trust in 1997.[45] Its definition of Islamophobia indeed allows not only for the identification of the phenomenon, but also the adoption of a balanced reaction to its detection. This approach underlines the elements of dialogue and mediation that would help eliminate hostility and discrimination. For these reasons, it is worth recalling what the Runnymede Trust report considered as Islamophobia.[46] In the report the term "Islamophobia", which was reportedly coined at the end of the 1980s and used publicly for the first time in 1991,[47] referred to:

> (...) unfounded hostility towards Islam. It refers also to the practical consequences of such hostility in unfair discrimination against Muslim individuals and communities, and to the exclusion of Muslim from mainstream political and social affairs. (Runnymede Trust, 1997: 10)

Most notably, the approach proposed by the report allowed to differentiate between a closed, phobic view of Islam and an open view. Islamophobia is associated with a closed view of Islam. In particular:

[45] A very detailed account of the activities of the Runnymede Trust is contained in Allen (2010: Chapter 4). In 1992 Runnymede set up a Commission to consider antisemitism in contemporary Britain. Its report entitled A Very Light Sleeper, published in 1994, carried as one of its recommendations the proposal that Runnymede should set up a broadly similar commission to consider Islamophobia. Early in 1997 the Commission on British Muslims and Islamophobia, under the chairmanship of Professor Gordon Conway, issued a consultative document. The final report, Islamophobia: A Challenge for Us All, was launched in November 1997 by the Home Secretary, Jack Straw. See https://www.runnymedetrust.org/projects/commissionOnBritishMuslims.html.

[46] Islamophobia: A Challenge For Us All, Report of the Runnymede Trust Commission on British Muslims and Islamophobia, was published in 1997. Web-site: https://www.runnymedetrust.org/companies/17/74/Islamophobia-A-Challenge-for-Us-All.html. Accessed on May 19, 2020.

[47] Allen (2010: 5–13) provides a more detailed chronology of the use of the term Islamophobia, dating back to 1925.

- Islam is seen as a single monolithic bloc, static and unresponsive to new realities.
- Islam is seen as separate and 'Other'—

  (a) not having any aims or values in common with other cultures;
  (b) not affected by them;
  (c) not influencing them.

- Islam is seen as inferior to the West—barbaric, irrational, primitive, sexist.
- Islam is seen as violent, aggressive, threatening, supportive of terrorism, engaged in a clash of civilisations.
- Criticisms made by Islam of 'the West' are rejected out of hand.
- Hostility towards Islam is used to justify discriminatory practices towards Muslims and the exclusion of Muslims from mainstream society.
- Anti-Muslim hostility is accepted as natural and 'normal' (Runnymede Trust, 1997: 10).

The report suggests that such a phobic view of Islam should be replaced by an open view of Islam, proposing a series of dichotomies representing the two different approaches. The idea of Islam as a monolithic culture should be replaced by its understanding as a diverse one; that of being separate and other from other cultures, with considering Islam as interdependent with others; Islam should not be seen as inferior to the West, but as different; not an enemy of the West, more a partner; Islamic religion is a true faith, not a manipulative ideology aimed at militarily conquering the West; therefore, criticisms of the West by Muslims should not be rejected straightforwardly but considered within a balanced debate; hostility towards Muslim individuals and communities shall not be considered normal, but should be stigmatised and any discriminatory practices following from such hostility should be criticised (Table 8.1).

In a similar vein, Allen (2020) adopts a very tangible definition of Islamophobia:

> ...as any malicious or violent act directed at Muslims (or those perceived to be Muslim) or any material entity associated with them (including mosques) where there exists evidence that the motivation, content or perpetrator focused on a perceived Muslim identity or other symbol of Muslims or Islam. (Allen, 2020: loc.838 kindle ed.)

Table 8.1 Closed and open views of Islam

Closed and open views of Islam

Distinctions	*Closed views of Islam*	*Open views of Islam*
1. Monolithic/diverse	Islam seen as a single monolithic bloc, static and unresponsive to new realities	Islam seen as diverse and progressive, with internal differences, debates and development
2. Separate/interacting	Islam seen as separate and other—(a) not having any aims or values in common with other cultures (b) not affected by them (c) not influencing them	Islam seen as interdependent with other faiths and cultures—(a) having certain shared values and aims (b) affected by them (c) enriching them
3. Inferior/different	Islam seen as inferior to the West—barbaric, irrational, primitive, sexist	Islam seen as distinctively different, but not deficient, and as equally worthy of respect
4. Enemy/partner	Islam seen as violent, aggressive, threatening, supportive of terrorism, engaged in 'a clash of civilisations'	Islam seen as an actual or potential partner in joint cooperative enterprises and in the solution of shared problems
5. Manipulative/sincere	Islam seen as a political ideology, used for political or military advantage	Islam seen as a genuine religious faith, practiced sincerely by its adherents
6. Criticism of West rejected/considered	Criticisms made by Islam of 'the West' rejected out of hand	Criticisms of 'the West' and other cultures are considered and debated
7. Discrimination defended/criticised	Hostility towards Islam used to justify discriminatory practices towards Muslims and exclusion of Muslims from mainstream society	Debates and disagreements with Islam do not diminish efforts to combat discrimination and exclusion

(continued)

Table 8.1 (continued)

Closed and open views of Islam		
Distinctions	*Closed views of Islam*	*Open views of Islam*
8. Islamophobia seen as natural/problematic	Anti-Muslim hostility accepted as natural and 'normal'	Critical views of Islam are themselves subjected to critique, lest they be inaccurate and unfair

More clearly, the APPG on British Muslims proposed the following definition, "Islamophobia is rooted in racism and is a type of racism that targets expressions of Muslimness or perceived Muslimness" (All Party Parliamentary Group on British Muslims, 2018: 11).

Even more complex is the relation between Muslim organisations and Islamophobia. Allen points out that Muslim organisations, especially political Islam, represent a reaction to "the first decade" of Islamophobia, at least in the British case (2010: 4). However, other authors, as noticed above, have identified in the bonding social capital of Muslim political organisations the breeding environment for violent and even terrorist groups. This certainly does not help integration and fosters the marginalisation of Muslims. As Jackson notes, there has been a tendency to adopt a "community cohesion" discourse to discriminate against Islamic Communities (2018). This revolved around the notion of self-segregation with different cultural communities unable to communicate with each other (Jackson, 2018: 31).

From this point of view, the outcomes of a project financed by the European Union[48] taking place between February 1997 and April 1999 seem to substantiate the hypotheses that, even "before" the events of September the 11th, political Islam was both a reaction to and a further reason for discrimination against Muslim people and culture. Thus, it seems to establish that the relation between Islamophobia and political Islam is a dialectical one. The title of the project is indeed very revealing in this respect, as it was called: "Muslim voices: The enemy within".

[48] See Glavanis, M. P. (1999). "Muslim voices" in the European Union: The stranger within-community, identity & employment project no. ERB-SOE2-CT96-3024. CCSR/University of Manchester, Mimeo.

One of the main merits of this project was to provide evidence of the complicated relation between Islamophobia and political Islam at a very early stage of the development of both phenomena. While highlighting the various manifestations of Islamophobia in eight European countries, (the UK, Belgium, France, Italy, Greece, Switzerland, the Netherlands and Germany), the project also detected early attempts at political organisation, and the impact of bonding social capital on local communities.

As seen above, Muslim migrants in the EU as well as other OECD countries tend to be overqualified for their job placement and are inserted in low-skill sectors of the economy like manufacturing, agriculture and personal and social services (Chapter 7). However, the forms of discrimination that they experience exceed the general kind of discrimination faced by foreign-born labourers in hosting labour markets. It is a more pervasive form of discrimination directed at their distinct religious faith and is highly politicised in the international arena (Glavanis, 1999: 71).[49] It can therefore be considered "Islamophobia".

Islamophobia manifests itself in various ways and in different contexts. The term is used by the authors of the "Muslim voices" project to define "a fear or hatred of Muslim peoples and Islam which manifests itself in various forms of discrimination" (Glavanis, 1999: 71).[50] This is the definition provided by The Runnymede Trust.[51]

Thanks to the results of the "Muslim voices" research it is possible to identify the different instances of Islamophobia predating the events of September the 11th 2001. Here are some of the most common manifestations of this phenomenon: Islam was perceived as a threat both internationally and within national borders; religious hatred enhanced xenophobia and racism; Muslims were singled out as "the others"; the public debate would discard the Muslim viewpoint. Thus, Islamophobia predates the events of September the 11th. Even before this date there was a widespread perception in political circles, both national and international, that Islam had replaced communism as a threat for the western world. Islam was considered to be the "enemy" both without and within.

[49] Ibid., p. 71.

[50] Ibid.

[51] Islamophobia: A challenge for us all, Report of the Runnymede Trust Commission on British Muslims and Islamophobia, was published in 1997. http://www.runnymedetrust.org/projects/commissionOnBritishMuslims.html. As accessed on May 19, 2020.

In turn, Muslim communities were identified as "the other" and perceived as different from other migrant communities. The marginalisation of Muslim perspectives in academic life echoed the lack of interest in the public debate as to the contribution of Muslims to the general social, political or cultural life. Finally, there was a specific normalisation of anti-Muslim attitudes in the countries considered by the project.

As a consequence of similar discriminatory practices, a variety of "Muslim voices" emerged across Europe. "Muslim voices" were at the forefront of the requests for equality of access to social, economic and cultural resources by Muslim communities. They were articulating the need for the recognition of a "political Islam" not only at the national but also at the level of European institutions.

Not all Muslim organisations had an eminently political nature as they had complex structures including cultural, social, religious and political aims. However, this network of social organisations started to represent Muslims and their leaders in different European countries with mixed results in terms of integration. As already underlined indeed, the multiplication of "Muslim voices" did not necessarily contribute to the creation of bridging or linking social capital.

Allen (2020) underlines as early Muslim associations in the UK, up to Kalim Siddiqui's The Muslim Manifesto: A Strategy for Survival,[52] published in 1990, did not consider, at least openly, their activities as a reaction to anti-Muslim attitude in society (2010: 11–12). On the other hand, the Muslim Manifesto, and even more, The Muslim Parliament a few years later, are two distinct instances of Muslim "voices" attracting very negative media and public opinion coverage, akin to Islamophobia.

Conclusion

The relation between social capital and migration is a complex one and one which has received a great deal of attention both in the scholarly debate and in policy-making circles. That is a formidable reason not to ignore it and to address it in the context of a study on international migration. If the British government is nowadays set on including some measures of social capital among its official indicators of integration, others in the academic and public debate are increasingly pointing

[52] Siddiqui, K. (1990). The Muslim manifesto: A strategy for survival. The Muslim Institute.

at social capital as at least a concurring variable in the radicalisation of immigrant Muslim communities and even in terrorism.

Of course much of the contradictory considerations around the role of social capital in immigrant communities are derived from different definitions of the notion and of the way it is measured. It is indeed different whether one adopts a bonding, bridging or linking definition of social capital in terms of producing integrating or disintegrating tendencies for immigrant communities. However, the debate is still open on whether the notion of social capital is a neutral one and therefore can entail also negative outcomes, or always positive ones, i.e. always increasing social welfare. It is also still debated whether social capital can be assessed by looking at the number of associations in a given society or has to be measured in terms of trust and reciprocity, thus discarding the associationist argument.

It is therefore difficult to conclude on whether more bonding social capital, in-group solidarity, the increase in immigrant associations as well as of political Islam represent an asset and for whom. What is scarcely debatable, however, is the fact that discrimination against Muslim communities and culture has been on the rise, even before September the 11th, and this is a fact that cannot be denied.

What is also scarcely debatable, as we will see in the next chapter, is the fact that migrants have increasingly been blamed for many of the problems of our societies and economies. This happened to the extent that the discourse of new, emerging populist parties, especially right-wing ones, targeted migrants to successfully gain electoral support.

In the next chapter, we will therefore explore the relation between migration and populism.

References

Abadie, A. (2006). Poverty, political freedom, and the roots of terrorism. *The American Economic Review, 96*(2), 50–56.

Agrawal, A., & Ostrom, E. (2001). Collective action, property rights, and decentralization in resource use in India and Nepal. *Politics & Society, 29*(4), 485–514.

All Party Parliamentary Group on British Muslims (2018), Islamophobia defined. The inquiry into a working definition of Islamophobia. https://static1.squarespace.com/static/599c3d2febbd1a90cffdd8a9/t/5bfd1ea3352f531a6170ceee/1543315109493/Islamophobia+Defined.pdf.

Allen, C. (2010, October 28). *Islamophobia*. Routledge. ISBN 9780754651406.

Allen, C. (2020). *Reconfiguring Islamophobia a radical rethinking of a contested concept*. London: Palgrave.

Amin, A. (2005). Local community on trial. *Economy and Society, 34*(4), 612–633.

Asal, V. H., Nagar, N., & Karl Rethemeyer, R. (2014). Building terrorism from social ties: The dark side of social capital. *Civil Wars, 16*(4), 402–424. https://doi.org/10.1080/13698249.2014.981942. To link to this article: https://doi.org/10.1080/13698249.2014.981942

Body-Gendrot, S. (2007b). Order and disorder in the urban landscape. In R. Burdett & D. Sudjik (Eds.), *The urban age book.*

Body-Gendrot, S. (2012, May 29). Globalization, fear and insecurity: The challenges for cities North and South. Springer.

Body-Gendrot, S., de Wenden, C., & Wihtol de Wenden, C. (2014, November 20). *Policing the inner city in France, Britain, and the US*. Springer.

Bourdieu, P. (1980). Le capital social: Notes provisoires. *Actes De La Recherche in Sciences Sociales, 31*, 2–3.

Bowles, S., & Gintis, H. (2002). Social capital and community governance. *The Economic Journal, 112*(483), F419–F436.

Brehm, J., & Rahn, W. (1997). Individual-level evidence for the causes and Consequences of Social capital. *American Journal of political Science, 41*(3), 999–1023.

Carr, J. (2016). Experiences of Islamophobia living with racism in the neoliberal era. London: Routledge.

Chenoweth, E. (2010). Democratic competition and terrorist activity. *Journal of Politics, 72*(1), 16–30.

Coleman, J. S. (1988). Social capital in the creation of human capital. *American Journal of Sociology, 94*(Supplement: Organizations and Institutions: Sociological and Economic Approaches to the Analysis of Social, Structure), S95–S120.

Drever, A. I., & Hoffmeister, O. (2008). Immigrants and social networks in a job-scarce environment: The case of Germany. *International Migration Review, 42*(2), 425–448.

Engbersen, G., Van San, M., & Leerkes, A. (2006). A room with a view: Irregular immigrants in the legal capital of the world. *Ethnography, 7*(2), 209–242.

Fukuyama, F. (2001). Social capital, civil society and development. *Third World Quarterly, 22*(8), 7–20.

Gijsberts, M., Van Der Meer, T., & Dagevos, J. (2012). “Hunkering down” in multi-ethnic neighbourhoods? The effects of ethnic diversity on dimensions of social cohesion. *European Sociological Review, 28*, 1–11.

Glaeser, E. L., Laibson, D., & Sacerdote, B. (2002). An economic approach to social capital. *Economic Journal, 112*(483), 437–458.

Glavanis, M. P. (1999). "Muslim voices" in the European Union: The stranger within-community, identity & employment, Project No. ERB-SOE2-CT96-3024. CCSR/University of Manchester, Mimco.

Halliday, F. (1999). *Islam and the myth of confrontation*. I.B Tauris.

Putnam, R. D. (1994). Institutional change in Italy: The first two decades. In J. W. Harbeson, R. F. Hopkins, & D. G. Smith (Eds.), *Responsible governance: The global challenge* (pp. 121–176) (Essays in honor of Charles E. Gilbert). New York: University Press of America.

Jackson, L. B. (2018). *Islamophobia in Britain the making of a muslim enemy*. London: Palgrave.

Kindler, M., Ratcheva, V., & Piechowska, M. (2015). *Social networks, social capital and migrant integration at local level. European literature review* (IRiS Working Paper Series, No. 6/2015). Birmingham: Institute for Research into Superdiversity.

Knack S., & Keefer P. (1997). Does social capital have an economic payout? A cross-country investigation. *The Quarterly Journal of Economics, 112*(4), 1251–1288.

Kumar, D. (2012). *Islamophobia and the politics of empire*. Haymarket Books.

Lancee, B., & Dronkers, J. (2011). Ethnic, religious and economic diversity in Dutch neighbourhoods: Explaining quality of contact with neighbours, trust in the neighbourhood and inter-ethnic trust. *Journal of Ethnic and Migration Studies, 37*(4), 597–618.

Laurence, J. (2011). The effect of ethnic diversity and community disadvantage on social cohesion: A multi-level analysis of social capital and interethnic relations in UK communities. *European Sociological Review, 1*, 70.

Laurence, J. (2013). "Hunkering down or hunkering away?" The effect of community ethnic diversity on residents' social networks. *Journal of Elections, Public Opinion & Parties, 23*(3), 255–278.

Laurence, J., & Heath, A. (2008). *Predictors of community cohesion: Multi-level modelling of the 2005 citizenship survey*. Department for Communities and Local Government.

Li, Q. (2005). Does democracy promote or reduce transnational terrorist incidents? *Journal of Conflict Resolution, 49*(2), 278–297.

Magouirk, J., Atran, S., & Sageman, M. (2008). Connecting terrorist networks. *Studies in Conflict and Terrorism, 31*(1), 1–16.

Moddod, T. (2005). *Multicultural politics: Racism, ethnicity and Muslims in Britain by Tariq Modood*. Edinburgh University Press.

Piazza, J. A. (2006). Rooted in poverty? Terrorism, poor economic development, and social cleavages. *Terrorism and Political Violence, 18*(1), 159–177.

Piazza, J. A., & Walsh, J. I. (2009). Transnational terror and human rights. *International Studies Quarterly, 53*, 125–148. SOCIAL CAPITAL AND TERRORISM 379.

Putnam, R. (2000). *Bowling alone: The collapse and revival of American community*. Simon & Schuster.
Putnam, R. (2002). Democracies in flux, Oxford Scholarship online. www.oxfordscholarship.com. As accessed on 19 February 2007.
Runnymede Trust Commission on British Muslims and Islamophobia. (1997). *Islamophobia: A challenge for us all* (Final Report). http://www.runnymedetrust.org/projects/commissionOnBritishMuslims.html. As accessed on 12 October 2008.
Sageman, M. (2004). *Understanding terrorist networks*. University of Pennsylvania Press.
Scott Helfstein, S. (2014). Social capital and terrorism. *Defence and Peace Economics, 25*(4), 363–380. https://doi.org/10.1080/10242694.2013.763505
Sennet, R. (1996). Uses of disorder: Personal identity and city life. Faber and Faber.
Sheehi, S. (2011). *Islamophobia: The ideological campaign against Muslims*. SCB Distributors.
Tabellini, G., (2005) "Culture and Institutions: Economic Development in the Regions of Europe", IGIER working paper n°292.
Talani, L. S. (2010). From Egypt to Europe. London: I.B. Tauris.
Van Eijk, G. (2012). Good neighbours in bad neighbourhoods: Narratives of dissociation and practices of neighbouring in a "problem" place. *Urban Studies, 49*(14), 3009–3026.
Van Meeteren, M., Engbersen, G., & Van San, M. (2009). Striving for a better position: Aspirations and the role of cultural, economic, and social capital for irregular migrants in Belgium. *International Migration Review, 43*(4), 881–907.
Wolfreys, J. (2018). *Republic of Islamophobia. The rise of respectable racism in France*. Oxford University Press.
Worms, J. (2002). France: Old and new civic and social ties in France. In R. D. Putnam (Ed.), *Democracies in flux: The evolution of social capital in contemporary society* (pp. 137–188). Oxford University Press.

CHAPTER 9

Populism and Migration

This chapter assesses the relationship between populism and migration. More specifically, it focuses on the origins of populism. It explores whether populism is rooted in hostility towards foreign cultures and citizens, cultural hostility more broadly defined or, instead in pure economic considerations with hostility towards migrants being only a manifestation of discontent.

As the debate on these issues is still very open, this chapter will first review the different approaches to populism and the relationship between populism and migration. It will then adopt a political economy approach to populism in general, and Italian populism in particular, by focusing on the political economic consequences of the global financial crisis and of the eurozone crisis on the Italian economic and political landscape.

Based on Frieden (2017, 2018b, interview with the author),[1] this chapter traces the origins of the populist phenomenon in political discontent. The latter is in part due to failures of compensation—meaning the lack of sufficient social safety provisions for the losers of economic trends. For populism to happen, however, the crisis of compensation needs to be accompanied by failures of representation—which are the inability of

[1] Frieden, J. (2017, December). *The political economy of the globalization backlash: Sources and implications*. Jeffry Frieden Harvard University.

L. S. Talani, *The International Political Economy of Migration in the Globalization Era*, International Political Economy Series,
https://doi.org/10.1007/978-3-030-79321-0_9

traditional parties to represent the interests of an increasingly large section of the electorate especially the discontented one (Talani, 2018, Harvard University lecture).

This contribution hypothesises that the global economic crisis acted as a catalyst for the contradictions of globalisation to explode. This assumption is predicated on the case of Italy. Here, the consequences of the global financial crisis were exacerbated by the characteristics of the eurozone leading to the eurozone crisis. This is why originally the Five Star Movement and the Lega were against the euro.

The Italian manufacturing sector suffered a backlash from the establishment of the euro that brought about political discontent. The failure of compensation, due to austerity measures imposed by the eurozone crisis, and the failure of representation due to the disappointment with the governments of Monti, Letta and Renzi, pushed voters towards populist movements like the Five Star Movement and the Lega.

Italian populism, at least in the European context, is one of the most successful. While the global rise of populism in the last decade is well documented, in the Italian case, especially with respect to the Five Star Movement, populism managed, in the space of a few years, to achieve an astonishing electoral success and gain power with two different coalitions. More incredibly, the Five Star Movement managed to change the colour of the coalition in power, from centre right to centre left, while keeping the same Prime Minister, Giuseppe Conte, without passing through an electoral turn.

Before assessing the Italian case, it is necessary to address the theoretical discussion on the relationship between populism and migration.

Populism and Migration: Between Culture and Elections

Populism is a contested notion in the relevant literature. Although the concept of populism is certainly not new, the term began a new life with the recent election of many seemingly populist governments around the globe. Such elections include the Italian populist coalition between the Lega and the Five Star Movement (5SM) in 2018, and the coalition between the 5SM and the Democratic Party in September 2019.

Leaving aside whether or not the Trump administration or the successful implementation of the Brexit project are populist, the focus of this section is on the definition of populism and its relation to migration. In particular, it will be discussed whether populism is inherently anti-migrant and xenophobic or, instead, its anti-migrant component is acquired in a second moment to attract more votes and win power.

There are many definitions of populism both in the past and current debates. There is a "popular agency" approach to populism which minimally describes populism as a form of life organisation based on popular political engagement. This mostly refers to the experience of the populist party in late nineteenth-century America (Mudde, 2017: 3).[2]

The Laclauan approach to populism is currently en-vogue in critical studies. It emphasises the role of populism as an emancipatory force to overcome the limits of liberal democracies and substitute them with radical democracy. This would happen through the mobilisation potential of populism and its ability to allow marginal sections of society to participate in political power, eventually overcoming the status quo (Mudde, 2017: 3).

In the economist jargon, based on Dornbusch and Sachs understanding of the concept, populism is equated to irresponsibility in economic policy-making, especially with respect to an extremely relaxed fiscal policy aimed at pleasing the electorate and keeping power. This is also known as the socio-economic approach to populism (Mudde, 68
2017: 3).

More recently, populism has been defined as a political strategy used by charismatic leaders, like Ghaddafi for example, to govern by relying directly on popular consensus. As such, populism cannot survive the death of the populist leader. Finally, some scholars consider populism to be just a political style, used to mobilise the masses (Mudde, 2017: 4).

One of the most renowned and discussed definitions of populism is the one proposed by Cas Mudde, whose contribution to the debate is so substantial that it can hardly be ignored.[3] His understanding of the concept is notoriously an "ideational one", i.e. one that considers populism a true ideology, albeit a thin one, and not simply a political strategy or, even less, a style of leadership.

[2] Lawrence Goodwyn's Democratic Promise: The Populist Moment in America.

[3] Mudde, C. (2017). *Populism: A very short introduction* (Very Short Introductions) (p. 3). Oxford University Press. Kindle Edition.

Although not a consensual definition of populism, the ideationalist approach has recently gained a lot of currency and has been applied to understand the rise of populism in the last years both in the Americas and in Europe.

In the words of the author himself:

> More concretely, we define populism as a thin-centered ideology that considers society to be ultimately separated into two homogeneous and antagonistic camps, "the pure people" versus "the corrupt elite," and which argues that politics should be an expression of the volonté générale (general will) of the people. (Mudde, 2017: 5–6)

The thinness of the populist ideology allows the phenomenon to change shape and nature according to the different circumstances where it appears, and thus create some "subtypes of populism": the right-wing one, the left-wing one, the clientelistic one and more variants. Despite its ability to adapt, populism is still an ideology containing some hard-core elements which constitute its ontology. These are: the people, the elite and the general will (Mudde, 2017: 8–9).

In Mudde's conceptualisation, the notion of "the people" can change but it usually refers to the people as sovereign, as the common people and as the nation (Mudde, 2017: 9). "The people" are morally and ontologically distinct from "the elite", and therefore, even if they acquire power, they cannot become "the elite" in the ideationalist approach to populism. This allows populist leaders to maintain their anti-establishment position even when they are in power. The establishment, from this perspective, is not represented by the elected populist leaders, but by some obscure illegitimate forces acting behind the back of the people (or the nation). The elite, in turn, can be defined in economic terms, most often, or sometimes even in ethnic ones. Finally, core to the populist ideology is Rousseau's notion of "the general will". The French philosopher, and the populist ideology after him, distinguished between "la volonte' general" (the general will) and "the volonte' de tous" (the will of all). This is a very well-known distinction in political theory between the common interest, and the sum of the particular interests of the individuals. The idea here is that, for the populist ideology, representative democracy cannot implement "the general will" or achieve the common interest for that matter (Mudde, 2004: 542–563; 2017: 16).[4] Only direct democracy of the people, in all its forms, including the most technological ones, can achieve the realise the general will. It does not seem a mere chance here that the IT platform used by the Five Star Movement to run its technological project is indeed called Rousseau (Casaleggio, 2011, 2012, 2016; Gerbaudo, 2018, 2019a, b).

This latter element, however, easily leads to the dark side of populism, i.e. its authoritarian tendencies. Indeed, the will of the people does not easily encompass criticism, as the people cannot be wrong and when a similar idea becomes consensual, then criticism becomes more and more ostracised up to the point that the system becomes illiberal and authoritarian. The will of the people is, in a word, absolute (Mudde, 2004: 542–563; 2017: 18).

Norris and Inglehart share with Mudde the idea that populism can lead to authoritarian outcomes, but not his ideational definition.[5] In their opinion, populism is not an ideology, not even a thin one, but just a discourse aimed at bringing down the established sources of power and elected representatives in favour of the rule of the people. In the words of the authors:

> Populism is understood in this book minimally as a style of rhetoric reflecting first-order principles about who should rule, claiming that legitimate power rests with 'the people' not the elites. It remains silent about second-order principles, concerning what should be done, what policies should be followed, what decisions should be made. (Norris & Inglehart, 2019: Kindle Locations 372–373)

Any alternative definition of populism is rejected explicitly, including the definitions discussed above such as the socio-economic approach, the style of leadership approach, the definition referring to a party with a popular social base and, importantly the ideological one, as proposed by Mudde. The latter, in the opinion of the authors, includes unnecessary elements which are discarded in their definition of populism as a rhetoric. In particular, according to Norris and Inglehart, it is not a constituent

[4] Mudde, C. (2004). The populist Zeitgeist. *Government and Opposition, 39*(4), 542–563.

[5] Norris, Pippa, Inglehart, Ronald. Cultural Backlash (Kindle Locations 2039–2043). Cambridge University Press. Kindle Edition.

part of populism the idea, proposed by Cas Mudde, of separating society in two homogeneous and antagonistic groups, ‘the pure people’ versus ‘the corrupt elite’ and the notion that politics should express the general will of the people.[6]

Devoid of any ideological content, the notion of populism proposed by Norris and Ingelhart can acquire any shape, from the conservative to the socialist and from the authoritarian to the progressive (Norris & Inglegart, 2018, Harvard seminar). This rhetoric makes two core claims about how to govern society: (a) it attacks “the establishment” and, with it, the elected representative of liberal democracies; (b) it claims that the only source of democratic legitimacy is given by the ‘people’. These are the only two components of the populist discourse recognised by the authors (Norris & Inglehart, 2018, 2019 Harvard seminar).

These two elements are also what make the populist narrative very compelling as it recognises as the only legitimate source of authority in a democracy the will of the majority, thus delegitimising any established representative institutions and minorities. However, again, the dark side of populism is that leaders using this discourse can easily justify on its basis authoritarian outcomes by claiming that they, and only they, represent the will of the people.

As clearly stated by the authors (Norris & Inglehart, 2018, 2019 Harvard seminar), this is not a necessary outcome of populist discourse, nor are all authoritarian leaders necessarily populist. However, there is a tendency in the rhetoric style associated with populism to undermine the legitimacy of democratically elected institutions, leaving the door open to the consolidation of authoritarian systems. In turn, Norris and Inglehart (2018, 2019) define the notion of “authoritarian” not as a political regime, but as a cluster of values which gives primacy to the idea of collective security of the group as opposed to the freedom of the single individual. An authoritarian system of values is based on the following three elements: (1) the security element, articulated in various ways against foreigners, migrants, terrorists and anyone else defined as ‘the Other’; (2) the conservative element, articulated in a way as to preserve the in-group values from external attacks; (3) the identification of the leader as the protector of the group and its values.

[6] Muller, J.-W. (2016). *What is populism?* University of Pennsylvania Press.

In this "authoritarian conception", the group, driven by a "politics of fear", is a "tribe" that needs a leader to protect it from external attacks to its way of life in exchange for loyalty and obedience (Wodak, 2015).[7] Here the idea of tribe can be represented by "the nation", the community of citizens, but also by other signifiers such as socio-economic status, race, sex, gender, ethnicity, location or generation. A tribe is defined as a cohesive community of values, perceived as being threatened by an out-group, often a scapegoat, and therefore needing to be protected by a leader (Tajfel & Turner, 1985).

What ensues is a "cultural backlash" against the perceived threatening groups (Norris & Inglehart, 2019). The sequence of events leading to an "authoritarian" cultural backlash led by a populist leader is as follows. First, the social group in question needs to have experienced in the course of the years a significant deterioration of its socio-economic condition. Second, this deterioration of socio-economic conditions leads to a change in cultural values. Third, the in-group cultural values become more conservative and prone to the authoritarian backlash. Fourth, the existence of an increased social diversity accelerates the conservative backlash. Fifth, the conservative backlash brings votes to authoritarian populist leaders. Sixth, this translates into seats for the authoritarian leaders in democratic elections. Seventh, the democratic system is threatened by the cultural backlash (Norris & Inglehart, 2018, 2019).

This thesis is demonstrated by looking at how the cultural backlash against socially liberal and egalitarian values prevalent in progressive urban areas has led to the election of populist and sometimes also authoritarian leaders around the world (Norris & Inglehart, 2018, 2019). As western societies become more liberal, those who are left behind by this social revolution are, allegedly, expected to react. Those who feel more estranged are, according to Norris and Inglehart, the older generations, especially, in their words:

> …non-college graduates, the working class, white Europeans, the more religious, men, and residents of rural communities. These groups are therefore most likely to feel that they have become estranged from the silent revolution in social and moral values. (Norris & Inglehart, 2019: Kindle Locations 624–628)

[7] Wodak, R. (2015). *The politics of fear: What right-wing populist discourses mean*. Sage.

Thus, especially older, white, uneducated men, who used to be the dominant groups until recently in Western societies are likely to become increasingly hostile to and react strongly against the new, inclusive and progressive liberal values. But what kind of reactions are envisaged? Among the many possible reactions to a similar threat to their cultural value and social position, Norris and Inglehart (2019) identify, as the one that has been widely adopted, the so-called populist authoritarian reflex.

What is most important in Norris and Inglehart's analysis is that the populist authoritarian reflex is not provoked by the growing ethnic diversity of western societies, the increase in migration and refugee flows, or by anti-immigrant attitudes due to the perceived cultural threat posed by foreigners. In the opinion of the supporters of the "cultural backlash" explanation for populism:

> The authoritarian reflex is not confined solely to attitudes toward race, immigration, and ethnicity, but also to the rejection of the diverse lifestyles, political views, and morals of ' out-groups ' that are perceived as violating conventional norms and traditional customs, including those of homophobia, misogyny, and xenophobia. Moreover, these sentiments are strongest among those groups, such as homogeneous rural communities and older citizens, who feel the most threatened by the spread of multicultural diversity. (Norris & Inglehart, 2019: Kindle Locations 694–700)

Hence, they propose and support the argument that voting for populist parties is characteristic of older generations of less educated voters, mostly males, more religious and belonging to the ethnic majority.

Importantly however, this does not appear to be true in the case of Italy, where the vote for the Five Star Movement was mostly concentrated among the young, technologically literate and educated generations. There is a certain degree of consensus in the literature that the Five Star Movement displayed many of the characteristics of a populist, anti-establishment party whose attitude was not anti-immigrant at the onset (Bickerton & Accetti, 2018; Bobba & McDonnell, 2015; Corbetta, 2017; Tarchi, 2014). However, more than relying on the resentment of older white men estranged by the liberal revolution of western societies' values, it fostered the collective identity of young, digital voters left behind by the global financial and economic crisis of 2008 and 2009 (Canestrari &

Biondo, 2018, 2019[8]; D'Arma, 2015; Gerbaudo, 2018, 2019a, b; Lo Scalzo, 2019; Mosca & Tronconi, 2019[9]).

Indeed, the argument that the Five Star Movement gained much consensus from creating a strong collective identity of young generations of technologically educated voters left behind by the crisis of 2008 and 2009 is well established and convincing. Chronologically, the Five Star Movement first appeared on September 9, 2009. It has been one of the most successful populist parties in the world, not only winning the elections both in 2013 and in 2018, but also managing to keep in power after the crisis of August 2019 and the demise from the government of Salvini's Lega. It achieved this result by shifting its alliance to the opposite political party, the Democratic Party of Zingaretti. The populist credentials of the Five Star Movement are undisputed in the literature, and so are its authoritarian tendencies especially as related to the use of the digital platform Rousseau (Casaleggio, 2011, 2012, 2016; Canestrari & Biondo, 2018, 2019; Mosca & Tronconi, 2019).

With the results of the 2013 elections, the Five Star Movement became the first party in Italy, with 25.5% of the votes for the House of Commons.[10] The Democratic Party came second, with 25.4% of the votes, essentially making the Italian political system a tripolar one, with the centre-left coalition achieving 29.5%, the centre right coalition 29.1% and the Five Star Movement 25.5%. The success of the Five Star Movement was based on its almost standard populist stances, based on a very thin ideology, a clearly personalistic leadership by ex-comedian Beppe Grillo, its evident anti-establishment sentiments and its relaying on the support of "the people"(Conti & Memoli, 2015; Corbetta, 2017; Mosca & Tronconi, 2019; Tronconi, 2018). Importantly, while remaining pretty elusive on migration before the 2013 elections, during the electoral campaign of 2013 and even more in its aftermath, its position became

[8] Gerbaudo (2019): "The digital party" digital democracy-Rousseau Gerbaudo 2018, Il partito piattaforma, Gerbaudo (2019), One person one click!; D'Arma (2015) Digital democracy..charismatic leader populism; Canestrari, N., & Biondo, M. (2018). La supernova, 2019 Il Sistema casaleggio; Lo Scalzo 2019 on the Rousseau platform; Casaleggio 2011 Siamo in Guerra, 2012 Tu sei rete, 2016 Aforismi.

[9] Mosca, L., & Tronconi, F. (2019). Beyond left and right: The eclectic populism of the Five Star Movement. *West European Politics, 42*(6), 1258–1283, https://doi.org/10.1080/01402382.2019.1596691.

[10] https://www.repubblica.it/static/speciale/2013/elezioni/camera/riepilogo_nazionale.html.

more and more anti-migrant to the extent that in 2014 it even signed an agreement with the UK Independence Party (UKIP) to form a joint political group in the European Parliament (Mosca & Tronconi, 2019).

This is in line with Mudde's understanding of populism as a thin-centred ideology (Mudde, 2004: 544). Eventually this allows the party to gather votes from all sides of the electorate, adapting to the changing sentiments of the people, like a classic "catch-all party". Indeed, although the Five Star Movement was not born as an anti-migration party, it moved towards more anti-migration positions, which allowed it to gather a greater consensus and win the 2018 elections by a landslide. On the 4th of March 2018, the Five Star Movement emerged from the polls as by far the biggest Italian party, with 32.7% of the votes for the House of Commons. The Democratic Party, coming second, only achieved 18.7% and the Lega came in third with 17.4%.[11] This electoral success paved the way for the formation of the first populist government coalition between the 5SM and the Lega in May 2018 led by Giuseppe Conte. This was then followed, in September 2019, by the second 5SM government, this time in coalition with the Democratic Party, again led by Giuseppe Conte.

Thus the Five Star Movement was an incredibly successful populist party which did not seem to rely on the support of the cultural communities identified by the cultural backlash approach. The electoral base of the Five Star Movement, in the electoral turnout of 2018, was very far from the white, old, uneducated men predicted by the theory. To start with, both men and women voted in almost exactly the same percentage for the Five Star Movement, 32.8% of male total valid votes and 32.9% of female total valid votes. More to the point, the Five Star Movement achieved the highest percentage of votes among the youngest generations, with 35.3% of the valid votes of electors between 18 and 34 years, 35.4% of the valid votes of electors aged between 34 and 49; 34% of those between 50 and 64 and only 27.1% of the valid votes of those over 65.[12] This is exactly the opposite demographics predicted by Norris and Inglehart (2018, 2019). Even more surprisingly, in the Italian case, it is mostly very educated people that voted for the Five Star Movement. In fact, 29.3% of electors with a degree or higher and 36.1% of those with

[11] https://elezioni.repubblica.it/2018/cameradeideputati.

[12] https://st.ilsole24ore.com/art/notizie/2018-03-06/genere-eta-professione-identikit-nuovi-elettori-cinque-stelle-190100.shtml?refresh_ce=1.

secondary education voted for the 5SM. Finally, in social terms, retired people voted the least for the Five Star Movement (26.4% of them); while the highest percentage was from among the unemployed with 37.2% of their valid votes being for the 5SM.[13]

Overall, it seems fair to claim that the electoral base of the Five Star Movement is composed of young, educated men and women who have however had problems in finding employment at their level of education in the post global and economic crisis environment. This is also confirmed by analyses showing an inverse relationship between the number of votes to the 5SM in the 2018 elections and the unemployment rate by Italian region.[14]

Building on these observations, it seems possible to claim that the cultural backlash theory is not holding in the case of Italy and the spectacular success of the populist wave in 2018, while considerations relating to economic and job insecurity seem to prevail.

Dani Rodrick (2017) famously proposes fears of economic insecurities as an alternative view of the origins of contemporary populism. However, the connection to the experience of white, blue-collar, well paid, male communities losing their centrality in the political economy of their countries still seem to prevail in his narrative of the populist phenomenon.[15]

The next section will be devoted to an examination of the political economy origins of populism.

The Political Economy Origins of the Populist Phenomenon

Scholarly interventions on the economic causes of populist phenomena, especially in the US debate, tend to focus on the impact of globalisation on import competing production (Autor et al., 2013; 2016a; 2016b; Rodrik, 2017). For example, Autor et al. (2013) notice how rising import competition from China, as related to globalisation, has contributed to

[13] https://st.ilsole24ore.com/art/notizie/2018-03-06/genere-eta-professione-identikit-nuovi-elettori-cinque-stelle-190100.shtml?refresh_ce=1.

[14] https://www.money.it/voti-movimento-5-stelle-regioni-disoccupazione-pil.

[15] Rodrik, D. (2017). *Populism and the economics of globalization*. https://drodrik.scholar.harvard.edu/files/dani-rodrik/files/populism_and_the_economics_of_globalization.pdf.

the polarisation of US politics. Looking at the results of the US Congress elections of 2002 and 2010 and at the presidential ones of 2000, 2008 and 2016, they identify an ideological realignment especially in those local labour markets majorly exposed to trade with China. In particular, they noticed how electoral districts with a majority of white voters who already vote for the Republican party were increasingly moving to the extremes of the party when exposed to more import competition from China. On the other hand, those districts with a more ethnically diverse population, who were already voting democrat, moved further towards liberal democratic ideas when their trade-exposure increased.

The presidential elections, however only, showed more votes for the Republicans in import competing districts. This supports the idea that adverse economic conditions favour extremist politicians (Autor et al.,).

Although this is a significant contribution to the debate, the relation between economic difficulties and voting for anti-establishment parties, i.e. populist ones, is not clear.

Burgoon (2001) identified the relation between globalisation and the failure of welfare state compensation already in 2001. Back then there were three perspectives dominating the debate: that globalisation, i.e. market liberalisation, had little impact on welfare; that openness constrained welfare expenditure and, finally, that globalisation required some welfare. Burgoon's analysis predicts that different kinds of trade openness produced different effects on the welfare system, either retrenching or expanding different components of it (Burgoon, 2001). However, the connection between globalisation, economic insecurity and populism is not explicitly made.

On the contrary, Dani Rodrik (2017) is very explicit in linking populism directly to globalisation. In Rodrik's conceptualisation, the definition of populism does not only include an anti-establishment orientation and the claim to represent the will of the people against the elite, but also opposition to globalisation and liberal economics. The authoritarian element is, instead, not necessarily there (Rodrik, 2017). For this author, populism has different versions, but the most relevant are its right-wing variant and its left-wing one. The first is represented by Trump, the second by Latin American populism, a la Chavez. In both cases, what is most interesting is their opposition to globalisation, which is also at the root of their differentiation. When, indeed, globalisation manifests itself mostly under the form of growing migratory and refugee inflows, populism takes the form of an anti-migrant movement as it is easier to mobilise the electorate alongside the lines of national/ethnic considerations. According to Rodrik, this is the case for most Western forms of populism. When, on the other hand, globalisation mostly takes the form of foreign direct investment or other financial flows from abroad, such as in the case of Latin America or even the south Europe, then populism is more of a left-wing kind and people are mobilised alongside social and class lines (Rodrik, 2018, Harvard module). However, in any case, populism has an economic origin and not a cultural one, as claimed by Norris and Inglehart. In Rodrik's words:

> The economic anxiety and distributional struggles exacerbated by globalization generate a base for populism, but do not necessarily determine its political orientation. The relative salience of available cleavages and the narratives provided by populist leaders is what provides direction and content to the grievances. Overlooking this distinction can obscure the respective roles of economic and cultural factors in driving populist politics. (Rodrik, 2017: 2)

Moreover, starting from a sceptical approach as to the ability of globalisation to produce positive outcomes for everyone, Rodrik openly advocates for a rebalancing of the process. In particular, Rodrik argues that we need to rebalance the inequalities brought about by globalisation in three specific realms: from capital to labour, from global governance to national governance and geographically from core countries in the global political economy to marginal ones. If these three rebalancing acts do not take place, globalisation can lead to discontent which, as seen above, in electoral terms can bring about the prevalence of populist sentiments and eventually even authoritarian outcomes (Rodrik, 2017, 2018, Harvard module).

More specifically, in Frieden's analysis (2018b, interview with the author), political discontent is caused by failures of compensation, which in turn are the consequence of the lack or loss of adequate social safety nets to accommodate the needs of the losers of globalisation. However, to give rise to sizeable populist movements the economic element needs to be matched by a political one which, as Frieden (2017) defines it, is the failure of representation. This phenomenon arises when traditional political parties are unable to represent the grievances of a large share of the electorate harmed by economic trends.

Like other political economists discussed above, Frieden's conceptualisation of political discontent proposes that the economic trend responsible for the creation of inequalities is globalisation.

However, the hypothesis that globalisation is at the origins of the populist phenomenon is debatable. The problem is one of time-lag. In fact, if globalisation is an economic trend that can be dated back more or less to the early/mid-1980s, the rise of populist governments, such as Brexit in the UK (Ballard-Rosa et al., 2017a, b; Becker et al., 2018), Orban in Hungary, Italian populism and, to a certain extent also Trump in the US (Autor et al., 2013, 2016a, 2016b), did not occur until the early/mid-2010s.

Hence, this chapter suggests that the global economic crisis acted as a catalyst for the contradictions of globalisation to explode. This hypothesis is substantiated by recent studies by Guiso et al. (2017, 2018, 2019).[16] Relying on data from voting behaviour in different European countries, they conclude that economic insecurity drives the demand for populism. This happens both directly, as economic insecurity destroys trust in politics, and indirectly, as it also increases hostility towards immigrants. In turn, the appeal of populist parties is the consequence of economic crisis. This is contrary to what Frieden believes, connecting the supply of populism to a crisis of representation of traditional parties, and thus providing a political dimension (Frieden, 2017). Finally, differently from Rodrik (2017), they relate the political orientation of the populist party not so much to the consequences of globalisation, but to the existence of a political space. What is most important however is that they connect economic insecurity, not so much to globalisation per se, but to the impact of austerity policies ensuing from the global financial crisis and the euro-zone crisis.[17]

Starting from the assumption that the support to populist parties is the consequence of the failure of institutions to address economic difficulties, they locate the source of the economic insecurities bringing the

[16] Guiso, L., Herrera, H., Morelli, M., & Sonno, T. (2017, February). ***Populism: Demand and supply*** (CEPR Discussion Paper No. DP11871). Available at SSRN: https://ssrn.com/abstract=2924731. Guiso, L., Herrera, H., Morelli, M., & Sonno, T. (2019, January). Global crises and populism: The role of Eurozone institutions. *Economic Policy, 34*(97), 95–139, https://doi.org/10.1093/epolic/eiy018.

[17] Guiso, L., Herrera, H., Morelli, M., & Sonno, T. (2019, January). Global crises and populism: The role of Eurozone institutions. *Economic Policy, 34*(97), 95–139, https://doi.org/10.1093/epolic/eiy018.

latest wave of populism success in the shocks experienced by the eurozone countries after the global financial crisis (GFC). In this analysis, populism is a phenomenon characterised by three stages: (1) an initial anti-elite rhetoric; (2) an offer of immediate protection; (3) the impossibility to then deliver such protection and the hiding of the costs related to it (Guiso et al., 2017).

As applied to the case of populism's success in the context of the post-eurozone crisis, we have the following explanation: the eurozone crisis, and the way in which the EU institutions addressed it through austerity measures, increased economic insecurity especially in peripheral countries; such increase in economic insecurity reduced the trust of the electorate in traditional politics; populist parties arose and proposed easy ways out of the crisis; as eventually these solutions cannot be implemented, a scapegoat for their failure is identified in immigrants.

Thus, the explanation of populism cannot be cultural. To the contrary, its cultural manifestations, i.e. its distrust of power and anti-immigrant feelings, are a consequence of the worsening of economic insecurity. In turn, once manifested, populism feeds on anti-establishment and anti-immigrant feelings.

Guiso et al. (2017) show also that where globalisation did not bring economic insecurity, populism did not gain consensus. On the contrary, membership of the eurozone exacerbates the difficulties of eurozone states to address economic shocks ensuing from globalisation in general and the global financial crisis and eurozone crisis in particular. As we will see in the next section, the reason why this happens is because of the lack of fiscal policy independence due to the straitjacket of the Stability and Growth Pact and because of the limited role of the European Central Bank as lender of last resort.

This would explain why Italian populist movements such as the Five Star Movement and the Lega were, originally, vehemently opposed to the Economic and Monetary Union (EMU) and the euro (Talani, 2017). They did, however, change their ideology once their electoral success made it more convenient to compromise with EU institutions, especially after taking power. Then, the political discourse shifted towards an anti-immigrant attitude, identifying immigrants as a new scapegoat for the grievances of their electorate.

To be sure, the position of the Five Star Movement, before taking hold of the Italian government in 2018, was very overtly against the European Economic and Monetary Union. Also the relation between the adoption of the euro, the eurozone crisis and the impoverishment of the Italian economy was made very clear in the public interventions of the movement before winning power.

In an article published on the Five Star Movement blog in 2014,[18] it is plainly clear that, according to the movement, it is indisputable that the disaster of the Italian economy started with the adhesion of the country to the EMU. Moreover, it is explicitly claimed that the Italian economic crisis after the global financial crisis of 2008 has been exacerbated by the decisions of the institutions in charge of running the euro. As a consequence of austerity, the Italian economy was in recession and the Italian manufacturing production was back to the level of 1980.

In the opinion of the Five Star Movement, not even in the course of the great depression, under the government of Mussolini between 1929 and 1939, did Italy experience such a dire economic result. Indeed, they claimed that during that period the Italian economy grew by 16%.[19] For the Italian populist movement there is no doubt that, following the 2008 global crisis, the reason for Italy's dire economic damage was the permanence of the country in the EMU and the way in which the EU institutions, the European Central Bank (ECB) and the EU Commission in primis, handled the economic crisis and the crisis of the eurozone. Indeed, before entering the currency union, Italy had a trade surplus with Germany and Italian small and medium enterprises were seen as formidable competitors to German manufacturers given the weakness of the Lira.

Furthermore, according to the Five Star Movement, after pegging the Lira to the Deutch Mark (DM) in 1996, when Italy re-entered the Exchange Rate Mechanism of the European Monetary System, Italy experienced a vicious cycle losing productivity and competitiveness. To exit this cycle Italy would have needed more investment in human capital, infrastructure and innovation, which could not happen because of austerity.

Hence, the only solution for Italy was to get out of the "EMU trap", take back control of its monetary policy tools and denominate its public debt in Lire, introducing capital controls until the situation would calm

[18] See https://www.ilblogdellestelle.it/2014/08/fuori_dalleuro_per_non_morire.html. As accessed on January 9, 2019.

[19] See above.

down. In 2014, for the Five Star Movement, although there is no easy way out of the euro, the only way to save the economy was to "bet everything on the Lira".[20]

A similar position was reiterated in many occasions by the movement before the electoral campaign leading to the general elections of March 2018. For example, in 2017, intervening in an Italian political programme called "di martedi'", the leader of the movement, Luigi di Maio, when asked about his position on the currency Union, explicitly stated that he would vote yes to the Italian exit from the euro.[21]

The anti-euro discourse of the Lega before obtaining power is well documented. In various occasions, before the electoral tournament in 2018, Matteo Salvini, leader of the Italian eurosceptic party La Lega, proposed his opinion with respect to the euro. For example, on the 25th of January 2018, in a television interview, he declared his intention to get rid of the EU currency in case he took office.[22]

However, after the result of the elections were clear, and the possibility of governing with the Five Star Movement became a reality, his position towards the euro changed. On the 13th of March 2018, while he was in Strasburg giving his last speech as a member of the European Parliament, Salvini stated: "The euro was, and still is an unsuitable currency, but at the moment there is no possibility for a solitary exit from the common currency for a country like Italy. Our experts are working on a plan B in case Bruxelles decides not to cooperate with us".[23] Moreover, when asked about the possibility of a referendum to exit the eurozone he said:

> The Italian constitution impedes the Italian population from voting on international treaties, sadly. The modification of the constitution will be

[20] See above.

[21] See https://www.youtube.com/watch?v=v2wV1kq5IO4. Di Maio: diMartedì, Published on January 14, 2017, Luigi Di Maio risponde ad Antonio Caprarica in materia di moneta unica as accessed on January 9, 2019.

[22] See Salvini: 'The euro will cease to exist!' Italian election frontrunner vows to destroy EU currency (25 January 2018), https://www.express.co.uk/news/world/909747/Italian-election-2018-Silvio-Berlusconi-Jean-Claude-Junker-Matteo-Salvini-euro-EU#t=20s. As accessed on January 10, 2019.

[23] See https://www.ilblogdellestelle.it/2014/08/fuori_dalleuro_per_non_morire.html. As accessed on January 9, 2019. See https://www.ilblogdellestelle.it/2014/08/fuori_dalleuro_per_non_morire.html as accessed on January 9, 2019.

> part of the program of the next 5 years. This is not something that can be solved by tomorrow morning.[24]

Was the initial position of the two populist parties against the euro justified? Is it possible to claim that at the origins of the success of populism is the crisis of the eurozone and the way in which the EU institutions addressed it, at least in Italy? This is discussed in the next section of this chapter.

Populism and the Crisis of the Eurozone: The Case of Italy

Based on the discussion above about the political economy origins of populism and discarding mere cultural explanations, we can hypothesise that the crisis of the eurozone was a catalyst for the contradictions of globalisation to explode, bringing especially peripheral countries to a situation of dire economic insecurity. We can also hypothesise that the way in which EU institutions reacted to the crisis, instead of allowing for a redistribution of resources and thus decreasing the pressures on the economy, exacerbated insecurity through austerity. This is the economic substratum on which populism took roots and flourished in the post global financial crisis era. In this section we will analyse how this was manifested in the case of Italy.

Although the origins of the Italian decline are profound and need to be traced back in a number of concomitant structural phenomena that affect the Italian economy, it is also true that the global financial crisis and the ensuing crisis of the eurozone revealed the asymmetries of the Economic and Monetary Union which certainly did not favour peripheral countries, including Italy.

The definition of the eurozone crisis as a crisis of competitiveness and not as a fiscal crisis is now relatively undisputed in the literature (Bourgeot, 2013; De Grauwe, 2013; Eichengreen, 2014; Gros, 2011; Guiso et al., 2019; Krugman, 2012). The way in which a crisis of competitiveness is manifested is through a current account crisis (Bourgeot, 2013). Indeed, looking at the current account balance of the peripheral members of the euro-area, those which experienced the eurozone

[24] See https://www.ilblogdellestelle.it/2014/08/fuori_dalleuro_per_non_morire.html. As accessed on January 9, 2019.

crisis, i.e. Portugal, Ireland, Italy, Greece and Spain (the so-called PIIGS or GIIPS), their performance followed exactly the opposite dynamic of Germany from the onset of the euro to the eurozone crisis (see Figs. 9.1 and 9.2). While Germany experienced an unprecedented increase in the level of export and consequently a huge surplus of the current account balance, the GIIPS went increasingly in a deficit of the current account.

It is important at this point to discuss what it is meant by a 'crisis of competitiveness' as, according to which definition is selected, the policy options to adopt differ.

The measure of competitiveness most adopted in the literature is represented by the Unit Labour Costs (ULC). In particular, the nominal ULC are calculated as the ratio between total wages (which are nominal wages) and real GDP. This gives the labour cost of the production of a unit of real GDP. The other possibility to obtain the value nominal Unit Labour Cost is to divide the average wage (total wages over number of employees) by the productivity of one unit of labour (total real GDP divided by the total units of labour of a country).

On the basis of the definition of competitiveness given above, this is deeply affected by the level of nominal wages, which in turn are a function of the level of inflation.

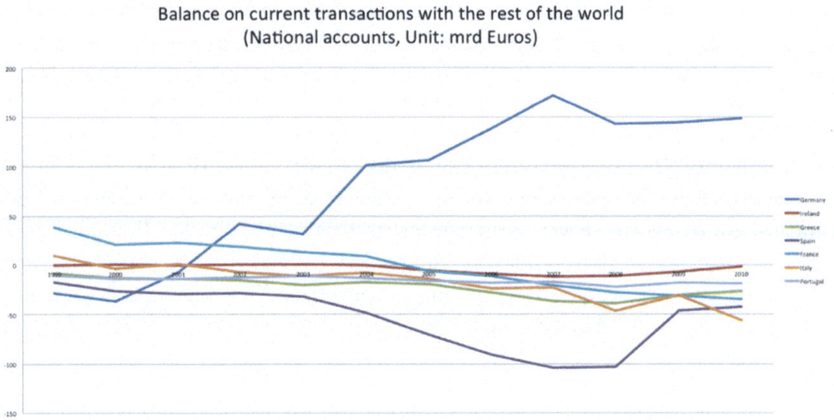

Fig. 9.1 Balance on current account transactions with the rest of the world: GIIPS and Germany, 1999–2010 (*Source* AMECO)

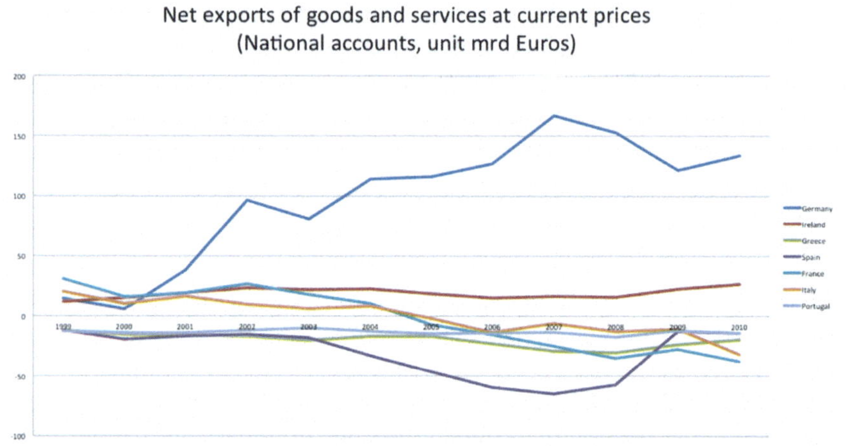

Fig. 9.2 Net exports of goods and services at current prices: GIIPS and Germany, 1999–2010 (*Source* AMECO)

On the other hand, the performance of a country in terms of competitiveness is linked to its productivity. Thus, the following indicators are relevant in assessing the motivations behind a fall in competitiveness: inflation rates, real costs of labour and productivity. On the basis of similar considerations, Bourgeot's contribution (2013) demonstrates how the divergences of nominal ULCs in the euro-area were not due to an increase in the level of real wages, also called wage drift, but to discrepancies in the levels of inflation and productivity. Indeed, looking at the performance of real wages, it emerges that the wage share of GDP was decreasing in countries at the periphery of the euro-area in the years between the establishment of the EMU and the crisis of the eurozone. For example, in Spain the wage share of GDP decreased by around 7% which is very similar to the 8% wage share decline experienced by Germany after the implementation of Agenda 2010 by Shroeder in 2003.[25] Notably, the

[25] In real terms ULCs represent the ratio between total real wage bill and real GDP (or as an equivalent the nominal wage bill and the nominal GDP). Real labour unit costs can therefore be expressed as a share of GDP and not just as variations or as an index number (unlike nominal ULCs). Likewise it can be calculated as the ratio between the average real compensation and real productivity. Their variations result from the development of real wages and of real productivity (and exclude inflation) (Bourgeot, 2013).

GIIPS countries, with the exception of Spain and Portugal, maintained always, before, during and after the euro-zone crisis, a wage share of GDP below the EU average, the euro-zone average and the level of Germany. Even in the case of Ireland, a peripheral country recording a significant increase of real wages in the period from 2002 to 2008, the wage share of GDP was always below the performance of Germany, the EU and eurozone (Fig. 9.3).

Bourgeot (2013) with Piccketty (2013) also underline how, in general, in the last forty years, real wages have been on a constant declining trend. The reasons for this are traced back to the weakening of the bargaining power of trade unions, as well as in the advent of globalisation. The same motivations are behind the constant increase in the share of profits over GDP (Bourgeot, 2013).

Thus, if we want to explain the increase in the Nominal Unit Labour Costs we need to refer either to an increase in the rate of inflation or to a decline in productivity.

In particular, it is undeniable that the inflation rates of the eurozone countries affected by the crisis of competitiveness were higher than those of Germany since before the establishment of the euro. It is therefore possible to conclude that competitiveness in the eurozone periphery has

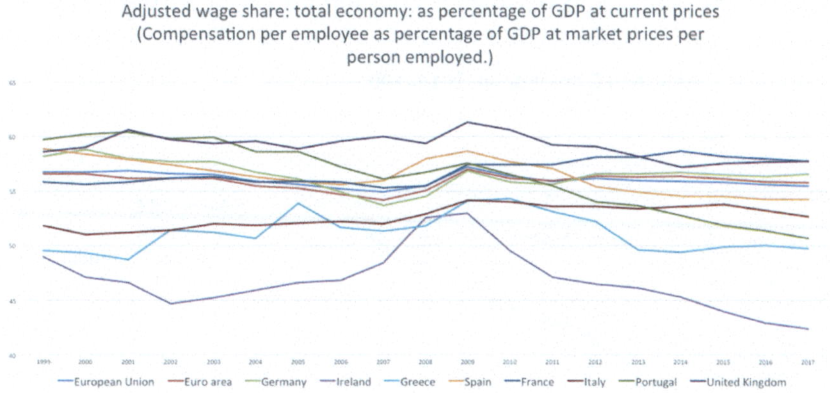

Fig. 9.3 Adjusted wage share: total economy as a % of GDP, 1999–2017 (*Source* The European Commission's Ameco online database [data retrieved in January 2016])

been profoundly impacted by higher inflation rates as reflected in the evolution of their national consumer price index (CPI).

Looking at the performance of Nominal Unit Labour Costs, real ULCs, labour productivity and the CPI for the peripheral countries of the eurozone in comparison with Germany it is possible to assess the relative role of each indicator on the decline of competitiveness. In particular, in the case of Italy whereas real wages, as reflected in the performance of real ULCs, and productivity did not change for the whole period of permanence in the Euro-zone, inflation, as reflected in the CPI, increased thus explaining the loss of competitiveness (Fig. 9.4).

According to Bourgeot (2013) this dynamic was particularly evident in the manufacturing sector, a very important sector when evaluating the degree of competitiveness of a country.

Summing up, a pattern of diverging competitiveness in a monetary union needs to be assessed as a consequence of divergent performances of the inflation rates in the absence of an increasing trend of the wage share of GDP or a declining productivity. Similar conclusions had been achieved by Walters (1990) in relation to the evolution of the Exchange Rate Mechanism of the European Monetary System (ERM of the EMS) when he posited that inflation divergences lead to divergences in real exchange rate and, consequently to a loss of competitiveness, the so-called "Walter's critique". This dynamic is even more valid in a fixed exchange

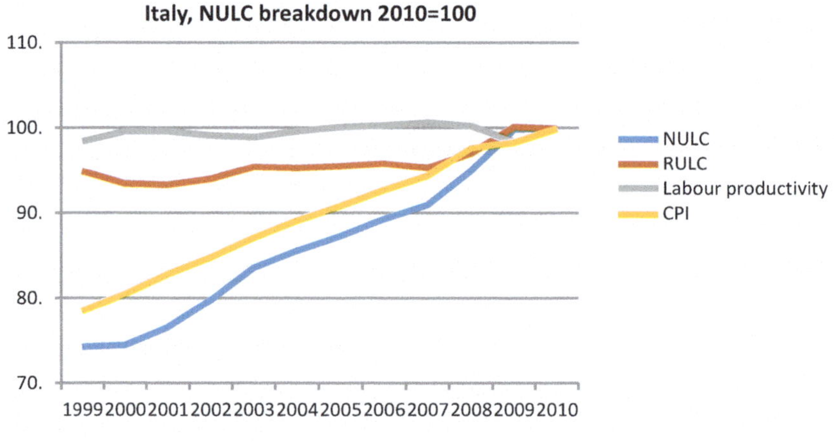

Fig. 9.4 Italy: NULC breakdown 2010 = 100 (*Source* AMECO)

rate system or in a currency union. Noticeably, in the Economic and Monetary Union (EMU), exactly like in the ERM of the EMS, the real exchange rate would diverge with respect to Germany, as there are inflation differentials between Germany and the countries of the eurozone periphery. Thus, the latter would lose competitiveness vis-à-vis Germany and this would explain the opposite performance of the current account in Germany and in the GIIPS as seen above.

As competitiveness was already an issue for the Italian capitalist system, the advent of the euro, in the absence of an Optimum Currency Area and of a common EU fiscal policy, exacerbated the difficulties the country was already experiencing. This was particularly detrimental in a period in which, after the global financial crisis, the country had been plunged in the worst depression recorded since the Great Depression of the 1930s (Alesina, 2018).

The way in which European institutions reacted to the eurozone crisis further aggravated what Frieden has termed the "failures of compensation" related to the inability of peripheral countries to compensate the losers of the Economic and Monetary Union and of the global financial crisis for their worsening economic conditions.

The asymmetries plaguing the EMU from the onset have been referred to as the so-called n−1 problem (De Grauwe, 2012a, b), a problem already existing in the previous European Monetary System, the ERM of the EMS, but which the loss monetary policy sovereignty by Member States made even more serious. In short, before the establishment of the currency union the problems of competitiveness affecting peripheral countries, the "n−1" countries, could be eased by devaluing their currencies, while the "1" country, i.e. Germany was setting the interest rates for the whole area belonging to the exchange rate agreement. However, after the adoption of a single currency and a single monetary policy, problems of competitiveness in the "n−1" countries could not be resolved any longer and were due to become progressively more serious.

In the lack of the so-called exchange rate and monetary tools to resolve situations of asymmetries, in a non-Optimum Currency Area where there is no common fiscal policy, the only possibility to react to a crisis of competitiveness is through the Central Bank acting as a lender of last resort to stop financial markets from speculating against the Member States of the currency union (Krugman, 2012).

De Grauwe (2012a, b) notices that it is an essential role of central banks to act as lender of last resort for both the government and the banking sector in a serious crisis of confidence. The case of the UK is indicative of this as, given that the country still enjoys a totally autonomous monetary policy, if attacked by financial markets the Bank of England would have immediately reacted by buying unsold Treasury bills. Unfortunately, in the euro-area the European Central Bank cannot perform such an essential role as this is prohibited by the Treaty establishing the Monetary Union. Therefore, during the eurozone crisis, the ECB could not intervene to stop financial markets if not in a "hidden" or disguised way and at a very late time (Krugman, 2012).

This explains why, when finally the governor of the ECB, Mario Draghi, was allowed to perform a role of "hidden" lender of last resort doing "whatever it takes" to save the euro area (Moravscik, 2012), most of the damage of the eurozone crisis had already been done. Indeed, a good amount of damage had been inflicted to the peripheral countries of the euro-area by European institutions themselves, focusing exclusively on a definition of the crisis as a fiscal crisis needing fiscal austerity measures to be resolved (Talani, 2017).

In fact, the austerity measures adopted in the wake of the eurozone crisis only depleted the ability of the GIIPS to react to the recession, without contributing to the achievement of a real fiscal union for the euro-area. Quite to the contrary, the austerity measures limited the capacity of Member States to intervene to rescue their economies without attributing such a role to any EU institutions thus curtailing any possibility to emerge relatively quickly from the difficult situation.

Among the various measures adopted by the EU institutions in the wake of the eurozone crisis, there was the establishment of an ad hoc solution lacking both institutional depth and democratic legitimacy. This was the European Financial Stability Facility (EFSF)[26] which then became the European Stability Mechanism (ESM)[27] in December 2010. The EFSF was established by the ECOFIN on the 9th May, 2010, when speculators started attacking the Greek sovereign debt. The fund was endowed with a total of €750 billion. This amount can be considered very limited and insufficient to guarantee against financial speculation especially if we

[26] See http://www.efsf.Europa.eu/about/index.htm. As accessed on December 15, 2010.

[27] For more details, see http://www.consilium.Europa.eu/uedocs/cms_data/docs/pressdata/en/ec/118578.pdf. As accessed on December 21, 2010.

compare it to the daily turnover of the London FOREX which is around US$5.3 trillion per day.[28]

The EFSF could also be financed by issuing bonds guaranteed by Euro Area Member States (EAMS) for an amount up to €440 billion which could be used for onward-lending to EAMS in difficulty. The bonds issued by the EFSF for the financing of the fund were awarded triple A credit ratings by Standard & Poor's, Fitch Ratings and Moody's. However, borrowing from the EFSF required abiding to heavy austerity conditions previously negotiated by the rescued Member State with the "troika" of the European Commission, the European Central Bank and the International Monetary Fund, approved by the EUROGROUP.

The EFSF did not manage to stop the run on the sovereign debt of the GIIPS[29] and was soon substituted by a more permanent institution, the European Stability Mechanism (ESM) deliberated by the European Council in December 2010 and inaugurated, after a ratification process that lasted two years, in October 2012.[30] With the entry into force of the EMS, the EFSF was phased out. Similarly to the ESFS, the ESM was meant to extend heavily conditional loans to the European Member States experiencing severe financial difficulties. The funding of the EMS was provided through money market instruments and medium and long-term debt with maturities of up to 30 years. These funds were to be guaranteed by the EAMS with capital provided in accordance with a contribution key annexed to the ESM Treaty.[31] De Grauwe (2012a, b) notices how the limitations of this fund to rescue countries in a crisis of competitiveness were underlined by the fact that countries in need had to ask for help also to the IMF.[32] Moreover, the EMS would always decide which conditions to impose to the countries willing to borrow from the institution in accordance with the IMF. Nowadays, with the COVID-19 crisis, the

[28] See more at: http://www.cityindex.co.uk/forex-trading/what-is-forex-trading.aspx#sthash.fSDJJL6w.dpuf. As accessed on January 20, 2016.

[29] See http://www.efsf.Europa.eu/about/index.htm. As accessed on December 15, 2010.

[30] The ESM Treaty entered into force on 27th September 2012. All seventeen euro area Member States had ratified by 3rd October 2012.

[31] See ESM website, available at http://www.esm.europa.eu/about/index.htm. As accessed on October 12, 2012.

[32] See ESM website, available at http://www.esm.europa.eu/about/index.htm. As accessed on October 12, 2012.

nature of one of the lines of intervention of the EMS has changed dramatically. Funds extended by the EMS to react to the COVID-19 crisis are no longer conditional, are not provided in accordance with the IMF and can be used only to support the health and care systems of the EAMS during the pandemic.[33]

Still relying heavily on austerity was another decision taken on the 2nd March, 2012 by the European Council to react to the eurozone crisis: the passing of the so-called 'Fiscal Compact' (officially the Treaty on Stability, Coordination and Governance, TSCG).[34] This required signatory Members States to remain committed to and include in their constitutions the rule of a balanced government budget or in surplus (Schweiger, 2014).

However, like similar austerity measures already adopted, this commitment to a balanced or in surplus budget failed to calm financial markets, the main reason being that the crisis was not a fiscal crisis but a crisis of competitiveness and therefore could not be resolved by fiscal austerity.

The only fiscal intervention that could have addressed the crisis, as often underlined in the literature (De Grauwe, 2012a, b), would be the establishment of a common fiscal policy. However, the Fiscal Compact was not in any way a fiscal union for a number of reasons. Partly because it was not signed by the UK, with the declared aim to avoid harming the City of London as publicly stated by the then Prime Minister David Cameron.[35] Most importantly, it was not a fiscal union because the Fiscal Compact did not include the most important features of a true Federal budget, such as a solidarity clause, a redistributive and a stabilisation function nor did it provide for a substantial increase in the size of the EU budget (Ackrill, 2000).[36] In reality, the Fiscal Compact reflected the same aims of the Stability and Growth Pact rendering them even stricter by making it compulsory for EAMS that signed it to enshrine in their

[33] See ESM web-site available at https://www.esm.europa.eu/content/europe-response-corona-crisis. As accessed on December 1, 2020.

[34] For the full text see http://www.european-council.europa.eu/media/639235/st00tscg26_en12.pdf.

[35] See https://www.theguardian.com/world/2011/dec/09/david-cameron-blocks-eu-treaty. As accessed on December 2, 2020.

[36] De Grauwe, P. (2012). Interview available at: http://aregan.wordpress.com/2012/03/20/interview-with-paul-de-grauw/. Accessed on October 18, 2012.

constitutional laws austerity and fiscal restraint. In this context, it is particularly striking that the Fiscal Compact did not include any reference to a solidarity mechanism that could be activated in case of the inability of one of the EAMS to react to the crisis. The joint declaration of the five Presidents of the EU on the 22nd of June 2015 detailing further steps in the eurozone integration process,[37] including the establishment of a EU Treasury, raised hopes that sooner than later a common fiscal policy could be adopted.[38] Indeed, the response to the COVID-19 crisis seems to move in the right direction.[39] However, the Fiscal Compact was not a step in the direction of an extension of solidarity to all the members of the euro-area or towards the establishment of a fiscal union (Talani, 2017).

Overall, the austerity measures devised by the EU institutions to react to the sovereign debt crisis not only did not stop market speculation but also worsened the difficulties faced by peripheral countries. Effectively, only the European Central Bank acting as a hidden "lender of last resort" had the ability to rescue the GIIPS from the speculation unleashed by financial markets against them. Although the ECB could not act automatically, as all other central banks can do, to rescue struggling economies by intervening directly in the primary markets to acquire unsold debt, it did resort to acting in secondary markets doing "whatever it takes" to stop the fury of the markets. This strategy did, indeed, work (De Grauwe, 2012a, b; Eichengreen, 2014; Giannone et al., 2011; Krugman, 2012).

On the other hand, the austerity measures adopted to counteract the sovereign debt crisis only produced deflation. Deflation is very dangerous in a situation of recession as it signals the contraction of demand and the entering into a vicious circle of less aggregate demand, less growth and less employment. Also on this occasion the intervention of the ECB was vital and consisted in establishing a quantitative easing (QE) programme on the 22nd January, 2015. Through such a programme the ECB provided the eurozone with 60 billion euros a month to help

[37] See Van Rompuy, H., Barroso, J., Juncker, J.-C., & Draghi, M. (2012, June 26). *Towards a genuine and economic monetary union*. European Council, Brussels, EUCO 120/12.

[38] See EU web-site http://europa.eu/rapid/press-release_IP-15-5240_en.htm. As accessed on October 22, 2015.

[39] See https://ec.europa.eu/info/strategy/recovery-plan-europe_en. As accessed on December 2, 2020.

stimulate aggregate demand, growth and employment.[40] It is particularly important to note that, although the recession in the eurozone had been there since the 2008/2009 global financial crisis, the ECB was allowed to start QE only when deflation hit the EU. This delay in increasing the monetary supply made it possible to shift the burden of the global financial crisis onto the weaker countries and the weaker societal strata of the EU.

It is on the basis of a similar shift in the burden of the global financial and eurozone crisis, that it is possible, as detailed in the first two sections of this chapter, to explain the exponential growth and rise to power of populist parties, such as the Five Star Movement in Italy.

Conclusion

In conclusion, there is no doubt that, at least in Europe, the adverse consequences of the global financial crisis were shifted to the weakest countries of the system and to the weakest social groups within them, especially women and the youth. This shift was effected through the two interrelated phenomena of the sovereign debt crisis and its ensuing austerity measures. The global financial crisis and the eurozone crisis exacerbated already existing geographical and social discrepancies. On the other hand, austerity, far from allowing for a resolution of the contradictions of the crises, substantially limited the resources available to the state to compensate the losers. This created what Frieden (2018a) calls "a crisis of compensation" which, in some countries, where it was accompanied by a "crisis of representation", led to the prevalence of populist parties in the elections.

Thus, it is possible to explain the rise and growth of populism in the eurozone periphery as a consequence of the economic difficulties experienced by the unemployed youth and both men and women with a high education level after the economic crisis of 2008/2009 and the sovereign debt crisis of 2010/2011.

The intervention of the European Central Bank, by underlying its will to do "whatever it takes" to stop market speculation against the sovereign debt of the peripheral countries, eventually resolved the eurozone crisis. However, the disparities and asymmetries which characterised the first

[40] See BBC, http://www.bbc.co.uk/news/business-30915210. As accessed on October 22, 2015.

ten years from the establishment of the EMU were far from being over. The only long-term solution for them would be the establishment of a common fiscal policy and, with it, of true solidarity between the members of the currency union.

Instead of adopting such a bold solidarity step, the EU institutions relied heavily on austerity and "internal devaluation" to react to the crises, thereby substantially reducing, instead of increasing, the ability of states to address existing social and geographical inequalities. This happened despite the rhetoric about more integration and the creation of a new supranational economic governance system, neo-functionalist style, which was proposed, perhaps disingenuously, by both academic and EU institutional circles (Caporaso & Kim, 2015; Schimmelfennig, 2014).

As widely anticipated, this austerity-oriented policy resulted in the popular discontent, political instability and threats of disintegration which characterise the recent populist phenomena in Europe.

In the case of Italy, there is a widespread agreement among the scholarly community that the crisis of 2010/2011 was mostly a crisis of competitiveness, affecting in particular the Italian export-oriented manufacturing sector. The economic backlash of the crisis led to a growing discontent of some sections of the Italian electorate, especially the young, urban unemployed enjoying high educational levels. These sections of the electorate failed to be rescued by the State due to the highly restrictive fiscal policy imposed by EU institutions as a way out of the crisis. At the same time, the Italian political system was experiencing a "crisis of representation" triggered by the disappointment of the electorate towards the traditional Italian political parties. In particular, Berlusconi and Forza Italia were considered responsible for the onset of the sovereign debt crisis and the PD, with the Monti, Letta and Renzi government, were blamed for the imposition of restrictive fiscal policies and austerity (Talani, 2018).

The combination of the "crisis of compensation" of the losers of the economic crisis and of the "crisis of representation" of the Italian party system led the Italian electorate to accept the populist, anti-establishment offers of the Lega and the Five Star Movement. Both parties, as seen in this chapter, indeed underlined how the roots of the Italian economic strains were to be found in the asymmetries of the eurozone and in the eurozone crisis. Once in power, however, the populist government based on the coalition of the Lega and of the Five Star Movement, given the impossibility for Italy to leave the eurozone for reasons which is not here the place to discuss, managed to shift the blame for Italian problems to migrants who are easy scapegoats not least because they do not have the right to vote.

References

Ackrill, R. W. (2000). *The common agricultural policy*. Sheffield Academic Press.

Alesina, A. (2018). Interview with the author, Harvard University, Mimeo.

Autor, David H., Dorn, D., & Hanson, G. H. (2013, October). The China syndrome: Local labor market effects of import competition in the United States. *American Economic Review, 103*(6), 2121–2168. https://doi.org/10.1257/aer.103.6.2121

Autor, D., Dorn, D., Hanson, G., & Majlesi, K. (2016a). *Importing political polarization? The electoral consequences of rising trade exposure* (NBER Working paper). https://www.nber.org/papers/w22637. As accessed on January 15, 2019.

Autor, D., Dorn, D., Hanson, G., & Majlesi, K. (2016b, November). *A note on the effect of rising trade exposure on the 2016 Presidential Election* (Working paper).

Ballard-Rosa, C., Malik, M., Rickard, S., & Scheve, K. (2017a, November 17–18). *The economic origins of authoritarian values: Evidence from local trade shocks in the United Kingdom.* Paper presented at the Annual Meeting of the International Political Economy Society, Austin. https://ncgg.princeton.edu/new_ipes/2017/BMRS%20_IPES_2017.pdf

Ballard-Rosa, C., Malik, M., Rickard, S., & Scheve, K. (2017b, October). *The economic origins of authoritarian values: Evidence from local trade shocks in the United Kingdom* (Working Paper).

Becker, S., Fetzer, T., & Novy, D. (2018, January). Who voted for Brexit? A comprehensive district-level analysis. *Economic Policy, 33*, No. 93, 179–180, https://doi.org/10.1093/epolic/eix017

Bickerton, C., & Accetti, C. I. (2018). 'Techno-populism' as a new party family: The case of the Five Star Movement and Podemos. *Contemporary Italian Politics, 10*, 1–19.

Bobba, G., & McDonnell, D. (2015). Italy—A Strong and Enduring Market for Populism. In H. Kriesi & T. Pappas (Eds.), *European populism in the shadow of the great recession.* ECPR Press.

Bourgeot, R. (2013, September 23). Labour costs and crisis management in the Euro Zone: A reinterpretation of divergences in competitiveness. *European Issues*, No. 289. http://www.robert-schuman.eu/en/european-issues/0289-labour-costs-and-crisis-management-in-the-Euro-zone-a-reinterpretation-of-divergences-in. As accessed on January12, 2016.

Burgoon, B. (2001). Globalization and welfare compensation: disentangling the ties that bind. *International Organization, 55*(3) (Summer), 509–551.

Canestrari e Biondo. (2018). La supernova.

Canestrari e Biondo. (2019). Il Sistema casaleggio.

Caporaso, J. A., & Kim, M.-H. (2015). States choose but not under circumstances of their own making: A new interpretation of the integration debate in light of the European financial crisis. In J. A. Caporaso & M. Rhodes (Eds.), *The Political And Economic Dynamics of the Eurozone crisis* (pp. 15–47). Oxford University Press.

Casaleggio. (2011). Siamo in Guerra.

Casaleggio. (2012). Tu sei rete.

Casaleggio. (2016). Aforismi.

Conti, N., & Memoli, V. (2015). The emergence of a new party in the Italian party system: Rise and fortunes of the Five Star Movement. *West European Politics, 38*(3), 516–534.

Corbetta, P. (2017). *M5s Come cambia il partito di Grillo.* Bologna: Il Mulino.

D'Arma 2015 Digital democracy. Charismatic leader populism.

De Grauwe, P. (2012a). Interview available at http://aregan.wordpress.com/2012/03/20/interview-with-paul-de-grauw/. Accessed on October 18, 2012.

De Grauwe, P. (2012b, March 20). Interview with Paul de Grauwe. *Euro-Irish Public Policy*. https://wp.me/pnUaW-pX

De Grauwe, P. (2013). 'Foreword'. In M. Marco (Ed.), *The Economics of the Monetary Union and the Eurozone Crisis*. London: Springer.

Eichengreen, B. (2014, March). *The Eurozone crisis: The theory of optimum currency areas bites back* (Notenstein White Papers, pp. 1–18). https://eml.berkeley.edu/~eichengr/Swiss%20Bank%20White%20Paper%20No%202%20Text-1.pdf

Frieden, J. (2017, December). The political economy of the globalization backlash: Sources and implications, Jeffry Frieden Harvard University.

Frieden, J. (2018a, January). *The political economy of the globalization backlash: Sources and implications*. Paper presented at the annual meeting of the American Economic Association. http://pelg.ucsd.edu/Frieden_2018.pdf

Frieden, J. (2018b). Interview with the author. Harvard University, Mimeo.

Gerbaudo. (2018). Il partito piattaforma.

Gerbaudo. (2019a). One person one click!.

Gerbaudo. (2019b). "The digital party" digital democracy-Rousseau.

Giannone, D., Lenza, M., Pill, H., & Reichlin, L. (2011, January). *Non-standard monetary policy measures and monetary developments* (ECB Working Paper Series No 1290). Brussels. https://www.ecb.europa.eu/pub/pdf/scpwps/ecbwp1290.pdf

Gros, D. (2011, May 25). *External versus domestic debt in the Euro Crisis* (CEPS Policy Briefs, No. 243). https://www.ceps.eu/node/5677

Guiso, L., Herrera, H., & Morelli, M., & Sonno, T. (2017, February). *Populism: Demand and supply* (CEPR Discussion Paper No. DP11871). Available at SSRN: https://ssrn.com/abstract=2924731

Guiso, L., Herrera, H., Morelli, M., & Sonno, T. (2019). Global crises and populism: The role of Eurozone institutions. *Economic Policy, 34*(97), 95–139. https://doi.org/10.1093/epolic/eiy018

Krugman, P. (2012). Revenge of the optimum currency area. *NBER Macroeconomics Annual, 27*, 439–448. https://www.nber.org/chapters/c12759

Lo Scalzo 2019 on the Rousseau platform.

Moravscik, A. (2012, May/June). Europe after the crisis. *Foreign Affairs*. https://www.foreignaffairs.com/articles/europe/2012-05-01/europe-after-crisis. As accessed on July 15, 2016.

Mosca, L., & Tronconi, F. (2019). *Beyond left and right: The eclectic populism of the Five Star Movement. West European Politics, 42*(6), 1258–1283. https://doi.org/10.1080/01402382.2019.1596691

Mudde, C. (2004). The populist Zeitgeist. *Government and Opposition, 39*(4), 542–563.

Mudde, C. (2017). Populism: A very short introduction (Very short introductions) (p. 3). Oxford University Press. Kindle Edition.

Norris, P., & Inglehart, R. (2019). *Cultural backlash: Trump, Brexit, and authoritarian populism*. Cambridge University Press. Kindle Edition.

Norris, P., & Inglehart, R. (2018). Seminar on populism held at Harvard University, Mimeo.

Piccketty, T. (2013). *Capital in the twenty-first century*. The Belknap Press of Harvard University Press.

Rodrik, D. (2017). *Populism and the economics of globalization*. https://drodrik.scholar.harvard.edu/files/dani-rodrik/files/populism_and_the_economics_of_globalization.pdf.

Rodrik, D. (2018). *The political economy and its future*. Harvard module followed by the author. https://drodrik.scholar.harvard.edu/classes/political-economy-and-its-future

Schimmelfennig, F. (2014). European integration in the Euro crisis: The limits of postfunctionalism. *Journal of European Integration, 36*(3), 321–337.

Schweiger, C. (2014). The EU-25 fiscal compact: Differentiated spillover effects under crisis conditions. *Perspectives on European Politics and Society, 15*(3), 293–304.

Tajfel, H., & Turner, J. C. (1985). The social identity theory of intergroup behaviour. In S. Worchel and W. G. Austin (Eds.), *Psychology of intergroup relations* (2nd edn, pp. 7–24). Chicago: Nelson Hall.

Talani, L. S. (2017). *The political economy of Italy in the Euro. Between credibility and competitiveness*. Palgrave Macmillan.

Talani, L. S. (2018). *The political economy of the populist vote in Italy*. Harvard University Lecture, April 2019, Mimeo.

Tarchi, M. (2014). L'Italia populista e il caso Beppe Grillo Ten years after. Populist Italy and the case of Beppe Grillo. *Quaderni di Sociologia, 65*, 31–49.

Tronconi, F. (2018). The Italian Five Star Movement during the crisis: Towards normalisation? *South European Society and Politics, 23*, 163–180. https://doi.org/10.1080/13608746.2018.1428889.

Walters, A. A. (1990). *Sterling in danger. The economic consequences of pegged exchange rates*. Fontana/Collins

Wodak, R. (2015). *The politics of fear: What right-wing populist discourses mean*. Sage.

Conclusion: Migration and the 'Dark Side' of Globalisation

The theoretical discussion contained in this book underlines the uneven and unequal nature of globalisation. Globalisation emerges from the analysis as a process ridden with contradictions, whose consequences increase social discrepancies and geographical marginalisation, leading to what the author of this book has defined as the 'dark side' of globalisation. Here, it is argued that globalisation manifests its inherent dark-side in relation to international migration due to a number of phenomena that escape traditional controls and regulations. Below we look at the main components of the 'dark side' of globalisation in relation to international migration.

The Loss of Political Control of International Migration

The first aspect of the 'dark side' of international migration in the age of globalisation is whether or not globalisation-induced migratory flows can be governed and by whom.

As detailed in Chapter 1, this book has focused on the macro-analysis of immigration policy i.e. the study of the conditions of the entry and exit of migrants from national territories and whether these conditions can still be controlled by the nation state. This focus on the role of the nation state in facilitating, hindering or regulating migratory flows is, indeed, at the centre of the enquiry of political scientists' approach to migration studies, differentiating it from the economists' one, which, on the

L. S. Talani, *The International Political Economy of Migration in the Globalization Era*, International Political Economy Series,
https://doi.org/10.1007/978-3-030-79321-0

contrary, utilises methodological individualism to ascertain the individual costs and benefits of the migratory process. The political scientists' understanding of the IPE of migration focuses on how macro-political entities, such as states or institutions, and socio-economic groups, influence the size and frequency of international migratory flows in the age of globalisation as well as the motivations of migration both at the point of origin and of arrival.

According to which IPE approach is selected, however, the question relating to the role of the state in allowing or stopping migration has a different answer. In the analysis proposed by realist scholars, there is no doubt that national institutions are totally in control of migratory dynamics in the globalisation era and any outcome of migratory policy, including the increase of irregular migration, is an intended consequence of specific policy provisions. On the contrary, for neo-institutionalists, liberal democracies do not have the ability to implement restrictive migratory policy as international migrants are attracted to their territories by legal systems that protect their rights. The enforcement of those liberal legal provisions by national, but also regional or even international courts, makes it impossible for western democracies to limit the rights of migrants with the aim of limiting migration. There is, however, the scope for international organisations to intervene to regulate and control the phenomenon.

Overall, both mainstream IPE theories remain confident that international migration in the age of globalisation can be governed by either national, for the realists, or supranational institutions, for liberal institutionalists. The two theories still attribute to the political dimension, whether at the national or at the supranational level, the ability to control globalisation thus, in a way, negating its 'dark side'.

On the other hand, the globalisation thesis, as discussed in Chapter 2, posits that the phenomena comprising what Mittelman (2011) terms the globalisation syndrome, including international migration, cannot be governed by political entities at any level. The forces unleashed by globalisation escape governance as they are structural necessities.

In particular, starting from a qualitative definition of globalisation, the globalisation thesis argues that the structural transformations of the global political economy lead to the structural need for populations to move both within regions and outside them. This is the consequence of the three paradoxes of globalisation and their impact on the motivations for migration:

- The paradoxof "marginalisation" and its impact in terms of increase of extra-regionalpermanent migrationand brain drain.
- The paradox of "regionalisation" and its consequences in terms of intra-regional temporary migration.
- The paradox of "securitisation" and its consequences in terms of irregular migration.

These paradoxes follow from the structural nature of globalisation and the emergence of a New Global Division of Labour and Power, and therefore the urge to migrate cannot be stopped by political entities. From this perspective, thus, migration cannot be controlled, regulated or governed neither by the state nor by supranational institutions. By imposing restrictive regulatory regimes to international migration, the only result that political institutions can obtain is to transform regular into irregular migration. Moreover, because of the paradox of regionalisation within globalisation and of the paradox of marginalisation, the population of the non-regionalised, marginal areas of the global political economy experience an increased incentive to migrate, thus adding two further elements to the dark side of globalisation: the increase of mass migration and brain drain.

The Marginalisation of Regions and Countries and the Increase of Mass Migration and Brain Drain

The phenomenon of globalisation -induced regionalism has been recognised by all different IPE approaches, although its nature and consequences are different according to which IPE perspective is considered.

As noticed in Chapter 4, realists consider regionalisation as unrelated to globalisation, whose novelty and even existence they deny. From their viewpoint, regionalism and regionalisation only respond to the interest of the hegemonic states. They do not lead to a global governance system and, if the hegemonic states lose interest in the process of regional integration, this would be reversed and dissolved.

Both institutionalists and transnationalists, on the other hand, believe that the process of regionalisation is intimately connected to that of globalisation. However, institutionalists consider it as a step towards the establishment of a cosmopolitan democratic governance system. Differently, transnationalists notice how the relationship between globalisation and regionalisation is dialectical and entails a number of contradictions and paradoxes. Some regions, the ones that are more integrated in the global political economy, will develop advanced forms of regionalisation and regionalism. Those which are unable to catch up with globalisation will, on the other hand, be progressively more marginalised and will not see their degree of internal integration progress.

In turn, this constant increase in their marginalisation and the lack of regional integration represent a push factor for mass migration and brain drain to the core areas of the global political economy, the so-called "triad".

The analysis in Chapter 3 shows that the MENA area, and the three countries considered in the chapter, i.e. Tunisia, Morocco and Egypt, are experiencing a lack of integration in the global political economy and are progressively more and more marginalised from its core.

As far as Tunisia is concerned, despite the widespread rhetoric about the Tunisian economic miracle at the end of the 1980s/beginning of 1990s, the data proposed in the book clearly shows how the country has been progressively more marginalised from the global political economy.

This inability to catch up with globalisation has led to the wave of discontent at the roots of the revolt against Ben Ali in 2011. However, marginalisation also fueled the constant increase of emigration from the country. This is a phenomenon which has maintained the same characteristics in the last decades, both before and after the events of 2011, thus confirming the economic nature of the Tunisian diaspora. Moreover, together with unskilled labour, it includes the exit of highly educated Tunisians, thus leading to the phenomenon of brain drain. The latter is, especially detrimental in the globalisation era as it makes it even harder for the country to acquire and retain the skills necessary to catch up with the highly technological content of globalisation.

Egypt is also on its way to further marginalisation as the figures reported in Chapter 3 relating to its share of Global FDI stock, Mergers and Acquisitions and production for exports show. Reliance on liberalisation and privatisation programmes, starting with Sadat's Infitah in 1974 until the more recent implementation of the IMF and World Bank restructuring and stabilisation measures, did not succeed in attracting foreign investment. Instead, foreign investment steadily declined in terms of the global share of FDI stock in the last decades and is by far less important for the economy of the country than migrants' remittances.

Indeed, Egypt seems to be a paradigmatic case of the failure of liberalisation policies to deliver integration into the global political economy, leading instead to marginalisation.

In turn, marginalisation has acted as a push factor for extra-regional permanent emigration of Egyptians whose numbers in the US and in Europe are steadily increasing. Further adding to the marginalisation of the country is the fact that many of the migrants leaving Egypt for good are highly educated and highly skilled ones. This leads to a vicious circle which does not seem to have been affected by the various regime changes happening after the 2011 revolts. Indeed, as detailed in Chapter 3, the profile of the Egyptian migrant has remained the same and configures the outlook of a mainly economic related migratory dynamic.

Similarly, the profile of the Moroccan migrant, as verified by a survey conducted by the author, compares very closely to that of the Tunisian and of the Egyptian one. Indeed, the typical Moroccan migrant is an economically motivated, young male who is not deterred to emigrate irregularly, as the answers to the questionnaires implemented in Morocco reveal. Also, they are very likely to be underemployed in the receiving countries. This happens despite the fact that Morocco did not experience the same political upheavals of the other two countries analysed and enjoyed regime stability in the last decades. Still, the emigration dynamics of the country follow the same path as the other two countries of the MENA area showing a constant increase of extra-regional permanent migration to the global north, especially the EU.

Thus, the analysis conducted in the book shows clear evidence of both marginalisation and lack of economic integration in the MENA area and reveals how this phenomenon is related to the steady increase in extra-regional permanent migration of both unskilled labour and, worryingly, skilled workers. In turn, this further reduces the chances of marginalised geographical areas to be included in the global political economy, clearly manifesting the dark-side of globalisation. As seen in the case of the three countries selected, migrants leaving them are not only aware but also willing to embark in migratory processes without proper documents, which brings us to a further element of the dark side, the irregularisation of international migration.

The 'Irregularisation" of International Migration and the Criminalisation of Migrants

The "irregularisation" of migration is, then, another negative consequence of globalisation on migration. This entails the creation of new inequalities in labour markets, the rise of the so-called 'modern slavery', as well as the death toll that the process of migrating through irregular means inevitably produces.

The discussion in Chapter 7 reports on the widespread consensus in the scholarly community that international migration is generally beneficial for the economic performances of host societies. International migration is considered in the literature a positive sum game for destination countries as it allows them to cover the gaps of the labour market, complementing the skills of the local labour force and overall enhancing the productivity and efficiencies of their economies. This happens despite the fact that they are often underemployed, are relatively less employed than the local population and have to accept working conditions below standards, which is often the case for both regular and irregular migrants.

The negative aspects of globalisation-induced migratory flows, in fact, come from their irregularisation which substantially contributes to the antagonisation, even criminalisation, of international migrants by receiving societies.

Due to the paradox of the securitisation of migration within globalisation and the impossibility of stopping international migration by using restrictive migration policies and deterrence, legal migration is very limited while irregular forms of entry and stay thrive.

In turn, there are clear winners and losers in the transformation of regular into irregular migration. On the one hand, irregular migration allows for an uber-flexibilisation of the labour force, both the local and the migrant one. Besides, irregular work reduces the size and function of the welfare state, limiting substantially state contributions. Consequently, both workers and citizens would appear as clear losers from the growing 'irregularisation' of migration.

Moreover, the irregularisation of migration intensifies the negative perception of migrants by host societies which, in turn, leads to more marginalisation and exploitation, including instances of 'modern slavery'. This is due to the fact that irregular migrants can only be employed in the underground economy of receiving countries. As such, they are perceived by host societies as acting at the margins of legality, if not in the illegal economy proper, as they are often smuggled into the country of arrival by organised crime and are therefore indebted with it. Since, as irregular workers, they pay less taxes, this makes local populations blame them for using public services, such as education or the health system, without contributing to it.

It must be noticed, with Reyneri (2013), that by no means the underground economy is created by irregular migrants nor are irregular jobs their prerogative. To the contrary, it is the underground economy that acts as a pull factor for irregular migration, as discussed in Chapter 6. In reality, the biggest majority of irregular workers are represented by locals. However, there is a stigmatisation attached to irregular workers from third countries which does not apply to the natives. The formers are also usually considered more prone to committing crimes when they are involved in the underground economy.

De facto, it is possible to identify a phenomenon of the ethnification of the jailed population. This has been connected to a process of the self-selection of irregular migrants. If they are willing to undergo the process of migration irregularly and to be involved in the underground economy, then it is also possible that they have a higher propensity to misbehave. They might therefore be more likely to engage in deviant or even criminal behaviour. This could be incentivised by the fact that illegal activities are often more profitable than legal ones. Also, irregular migrants might also be bound to engage in criminal activities by the same criminal organisations that smuggled them in. There is then the possibility that migrants do not realise that rules have to be respected in a country that allows them to enter and stay illegally as well as to work in the submersed economy. Finally, it should be noticed that in many countries, irregular migration is actually a crime, and therefore irregular migrants are criminals by definition, which certainly lowers their reluctance to misbehave.

From a different viewpoint, the securitisation paradox, which is often justified as a way to limit global terrorism, could paradoxically fuel terrorist tendencies, not only in first-generation, but also in second and third-generation migrants as discussed in Chapter 8. By facilitating instances of discrimination and Islamophobia, the paradox of securitisation could act as a catalyst for Islamic communities to grow more hostility towards host societies, that could be reflected in social unrest, radicalisation and even terrorism.

All the considerations above contribute to the antagonisation of irregular international migrants by host societies and to their criminalisation.

This could, in what is called a short-circuit of criminalisation, incentivise them to engage in illegal activities.

Clearly, the solution to this conundrum would be to legalise international migration, thus eliminating its dark side. However, this does not seem to be the solution chosen by nation states or supranational institutions. Quite to the contrary, the tendency is towards an even greater securitisation of the border. This is the case even in relation to the entry of refugees, who would have a right to ask for asylum as per UN Convention on refugees of 1951, but do not have a right to reach a country regularly. They are therefore likely to undergo irregular migratory processes until they reach the shores of a country where they can ask for asylum. This increases insecurity for refugees, who may die at sea, may be trafficked by organised crime and may be smuggled into destination countries where they are exploited to repay the debts incurred to pay for the journey. It also makes host societies much less safe as they have to deal with irregular migrants, criminal networks and exploitation of workers, modern slavery and illegal activities on their territories.

In the case of the EU, the recent debate, after the refugee crisis of 2014/2015, talks about a de-facto militarisation of the border. Not only did similar measures not restrain migrants nor refugees from starting the migratory process irregularly, but they also did not have an impact on reducing the number of victims at sea, which is still extremely high, as reported in Chapter 5.

In conclusion, the securitisation of migration policy overall only results in the opposite effect, which is in the increase of insecurity. It does so through the irregularisation of migrants and refugees which leads to extremely dangerous if not deadly journeys to reach destination countries; to the involvement of organised crime in the smuggling and exploitation of migrants; to the increased precarity of the working conditions of both the local and the migrant working force which can configure instances of modern slavery; to the necessary involvement of irregular migrants in the underground economy and their related marginalisation and criminalisation by host societies; to the ethnification of jailed populations and an increased incentive to misbehave and even commit crimes; to the growing hostility of migrant communities against receiving countries that could lead to social unrest or even terrorism, and vice-versa i.e. the growing hostility of 'native populations' to migrant communities which leads to Islamophobia and the rise of right wing populism.

All this is a consequence of the paradox of securitisation within globalisation. If international migration is a structural component of globalisation, political institutions cannot stop it. The policy gap is real. The implementation of restrictive policies only produces the irregularisation of international migration. In a nutshell, increasing securitisation increases insecurity.

But, if this is the case, why is the practice and discourse of securitisation so pervasive around the globe? Who are the winners and the losers of these insecurity enhancing securitisation measures?

If the losers, as seen here, are represented by international migrants and refugees and by local citizens, there are, however, also winners and these are populists and right-wing anti-migrant parties.

Populism and the Rise of Anti-migrant Parties

The debate about populism and populist right-wing parties has been revamped by the recent wave of success of similar parties in elections worldwide. Populism, in itself, does not need to be considered necessarily a negative consequence of globalisation, although the literature unanimously underlines the authoritarian tendencies of the populist ideology. However, the fact that populism is often accompanied by a very explicit anti-migrant discourse is certainly an element of the dark side.

Chapter 9 of this book has explored the relationship between both the rise of populism and globalisation and between populism and anti-migrant attitudes.

In particular, the analysis has sought to verify whether the recent waves of populism are inherently anti-migrant and xenophobic or, instead, they become so in a second moment to attract voters.

Although the notion of populism is still very contested in the relevant literature, its relationship to globalisation seems to be less debatable. Some authors (Norris & Inglehart, 2019) underline how populism represents the "cultural backlash" of those sectors of the economy negatively impacted by globalisation. The progression of the events leading to populism is, according to them, as follows. First, some socio-economic groups experience a deterioration of their socio-economic conditions due to globalisation. Then their in-group cultural values become more conservative, anti-egalitarian and illiberal as well as tendentially authoritarian and against cultural and social diversity. If those groups manage to elect their representatives in democratic elections, then the cultural backlash threatens the democratic system. In particular, according to this view, it is, especially older, white, uneducated men, living outside urban conglomerates and impoverished by globalisation, that have become more hostile towards inclusive and progressive liberal values.

Other scholarly interventions emphasise in particular the economic causes of populist phenomena focusing, especially on the impact of globalisation on import competing production or on the relationship between globalisation and the failure of welfare state compensation.

Dani Rodrick (2017) famously talks explicitly about a globalisation backlash to explain the origins of contemporary populism. Without stressing the cultural element, he highlights the connection to the experience of white, blue-collar, well paid, male communities losing their centrality in the political economy of their countries as a consequence of globalisation to give account of the populist phenomenon. He goes even further by claiming that the definition of populism includes opposition to globalisation and liberal economics. As far as the relationship between populism and anti-migrant attitudes is concerned, he distinguishes between right-wing and left-wing populism. While both are characterised by opposition to globalisation, their differentiation derives from the way in which globalisation manifests itself in different regions of the globe. In the global north, globalisation manifests itself mostly under the form of growing migratory and refugee inflows. Therefore, populism adopts an anti-migrant discourse to mobilise the electorate alongside the lines of national/ethnic considerations. In the global south, globalisation takes the form of foreign direct investment or other financial flows from abroad. Consequently, populism there is more of a left-wing kind as it is easier to mobilise voters alongside social and class lines. Thus, the economic backlash of globalisation produces political discontent not a cultural backlash. This idea is shared by Frieden (2018) for whom, however, political discontent is caused by the fact that globalisation reduces the ability of the government to accommodate the needs of the losers of globalisation itself. To produce populism, however, the failure of compensation needs to be accompanied by the failure of representation i.e. the loss of the traditional party system to represent the needs of the discontents.

Chapter 9 concludes that considering globalisation at the origins of the recent populist wave could be too vague given the time-lag between the start of globalisation and the electoral success of populist movements. Instead, it is suggested that the global economic crisis and the crisis of the euro-zone acted as a catalyst for the contradictions of globalisation to explode. In fact, where globalisation did not bring economic difficulties, the populist backlash did not appear. Instead, both the global financial crisis and the euro-zone crisis strongly impacted on the economic situations of some countries, especially when austerity made it much more difficult to compensate the losers through suitable fiscal policies. Therefore, the crises' role in favouring the growth of populism is very evident. From this perspective, the cause of populism cannot be cultural. It is an economic one and the cultural manifestation of populism, and, in particular, anti-immigrant feelings, are a consequence of the worsening of economic insecurity. On the other hand, however, once in power, populism exasperates its anti-migrant attitude to increase its popular support, especially because migrants cannot vote.

This has certainly been the case in Italy, where the two populist movements, the Five Star Movement and the Lega were, originally, focusing their political discourse on the attack of Euro and the way in which the EU institutions reacted to the Euro-zone crisis. Only after taking power by forming a government coalition in 2018, did their political narrative insist on an anti-immigrant sentiment, making immigrants the scapegoat for the grievances of their voters.

References

Frieden, J. (2018, January). *The political economy of the globalization backlash: Sources and implications*. Paper presented at the annual meeting of the American Economic Association. http://pelg.ucsd.edu/Frieden_2018.pdf.

Mittelman, J. H. (2011). *The globalization syndrome: Transformation and resistance*. Princeton: Princeton University Press.

Norris, P., & Inglehart, R. (2019). *Cultural backlash: Trump, brexit, and authoritarian populism*. Cambridge University Press. Kindle Edition.

Reyneri, E. (2013). Irregular immigration and the underground economy in Southern Europe: Breaking the vicious circle. In G. Brochmann & E. Jurado (Eds.), *Europe's immigration challenge. Reconciling work, welfare and mobility*. London: I.B. Tauris.

Rodrik, D. (2017). *Populism and the economics of globalization*. https://drodrik.scholar.harvard.edu/files/dani-rodrik/files/populism_and_the_economics_of_globalization.pdf.

Index

L. S. Talani, *The International Political Economy of Migration in the Globalization Era*, International Political Economy Series,
https://doi.org/10.1007/978-3-030-79321-0

K

L

The manufacturer's authorised representative in the EU is Springer Nature Customer Service Centre GmbH, Europaplatz 3, 69115 Heidelberg, Germany. If you have any concerns regarding our products, please contact ProductSafety@springernature.com

Printed and bound by CPI Group (UK) Ltd, Croydon, CR0 4YY
30/09/2026
02225700-0016